Elementary Applied CALCULUS

J.S. Ratti
University of South Florida

Mariner Publishing Company, Inc.
Tampa, Florida

To Janelle.

The author is grateful to the Houghton Mifflin Company for permission to use materials from *Introductory Calculus with Applications*, Second Edition, by J.S. Ratti and M.N. Manougian, © copyright 1977 by Houghton Mifflin Company.

Cover design and art by Joe Traina
This book was set in English Times by Mariner Publishing Company, Inc. and printed by Kingsport Press
Typespecialist: Scott J. Smith
Production Manager: John W. Hobbes

Printed in the United States of America

Library of Congress Cataloging in Publication Data

Ratti, J S
Elementary applied calculus.

Includes index.
1. Calculus. I. Title.
QA303.R26 515 80-26578
ISBN 0-936166-05-3

Preface

The objective of this book is to present some basic concepts of calculus and to relate these concepts whenever feasible to business as well as the social, physical, and life sciences. The text is written for the student. An intuitive approach is used to introduce new concepts, and formal rigorous proofs are excluded.

The text contains numerous solved examples in every section. The emphasis is placed upon the applications of differentiation and integration to several fields and particularly to business applications. Each section contains a large number of problems ranging from direct substitutions and applications to the more challenging ones. We also use the usual custom of placing stars (*) on those problems which are relatively difficult. Answers to selected problems are given. The problems are an integral part of the text, and the student is reminded that it is by doing mathematics that one learns mathematics.

It is not intended that the teacher cover the four appendices. This review material is included merely for the convenience of the reader. There are no problems in the appendices.

Each chapter ends with a review consisting of an oral and written part. The oral part contains the main definitions and concepts in that chapter, while the written part contains problems covering every section in that chapter. Short biographical sketches of some of the leaders of mathematics are included. It is hoped that this will encourage the student to read further on the lives of these and other famous men and women of mathematics.

Portions of this book have been taken from the book authored jointly with M. N. Manougian: *Introductory Calculus with Applications*, published by Houghton Mifflin Company.

I wish to thank my wife Janelle for her patience. I also thank the staff of Mariner Publishing Company, Inc., for doing an outstanding job during the production of this book.

Tampa, Florida

J.S.R.

Contents

Chapter One

Functions, Graphs and Limits

In this chapter our primary concern shall be the concept of a function. Before we introduce this concept, we shall examine the notion of a coordinate plane. This will help us to visualize (graph) a function which is given by an equation. The chapter ends with the discussion on limits and continuity.

1.1 THE COORDINATE PLANE

In this section, we shall assume that it is possible to identify real numbers with points on a line-called the *number line* or the *number axis.* We shall also assume that the reader is familiar with the concept of absolute value and the terminology of sets. If these notions are new to you or you wish to refresh your memory, we suggest that you review Appendices 1 and 2.

We shall show that points in a plane can be associated with pairs of numbers. A pair of numbers in which the order is specified is called an *ordered pair* of numbers. We denote an ordered pair by the symbol (a, b). The number a is called the *first component* and b is called the *second component* of the ordered pair (a, b). Note that the pair $(3, 7)$ is different from the pair $(7, 3)$. This is why we call $(3, 7)$ an ordered pair.

Let R denote the set of all real numbers. We denote by R^2 the set of all possible ordered pairs of numbers. That is,

$$R^2 = \{(x, y) \mid x, y \in R\}$$

Geometrically, R^2 may be interpreted as the set of all points in the Cartesian (or geometric) plane. The *Cartesian* (or *rectangular*) *coordinate system* is ordinarily used to set up a correspondence between ordered pairs of real numbers and points in the plane.

The formulation of this correspondence is attributed to the French mathematician-philosopher Descartes.[1]

We begin by constructing a line in the plane from left to right which we call the x-axis, as shown in Figure 1. Then a vertical line is drawn called the y-axis. The two

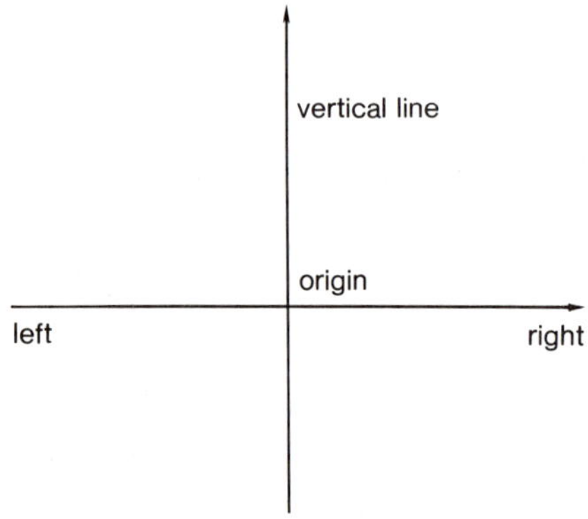

Figure 1

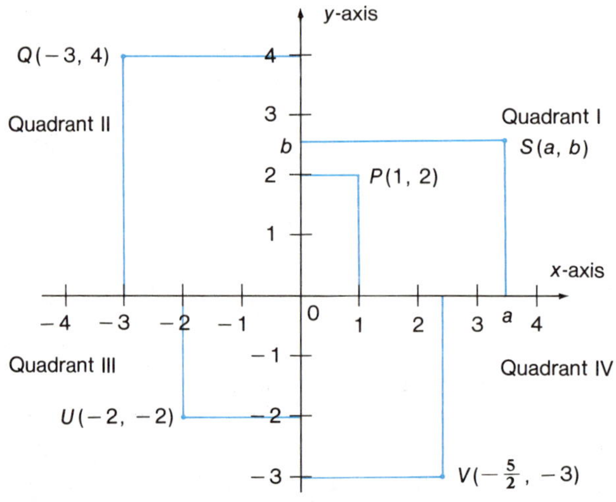

Figure 2

[1] RENE DESCARTES (1596-1650), was a Frence philosopher and mathematician. Descartes' coordinate geometry provided a vehicle whereby points in the plane could be viewed as ordered pairs of numbers and ordered pairs of numbers could be viewed as points in the plane. Thus, coordinate geometry unified algebra and geometry, enabling us to do geometry problems by algebra. The concept of the Cartesian product gives us an analytic way of viewing the geometric concept or dimension. Thus, one-dimensional space is represented by R, two-dimensional space by R^2 and so on.

lines are referred to as the *coordinate axes*, and their point of intersection is called the *origin*. A convenient unit of length is chosen, and a number scale is marked off on each of the axes. Usually the number zero is placed at the origin; on the x-axis the positive numbers are to the right, and on the y-axis the positive numbers extend upward. Figure 2 shows a Cartesian coordinate system including some points. The axes divide the plane into four parts called *quadrants*. The four quadrants are:

$$\begin{aligned}
&\text{Quadrant I:} && \{(x, y) \mid x>0, y>0\}\\
&\text{Quadrant II:} && \{(x, y) \mid x<0, y>0\}\\
&\text{Quadrant III:} && \{(x, y) \mid x<0, y<0\}\\
&\text{Quadrant IV:} && \{(x, y) \mid x>0, y<0\}
\end{aligned}$$

To avoid ambiguities, the coordinate axes are excluded from the quadrants.

The points P, Q, U and V are in the first, second, third and fourth quadrants, respectively. The ordered pair $(-3, 4)$ represents the *coordinates* of the point Q. The first component -3 is called the *x-coordinate* or *abscissa* of Q, while the second component 4 is called the *y-coordinate* or *ordinate* of Q.

It is apparent in Figure 2 that for each point S in the plane there are unique numbers $x = a$ and $y = b$ for which S can be labelled (a, b), and for each pair of real numbers (a, b) there is a unique point in the plane with coordinates (a, b).

DISTANCE

One of the oldest and most famous results in mathematics is the *Pythagorean theorem.* The ancient Greeks proved that the area of the square constructed on the hypotenuse of a right triangle is equal to the sum of the areas of the squares constructed on the other two sides (see Figure 3). If a, b, and c represent the lengths of the sides as indicated, then

$$a^2 = b^2 + c^2$$

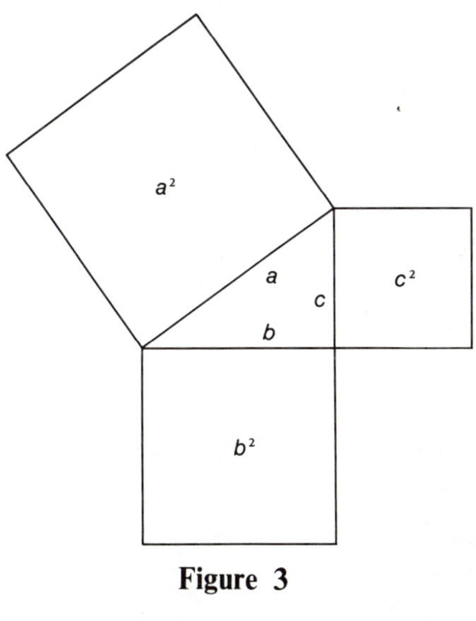

Figure 3

We shall use this result and the above coordinate scheme to find the length of the line segment between two points, say P and Q, in the plane. The length of the line segment between P and Q is called the *distance* between P and Q and is denoted by $d(P, Q)$.

We recall (see Appendix 2) that if a and b are two points on a number line (rigorously, the coordinates of two points), then the distance between a and b is defined by $|a - b|$.

We shall establish a formula for $d(P, Q)$ in terms of the coordinates of the points P and Q in the plane.

Consider two points on the x-axis, $A(x_1, 0)$ and $B(x_2, 0)$. We have defined the distance $d(A, B)$ between A and B to be

$$d(A, B) = |x_2 - x_1|$$

Consider the points $A'(x_1, b)$ and $B'(x_2, b)$, where $b \neq 0$ (see Figure 4). Since the second components of A' and B' are equal, the line segment $A'B'$ is parallel to the x-axis. The reader will note that the four points A, B, B', and A' are the vertices of a rectangle. Hence the line segment $A'B'$ may be assigned a length

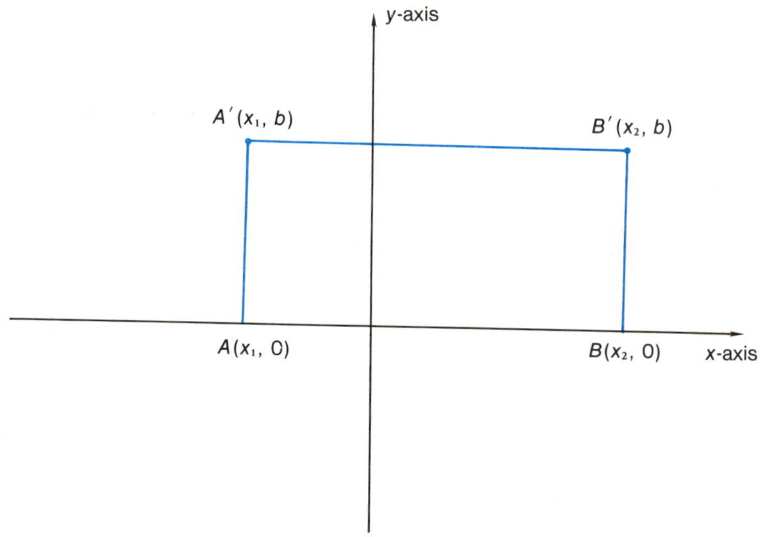

Figure 4

$$d(A', B') = d(A, B) = |x_2 - x_1|$$

Similarly (see Figure 5), we have

$$d(C', D') = d(C, D) = |y_2 - y_1|$$

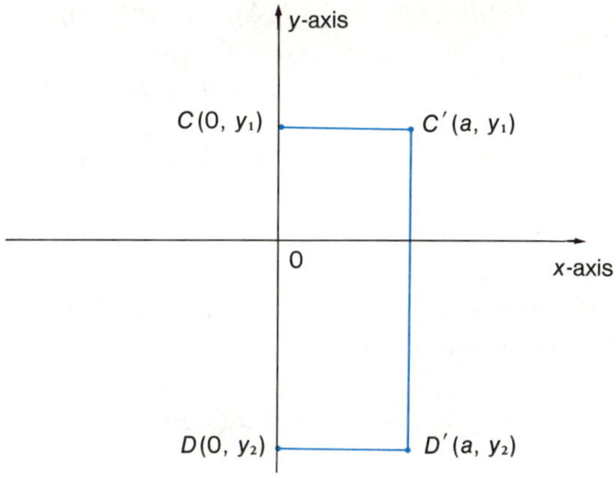

Figure 5

THEOREM 1. If $P(x_1, y_1)$ and $Q(x_2, y_2)$ are points in R^2, then

$$d(P, Q) = \sqrt{(x_2 - x_1)^2 + (y_2 - y_1)^2} \quad (1)$$

Proof. Let S be the point (x_2, y_1). Then the points P, Q, and S are the vertices of a right triangle as shown in Figure 6. Since the line segment PS is parallel to the x-axis

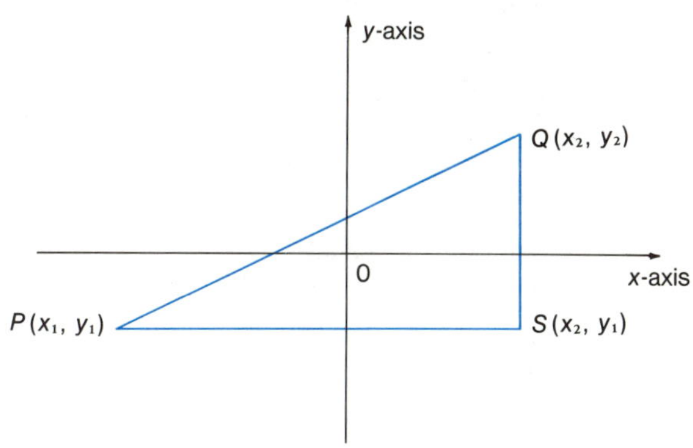

Figure 6

and the line segment QS is parallel to the y-axis, we have

$$d(P, S) = |x_2 - x_1| \quad (2)$$

and

$$d(Q, S) = |y_2 - y_1| \quad (3)$$

Hence, by the Pythagorean theorem,

$$\begin{aligned}[d(P, Q)]^2 &= [d(P, S)]^2 + [d(Q, S)]^2 \\ &= |x_2 - x_1|^2 + |y_2 - y_1|^2 \\ &= (x_2 - x_1)^2 + (y_2 - y_1)^2\end{aligned}$$

and

$$d(P, Q) = \sqrt{(x_2 - x_1)^2 + (y_2 - y_1)^2}$$

This proves Theorem 1.

It makes no difference whether the coordinates of P or those of Q are chosen as (x_1, y_1). The reader should show that

$$d(P, Q) = d(Q, P).$$

EXAMPLE 1. Find the distance between $P(2, -1)$ and $Q(5, 2)$.

SOLUTION. Using Equation (1), we have (see Figure 7)

$$\begin{aligned}d(P, Q) &= \sqrt{(x_2 - x_1)^2 + (y_2 - y_1)^2} \\ &= \sqrt{(5-2)^2 + [2 - (-1)]^2} \\ &= \sqrt{9 + 9} = 3\sqrt{2}\end{aligned}$$

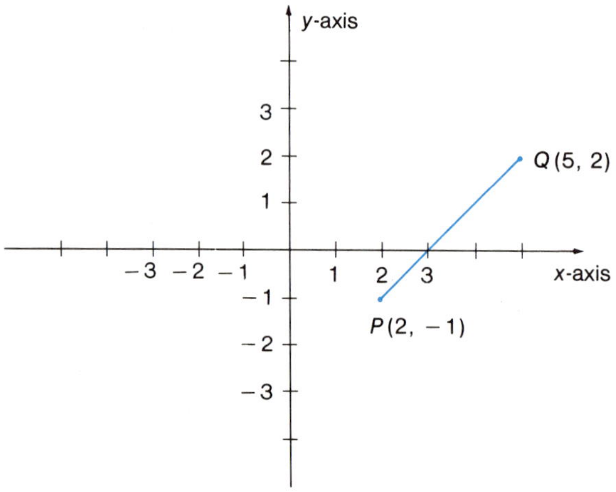

Figure 7

EXAMPLE 2. Show that the triangle with vertices $P(3, -2)$, $Q(4, 3)$, and $S(-6, 5)$ is a right triangle.

SOLUTION. Using the distance formula, we have

$$d(P, Q) = \sqrt{(4-3)^2 + (3+2)^2} = \sqrt{26}$$

$$d(Q, S) = \sqrt{(-6-4)^2 + (5-3)^2} = \sqrt{104}$$

and

$$d(P, S) = \sqrt{(-6-3)^2 + (5+2)^2} = \sqrt{130}$$

We observe that

$$(\sqrt{130})^2 = (\sqrt{104})^2 + (\sqrt{26})^2$$

or

$$[d(P, S)]^2 = [d(Q, S)]^2 + [d(P, Q)]^2$$

Hence, by the converse of the Pythagorean theorem, the triangle with vertices P, Q, and S is a right triangle with the line segment PS as its hypotenuse (see Figure 8)

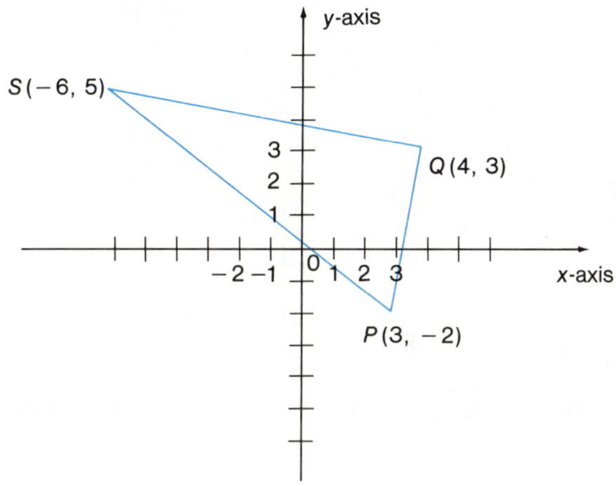

Figure 8

MIDPOINT FORMULA

Now we shall show how to find the coordinates of the midpoint of a line segment. Consider the line segment PQ in Figure 9. Let $M(x, y)$ be the midpoint of the line segment PQ. We wish to express x, y in terms of x_1, y_1, x_2, y_2. With the help of results in plane geometry, the reader will see that the triangles PMS and MQT are congruent.

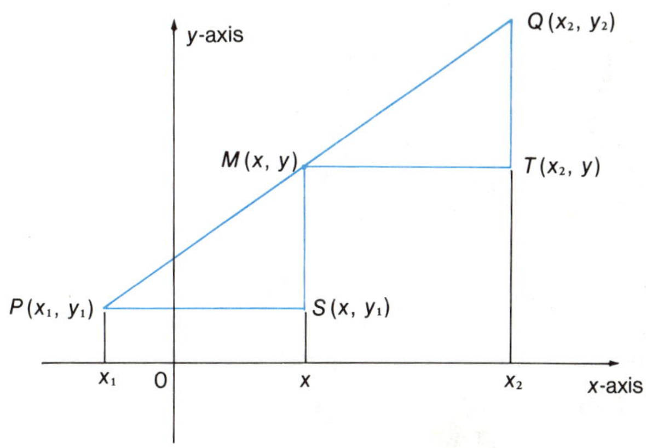

Figure 9

Therefore

$$d(P, S) = d(M, T) \quad \text{or} \quad |x - x_1| = |x_2 - x|$$

Further, since x is to the right of x_1 and x_2 is to the right of x, we have

$$x - x_1 = x_2 - x$$

Solving for x, we obtain

$$x = \frac{x_1 + x_2}{2} \tag{4}$$

Also

$$d(S, M) = d(T, Q)$$

or

$$y - y_1 = y_2 - y$$

$$y = \frac{y_1 + y_2}{2} \tag{5}$$

Hence the coordinates of the midpoint of the line segment PQ are given by (4) and (5).

EXAMPLE 3. Find the midpoint of the line segment with endpoints $A(-4, 2)$ and $B(3, 4)$.

SOLUTION. Let $M(x, y)$ be the midpoint of the line segment AB. Then

$$x = \frac{-4 + 3}{2} = -\frac{1}{2} \quad \text{and} \quad y = \frac{2 + 4}{2} = 3$$

Therefore the midpoint is $M(-\frac{1}{2}, 3)$.

EXERCISES 1.1

1. On a Cartesian coordinate system indicate the points with the following coordinates.
   (a) $(3, 2)$ (b) $(-3/2, 2)$
   (c) $(0, 5)$ (d) $(5, 0)$
   (e) $(-3, -3)$ (f) $(-2, -5)$
   (g) $(1, -3)$
2. Write down the coordinates of any five points on the x-axis. What do these coordinates have in common?
3. Do Problem 2 for the y-axis.
4. Plot the points $(1, 0)$; $(1, 1)$; $(1, 2)$; $(1, 3)$; $(1, -1)$; $(1, -2)$; $(1, -3)$ and $(1, y)$. Describe the line on which all these points lie.
5. Plot the points $(-3, -2)$; $(-2, -2)$; $(-1, -2)$; $(0, -2)$; $(1, -2)$ and $(x, -2)$. Describe the line on which all these points lie.
6. Plot the points $(-2, -2)$; $(-1, -1)$; $(0, 0)$; $(1, 1)$; $(2, 2)$ and (x, x). Describe the line on which all these points lie.

In Problems 7 through 14, find the distance between the points with the given coordinates. Also find the coordinates of the midpoint.

7. $A(1, 3)$, $B(4, 8)$ **8.** $A(2, 5)$, $B(-1, 4)$

9. $A(-3, -3)$, $B(-1, -6)$ **10.** $A(-2, 5)$, $B(-1, -2)$

11. $A(-2, 0)$, $B(-3, 1)$ **12.** $A(2, \frac{1}{2})$, $B(\frac{1}{2}, -3)$

13. $A(-1, 3)$, $B(-1, 7)$ **14.** $A(5, 2)$, $B(-3, 2)$

Show that ABC is a right triangle in Problems 15 through 20.

15. $A(-3, 2)$, $B(5, 5)$, $C(0, -6)$

16. $A(2, 0)$, $B(6, 3)$, $C(5, -4)$

17. $A(0, 0)$, $B(3, 2)$, $C(-4, 6)$

18. $A(4, 3)$, $B(1, 4)$, $C(-2, -5)$

19. $A(1, 2)$, $B(3, 4)$, $C(1, 6)$

20. $A(0, -3)$, $B(-1, -1)$, $C(-3, -2)$

In Problems 21 through 24, identify the triangle ABC as isosceles, equilateral, or right.

21. $A(-6, 6)$, $B(-2, 5)$, $C(-5, 2)$

22. $A(2, 1)$, $B(5, 5)$, $C(-2, 4)$

23. $A(1, -1)$, $B(-1, 1)$, $C(-\sqrt{3}, -\sqrt{3})$

24. $A(-4, 6)$, $B(0, 5)$, $C(-3, 2)$

25. Which of the following sets of points are collinear (lie on the same line)?

(a) $(0, 5)$, $(4, 1)$, $(12, -7)$

(b) $(-3, 2)$ $(2, 0)$, $(1, -2)$

(c) $(3, -5)$, $(\frac{1}{2}, 0)$, $(0, 1)$

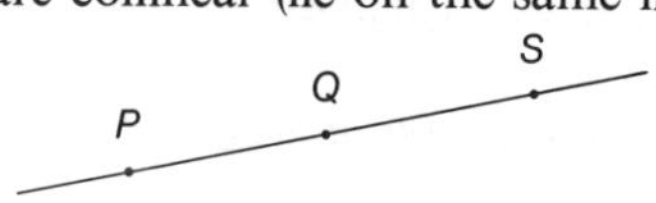

[*Hint:* If P, Q, and S are collinear, then $d(P, Q) + d(Q, S) = d(P, S)$.]

26. The distance from the point P to the point $(2, 0)$ is $\sqrt{45}$. If the abscissa of P is -4, find the ordinate of P.

27. Show that the points $A(3, -16)$, $B(-5, -1)$, $C(10, 7)$, and $D(18, -8)$ are the vertices of a square. Find the length of the diagonals.

28. If the vertices of a triangle are $A(1, 2)$, $B(-4, 5)$, and $C(3, 7)$, find the length of each of the medians (a *median* is a line segment drawn from a vertex to the midpoint of the opposite side).

29. A pilot is flying from Dullsville to Middale to Pleasantville. With reference to an origin, Dullsville is located at (2, 4), Middale at (8, 12), and Pleasantville at (20, 3), all numbers being in 100-mile units.

(a) Locate the positions of the three cities on a coordinate system.

(b) Compute the distance traveled by the pilot.

(c) Compute the direct distance from Dullsville to Pleasantville.

30. Let $P(-6, 2)$, $Q(8, -4)$, $R(4, 4)$ be the vertices of a triangle. Let A and B be the midpoints of the line segments PQ and PR, respectively. Show that

$$d(A, B) = \frac{1}{2} d(Q, R)$$

31. If the distance between $(a, 5)$ and $(2, 9)$ is 10 units, find a.

32. Find a relationship between numbers x and y for which the point $P(x, y)$ is equidistant from the points $A(2, 3)$ and $B(4, -5)$.

{*Hint:* Simplify $[d(P, A)]^2 = [d(P, B)]^2$}

1.2 GRAPHS

An equation in two variables x and y defines a set of ordered pairs. For example, consider the equation

$$y = 2x + 1 \tag{1}$$

An ordered pair (a, b) is said to *satisfy* equation (1) if, when a is substituted for x and b is substituted for y, the resulting statement is true. For example, the ordered pair (2, 5) satisfies equation (1), because the numbers $x = 2$ and $y = 5$ make equation (1) a true statement:

$$\begin{aligned} y &= 2x + 1 \\ 5 &= 2(2) + 1 \\ 5 &= 4 + 1 \\ 5 &= 5 \end{aligned}$$

An ordered pair that satisfies an equation is called its *solution.* The set of all solutions is called the *solution set* of the equation. Thus, the solution set is the collection of all ordered pairs that make the equation true.

Definition. The *graph* of an equation involving the variables x and y is the set of all points in the coordinate plane that correspond to the ordered pairs in the solution set.

EXAMPLE 1. Graph $y = 2x + 1$.

SOLUTION. We seek a visualization of the set of all (x, y) which satisfy the equation $y = 2x + 1$.

Now, if we let x assume a particular value, then it is easy to find a value for y that makes (x, y) a solution. For example, if we choose $x = -3$, then $y = 2(-3) + 1 = -6 + 1 = -5$. If $x = 0$, then $y = 2(0) + 1 = 1$. It is clear that the solution set of this equation is an infinite set (why?). We can determine some of the solutions by choosing particular values of x and computing the corresponding values of y from the equation. We list these solutions in the tabular form and plot the points $(-3, -5)$,

x	-3	-2	-1	0	1	2	3
y	-5	-3	-1	1	3	5	7

$(-2, -3)$, $(-1, -1)$, $(0, 1)$, $(1, 3)$, $(2, 5)$ and $(3, 7)$ in Figure 1(a). These points appear to lie on a straight line. In fact, we shall show in Section 1.3, that any equation which is equivalent to a *first degree equation* in two variables x and y,

$$ax + by + c = 0 \tag{2}$$

EXAMPLE 4. Graph $y = x^2$ (7)

SOLUTION. At the moment, to sketch the graph of (7), we have to plot several points corresponding to the solution set of (7). Later we shall see that calculus is tremendously helpful in deciding the shape of these figures. To find some of the points corresponding to the solution set of (7), we assign particular values to x and compute the corresponding values of y. In the tabular form, we have

x	-3	-2	-1	0	1	2	3
y	9	4	1	0	1	4	9

The graph of these points is shown in Figure 2(a). If we compute and plot additional points corresponding to (7) and connect them with a smooth curve, we obtain the graph of (7) shown in Figure 2(b).

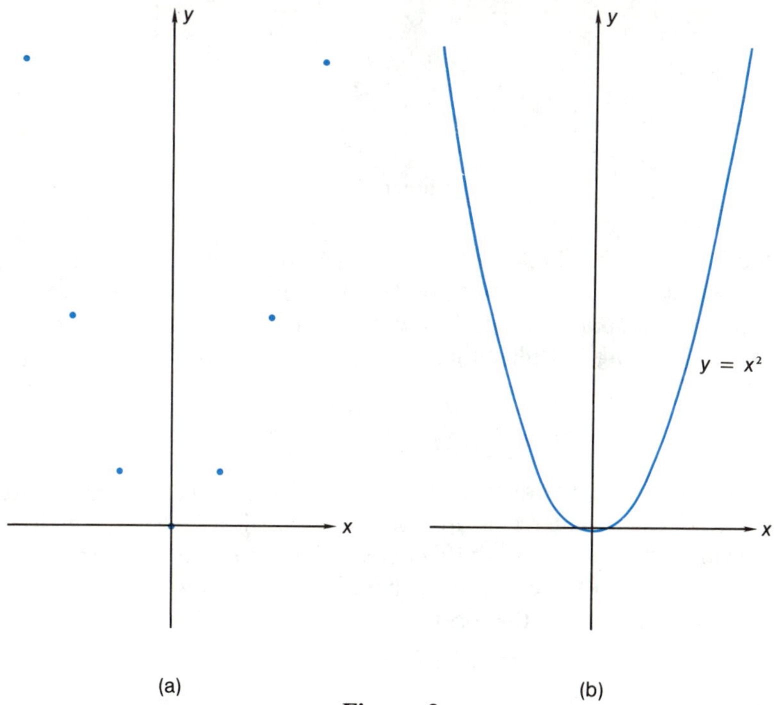

Figure 2

The graph of any equation of the form

$$y = ax^2 + bx + c \qquad (a \neq 0) \tag{8}$$

will have the same general shape as illustrated in Figure 2, although the specific location depends on the coefficients a, b, and c. Such graphs are called *parabolas*. In Figure 3, we have sketched the graphs of a few more parabolas:

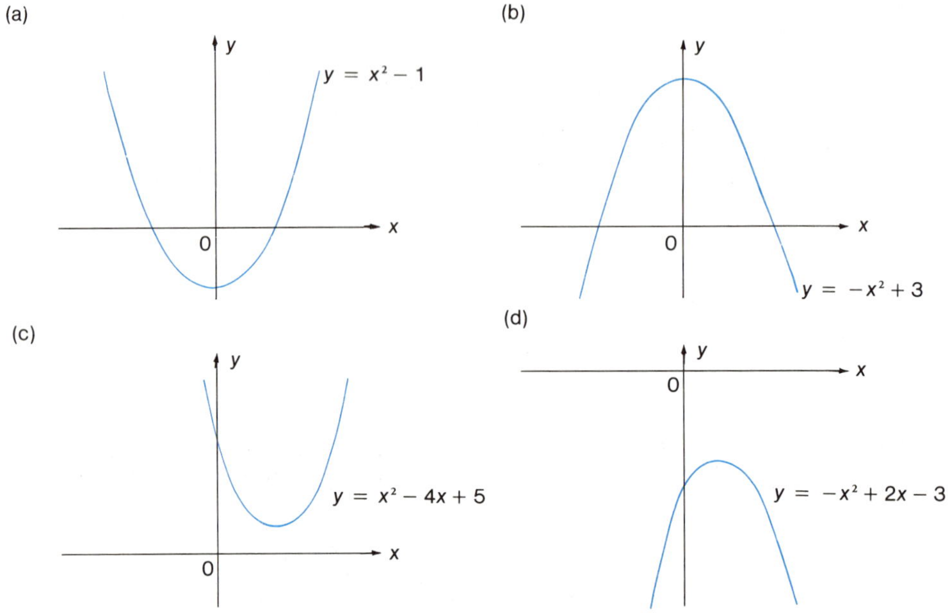

Figure 3

In Figure 3, we observe that the parabolas open upward if $a > 0$, and the parabolas open downward if $a < 0$. The lowest point of a parabola (if it opens upward) or the highest point on a parabola (if it opens downward) is called the *vertex* of the parabola. Later we shall see (using Calculus) that the coordinates of the vertex of the parabola given by (8) are

$$\left(-\frac{b}{2a}, c - \frac{b^2}{4a}\right)$$

The x-intercepts of a curve refer to the x-corrdinates at which the curve crosses or touches the x-axis. The x-intercepts are obtained by setting $y = 0$ in the given equation and solving for x. Similarly, the y-intercepts of a curve are obtained by setting $x = 0$ in the equation of the curve and solving for y.

So, to graph an equation of the form

$$y = ax^2 + bx + c \qquad (a \neq 0)$$

we shall need the following steps:

1. The graph is a parabola.
2. If $a > 0$, then the parabola opens upward. If $a < 0$, then the parabola opens downward.
3. The coordinates of the vertex are

$$\left(-\frac{b}{2a}, c - \frac{b^2}{4a}\right)$$

4. Find the x-intercepts by solving the quadratic equation $ax^2 + bx + c = 0$. If $b^2 - 4ac > 0$, then the parabola intersects the x-axis at two distinct points. If $b^2 - 4ac = 0$, then the parabola touches the x-axis. If $b^2 + 4ac < 0$, then the parabola does not cross the x-axis. That is, the parabola either lies above the x-axis (when $a > 0$) or lies below the x-axis (when $a < 0$).
5. The y-intercept is given by $(0, c)$.
6. Plot a few more points by assigning values to x and computing the corresponding values of y.
7. The graph is sketched by joining the points obtained from steps 3 to 6.

EXAMPLE 5. Sketch the graph of

$$y = x^2 - 4x + 3 \tag{9}$$

SOLUTION. In equation (9), $a = 1, b = -4,$ and $c = 3$. Since $a = 1 > 0$, the parabola opens upward. The coordinates of the vertex (lowest point) are

$$\left(-\frac{b}{2a}, c - \frac{b^2}{4a}\right) = \left(-\frac{-4}{2 \cdot 1}, 3 - \frac{(-4)^2}{4 \cdot 1}\right) = (2, -1)$$

The x-intercepts are given by

$$x^2 - 4x + 3 = 0$$

or

$$(x - 1)(x - 3) = 0$$

or

$$x - 1 = 0 \quad \text{or} \quad x - 3 = 0$$
$$x = 1 \quad \text{or} \quad x = 3$$

The y-intercept is (0, 3). We find additional points by writing the table below. The graph of (9) is shown in Figure 4.

x	-2	-1	0	1	2
$f(x)$	13	8	3	0	-1

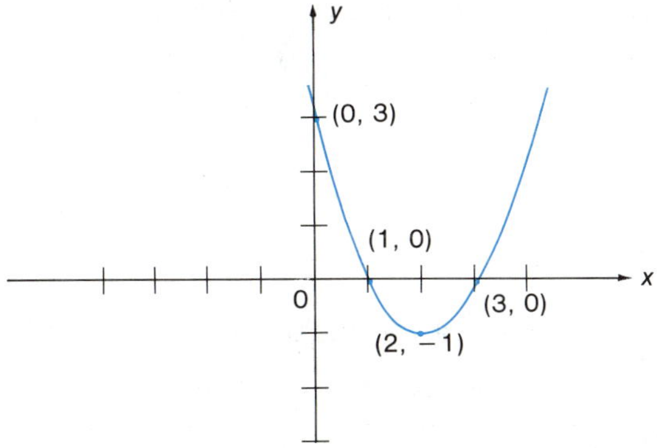

Figure 4

EXAMPLE 6. Graph $y = \dfrac{1}{x-3}$. (10)

SOLUTION. We notice that if $|x|$ is large, then the fraction $1/(x-3)$ will be close to zero. That is, for large values of $|x|$, the graph will almost touch the x-axis. Also, x cannot equal 3 (why?). If x is close to 3, the denominator of (10) will be close to zero. Hence, $|y|$ will be large. If x is to the right of 3, then y is positive; and if x is to the left of 3, then y is negative. Various ordered pairs that satisfy equation (10) are listed in the tabular form below, and the graph of (10) is sketched in Figure 5.

x	-100	-2	0	1	2.9	3.1	5	200
y	-0.01	-0.2	-0.33	-0.5	-10	10	0.5	0.005

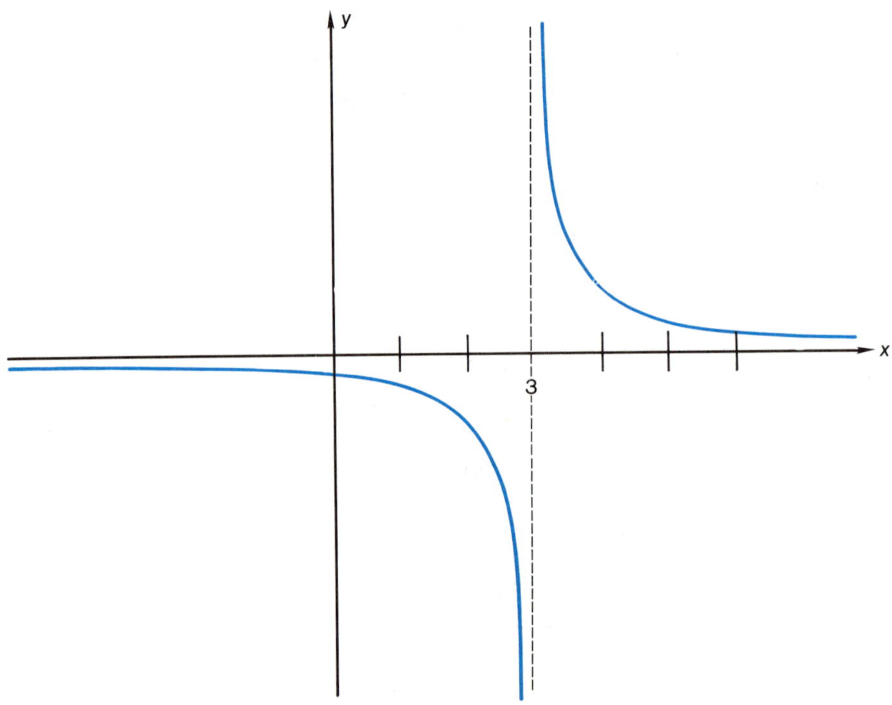

Figure 5

EXERCISES 1.2

In Problems 1 through 20, sketch a graph of the set of (x, y) which satisfies the given equation.

1. $y = x$

2. $y = -x$

3. $y = 3x + 5$

4. $x = 2y - 4$

5. $y = -2x + 1$

6. $2x + 3y = 6$

7. $3x - 4y = 12$

8. $x - 2y = -4$

9. $y = x^2 + 1$

10. $y = x^2 + 3$

11. $y = -x^2 + 3$

12. $y = 2x^2 + 4$

13. $y = -3x^2 + 5$

14. $y = -x^2 + x$

15. $y = -2x^2 - 3x + 2$

16. $y = x^2 + x - 2$

17. $y^2 = x$

18. $x = -y^2 + 1$

19. $y = \dfrac{1}{x}$

20. $y = -\dfrac{1}{x - 3}$

In Problems 21 through 24, specify the center and radius of each of the circles and sketch the graph.

21. $(x - 2)^2 + (y - 3)^2 = 25$

22. $(x + 1)^2 + (y - 3)^2 = 16$

23. $(x + 2)^2 + (y + 3)^2 = 11$

24. $(x - \frac{1}{2})^2 + (y + \frac{3}{2})^2 = \frac{9}{16}$

25. Change the equations in Problems 21 through 24 from standard form to the general form.

In Problems 26 through 30, change each of the general form equations to standard form for the circle and sketch the graph.

26. $x^2 + y^2 + 4x - 6y - 12 = 0$

27. $x^2 + y^2 - x = 0$

28. $x^2 + y^2 - 6x - 8y + 25 = 0$

29. $4x^2 + 4y^2 + 2x - 3y - 24 = 0$

30. $x^2 + y^2 - 4x - 6y + 14 = 0$

In Problems 31 through 38, find an equation of a circle that satisfies the conditions given.

31. Center at $(-2, 3)$ and radius 5.

32. Center at $\left(\frac{1}{2}, -\frac{1}{3}\right)$ and radius 2.

33. Center at $(-2, -3)$ and radius 4.

34. Center at $(3, -4)$ and passing through the point $(-1, 5)$.

35. Center at $(1, 2)$ and touching the x-axis.

36. Center at $(1, 2)$ and touching the y-axis.

37. Diameter is the line segment with endpoints $(7, 4)$ and $(-3, 6)$.

38. Center at $(3, 4)$ and touching the line $x = 6$.

In Problems 39 through 42, a graph G is described geometrically in terms of point $P(x, y)$. Find an equation of G in x and y.

39. $P(x, y)$ is on G if and only if its distance from the x-axis is equal to its distance from the y-axis.

40. $P(x, y)$ is equidistant from the two points $(1, 2)$ and $(3, -4)$.

41. The distance between $P(x, y)$ and $(2, 0)$ is equal to the distance between $P(x, y)$ and the y-axis.

42. $P(x, y)$ is equidistant from the line $x = 1$ and the point $(2, 3)$.

43. Suppose y varies directly[1] as x, and $y = 14$ when $x = 2$.
(a) Write an equation that relates y and x.
(b) Graph the equation.
(c) What is y when x is 6?
(d) What is x when y is 56?

44. Lisa's wages (w) vary directly with the number of hours (t) she works. If Lisa is paid \$140 for 40 hours of work, find an equation that relates w and t. If she is paid \$98, how many hours did she work?

45. The profits y of the ABC Widget Company vary directly with the sales x of widgets. Suppose the profits are \$4500 when sales are 7500. What are the sales when profits are \$2163?

---

[1] The variable y is said to *vary directly* with x if $y = kx$ for some k.

46. The social security payments (y) an indiviual receives vary directly as the cost of living index (x). Suppose Sam receives \$500 when the index is 0.1. What will Sam receive if the index is 0.12?

1.3 LINES

In this section, we shall investigate lines in a coordinate plane. We shall derive equations of lines and determine those equations whose graphs are lines. We begin by introducing the notion of the slope of a line in the Cartesian coordinate plane.

Definition. Let L be a line in the plane which is not parallel to the y-axis. If $P(x_1, y_1)$ and $Q(x_2, y_2)$ are two distinct points on L, then the *slope m* of L is given by

$$m = \frac{y_2 - y_1}{x_2 - x_1} \tag{1}$$

The slope of a line parallel to the y-axis is not defined (Why?).

Let us make a few remarks about the slope of a line:

(a) Since $\dfrac{y_1 - y_2}{x_1 - x_2} = \dfrac{y_2 - y_1}{x_2 - x_1}$, it follows that it is immaterial which point is labeled

P and which is labeled Q. For example, the slope of a line passing through $(-2, -3)$ and $(4, -5)$ is

$$m = \frac{-5-(-3)}{4-(-2)} = \frac{-5+3}{4+2} = \frac{-2}{6} = -\frac{1}{3}$$

or

$$m = \frac{-3-(-5)}{-2-4} = \frac{-3+5}{-6} = \frac{2}{-6} = -\frac{1}{3}$$

(b) The definition of slope is independent of the two points chosen on the line. To see this, suppose that on the line L two other points, $A(x_3, y_3)$ and $B(x_4, y_4)$, are chosen, Then, in Figure 1 we see that the triangles PQE and ABF are similar. Hence, their corresponding sides are proportional.

Consequently,

$$\frac{y_4 - y_3}{x_4 - x_3} = \frac{y_2 - y_1}{x_2 - x_1} = m$$

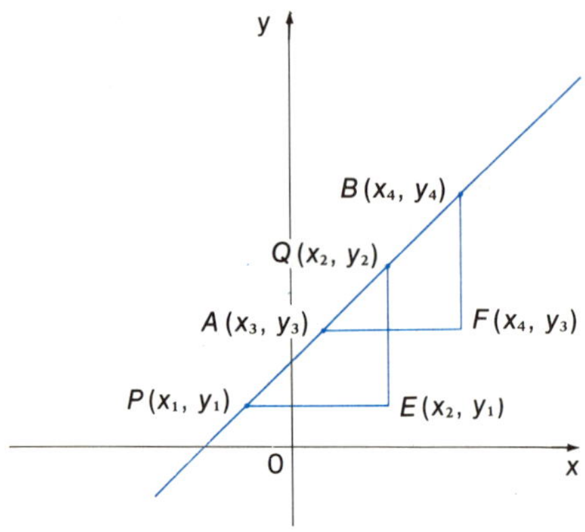

Figure 1

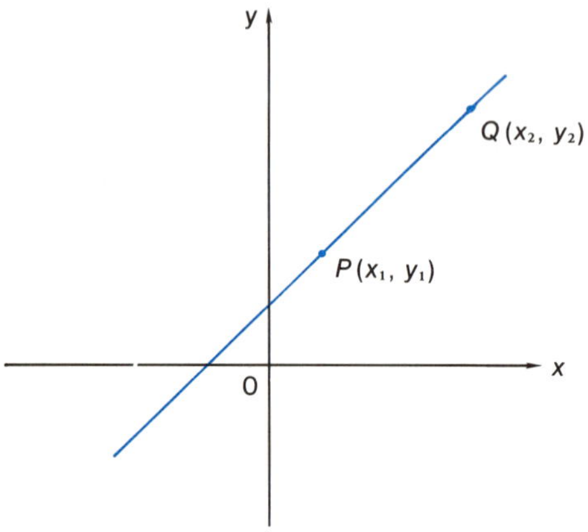

Figure 2

(c) In Figure 2, since $x_2 - x_1 > 0$ and $y_2 - y_1 > 0$, the line has a positive slope. In Figure 3, $x_2 - x_1 > 0$ and $y_2 - y_1 < 0$, the line has a negative slope. In Figure 4, the line parallel to the x-axis has a slope of zero.

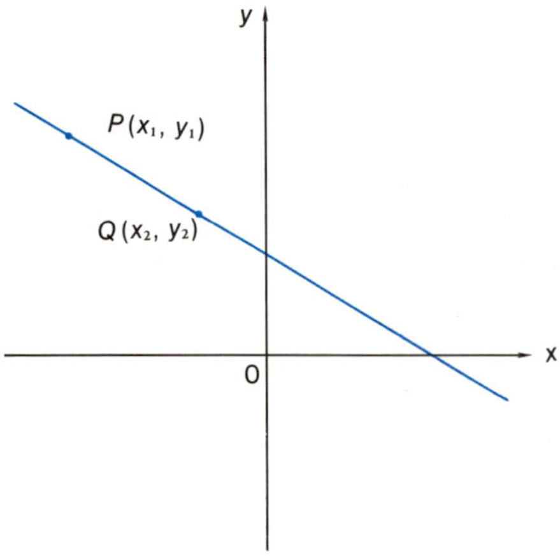

Figure 3

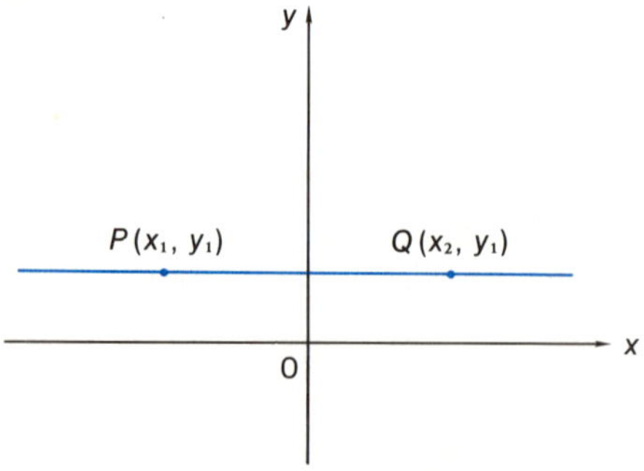

Figure 4

(d) The slope m is a measure of the inclination of the line L. In fact, those of you who are familar with trigonometry know that if a line L makes an angle θ with the positive horizontal axis, then the slope $m = \tan\theta$.

EXAMPLE 1. Sketch the line through the points $(2, -3)$ and $(-1, 4)$. Also, find the slope of this line.

SOLUTION. The line is sketched in Figure 5. The slope m of the line is

$$m = \frac{4-(-3)}{-1-2} = \frac{4+3}{-3} = -\frac{7}{3}$$

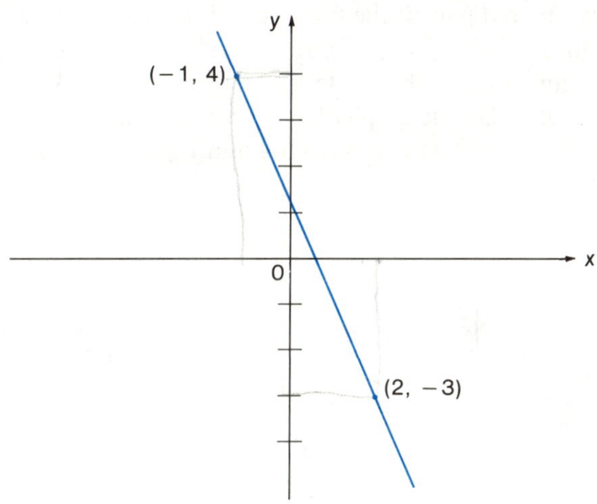

Figure 5

EXAMPLE 2. Given the line L with slope $m = 2$, if $P(-1, 3)$ and $Q(c, 1)$ are two points on L, find the number c.

SOLUTION. Using (1), we have

$$m = \frac{y_2 - y_1}{x_2 - x_1}$$

$$= \frac{1-3}{c-(-1)} = \frac{-2}{c+1}$$

Setting $m = 2$, we obtain

$$2 = -\frac{2}{c+1}$$

or

$$2c + 2 = -2$$

or

$$2c = -4$$

or

$$c = -2$$

EXAMPLE 3. Discuss the graphs of the following equations in the plane:
(a) $y = 2$ (b) $x = 4$

SOLUTION. (a) The equation $y = 2$ may be considered as an equation in two variables x and y by writing

$$0(x) + y = 2 \tag{2}$$

Thus any ordered pair of the form $(x, 2)$ is a solution of (2). Some of the solutions are $(-1, 2)$, $(0, 2)$, $(2, 2)$, and $(7, 2)$. It follows that the graph of $y = 2$ is a line parallel to the x-axis and 2 units above the x-axis. The graph is sketched in Figure 6.

(b) Similar to (a), the graph of $x = 4$ is a line parallel to the y-axis and 4 units to the right of the y-axis. The graph is shown in Figure 7.

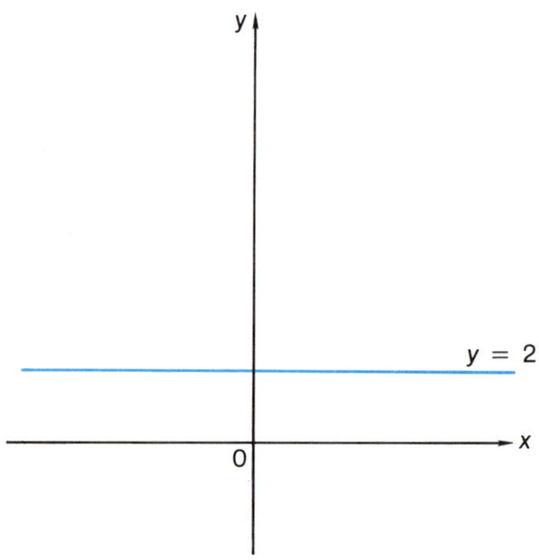

Figure 6

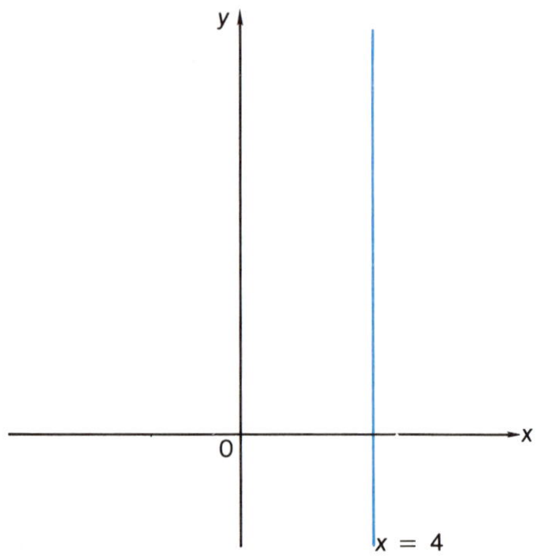

Figure 7

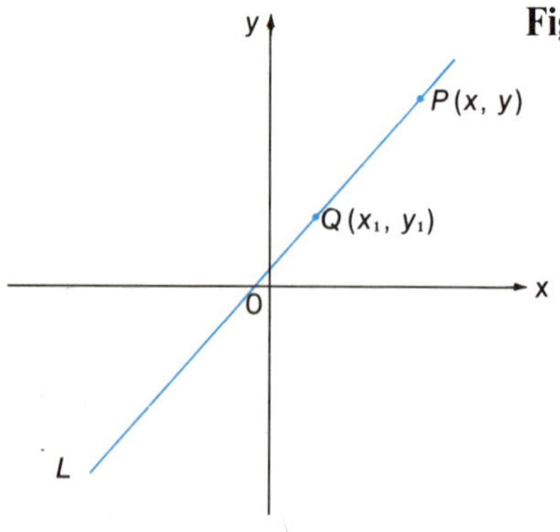

Figure 8

We shall now find an equation of a line L passing through the point $Q(x_1, y_1)$ and having slope m. If $P(x, y)$ with $x \neq x_1$ is any point, then P is on L if and only if the slope of the line passing through P and Q is m (see Figure 8). This is true if and only if

$$\frac{y - y_1}{x - x_1} = m$$

or

$$y - y_1 = m(x - x_1) \tag{3}$$

Equation (3) is called the *point-slope form* for the line.

EXAMPLE 4. Find an equation of the line passing through $(1, -2)$ and having slope $m = 3$.

SOLUTION. Substituting $x_1 = 1$, $y_1 = -2$, and $m = 3$ in (3), we have the equation

$$y - (-2) = 3(x - 1)$$

which is in the point-slope form. This equation may be simplified as

$$y + 2 = 3x - 3$$

or

$$y = 3x - 5$$

EXAMPLE 5. Find an equation of the line L passing through the points $P(-2, 1)$ and $Q(3, 7)$.

SOLUTION Using (1), we first find the slope m of L:

$$m = \frac{7 - 1}{3 - (-2)} = \frac{6}{3 + 2} = \frac{6}{5}$$

Now we use the point-slope form (3) and obtain

$$y - 7 = \tfrac{6}{5}(x - 3)$$

or

$$y = \tfrac{6}{5}x + \tfrac{17}{5}$$

In the point-slope form (3), we could have used the point P instead of Q, and we would still get the same answer.

It is readily seen from the examples above that equation (3) is equivalent to

$$y = mx + b \tag{4}$$

where $b = -mx_1 + y_1$ is a real number. In Equation (4), if we set $x = 0$, we obtain $y = b$. That is, the line intersects the y-axis at b. The number b is called the *y-intercept*. Similarly, the *x-intercept* of this line is $-b/m$ $(m \neq 0)$.

Equation (4) displays the slope and the y-intercept; it is called the *slope-intercept form* of the line. Thus the graph of an equation $y = mx + b$ is a straight line with slope m and y-intercept b.

In general, we have the following theorem.

THEOREM 1. The graph of the equation

$$ax + by + c = 0 \tag{5}$$

where a, b, and c are constants and not both a and b are zero, is a straight line.

Proof. We shall divide the proof into two cases.

CASE 1. Suppose that $b \neq 0$. In this case we divide both sides of (5) by b and obtain

$$\frac{a}{b}x + y + \frac{c}{b} = 0$$

Solving for y, we have

$$y = -\frac{a}{b}x - \frac{c}{b}$$

which is an equation of a line in the slope-intercept form. The slope of this line is $-\frac{a}{b}$ and its y-intercept is $-\frac{c}{b}$. In case $a = 0$, we have the horizontal line $y = -\frac{c}{b}$.

CASE 2. Suppose that $b = 0$ and $a \neq 0$. In this case equation (5) becomes

$$x = -\frac{c}{a}$$

which is a vertical straight line.

EXAMPLE 6. For the equation of the line

$$3x - 4y + 5 = 0 \tag{6}$$

find the slope, y-intercept, and x-intercept. Sketch the graph of (6).

SOLUTION. Equation (6) is equivalent to

$$y = \frac{3}{4}x + \frac{5}{4}$$

which is the slope-intercept form. Therefore, the slope m of (6) is $\frac{3}{4}$. The y-intercept is $\frac{5}{4}$. To find the x-intercept (the x-coordinate of the point where the line intersects the x-axis), we set $y = 0$ in (6) and obtain $x = -\frac{5}{3}$. Thus the x-intercept is $-\frac{5}{3}$.

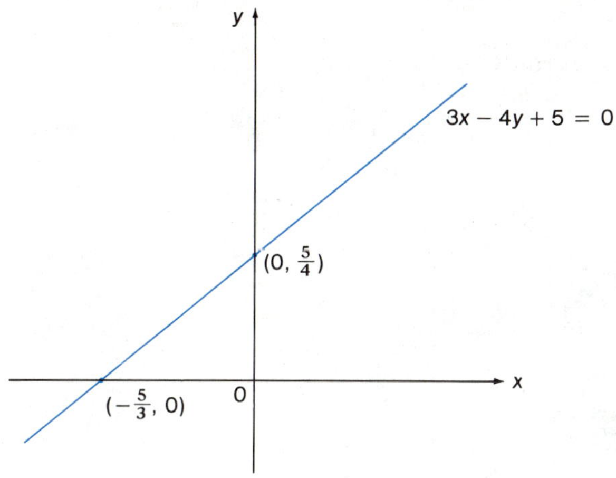

Figure 9

To sketch the graph of (6), since it represents a line, it is sufficient to find two points on the graph of (6) and draw a line through these two points. The two convenient points are the x and y intercepts. We have already found the x and y intercepts to be $-\frac{5}{3}$ and $\frac{5}{4}$, respectively. The graph is shown in Figure 9.

We shall now state two theorems from analytic geometry:

1. Let L_1 and L_2 be two distinct lines with slopes m_1 and m_2 respectively. The lines L_1 and L_2 are parallel if and only if $m_1 = m_2$.
2. Let L_1 and L_2 be two lines, neither of which is vertical, with slopes m_1 and m_2, respectively. The lines L_1 and L_2 are perpendicular if and only if $m_1m_2 = -1$.

EXAMPLE 7. Let L be the line $2x - 3y + 4 = 0$. Find an equation of the line through the point (4, 5) and (a) parallel to L, and (b) perpendicular to L.

SOLUTION. Writing the equation of L in the slope-intercept form, we have

$$y = \frac{2}{3}x + \frac{4}{3}$$

Thus the slope of L is $\frac{2}{3}$.

(a) The slope of the line parallel to L is also $\frac{2}{3}$. Thus, using the point-slope form, we have

$$y - 5 = \frac{2}{3}(x - 4)$$

This equation is equivalent to

$$2x - 3y + 7 = 0$$

which is the required equation.

(b) Since the slope of L is $\frac{2}{3}$, the slope of the line perpendicular to L is

$$-\frac{1}{(2/3)} = -\frac{3}{2}.$$

Again, using the point-slope form, we have

$$y - 5 = -\frac{3}{2}(x - 4)$$

This equation is equivalent to

$$3x + 2y - 22 = 0$$

which is the required equation.

EXAMPLE 8. Prove that the points $P(1, 5)$, $Q(-2, -4)$, and $R(-\frac{1}{3}, 1)$ are collinear.

SOLUTION. If m_1 is the slope of the line through P and Q, then

$$m_1 = \frac{5 - (-4)}{1 - (-2)} = \frac{9}{3} = 3$$

If m_2 is the slope of the line through Q and R, then

$$m_2 = \frac{1 - (-4)}{-\frac{1}{3} - (-2)} = \frac{5}{\frac{5}{3}} = 3$$

Since $m_1 = m_2 = 3$ and the lines through P and Q and through Q and R have a common point Q, these lines are identical. Hence the points P, Q, and R are collinear.

Two lines with different slopes (non-parallel lines) must intersect at exactly one point. A method of determining this unique point of intersection is the *method of substitution.* The general idea involved in the method of substitution is that the point of intersection (x, y) of two lines must be a solution to the equations of both lines simultaneously. That is, the variables x and y in one equation represent the same numbers as the variables x and y in the other equation. Therefore, we can replace (substitute) one of the variables in one of the equations by its equal from another equation. Let us see how the method works by considering an example.

EXAMPLE 9. Find the point of intersection of the lines determined by

$$2x + y = 1 \tag{7}$$

$$5x + 2y = 4 \tag{8}$$

SOLUTION. First we observe that the slope of line (7) is -2 and that of line (8) is $-\frac{5}{2}$. Hence, the lines are not parallel and they do have a unique point of intersection.

First, we solve equation (7) for y and obtain $y = 1 - 2x$. (Of course, one could proceed by solving for x. But in this case, solving for x will involve fractions.)

We then replace y in equation (8) by substituting $1 - 2x$ in its place, and obtain

$$5x + 2(1 - 2x) = 4. \tag{9}$$

Solving (9) for x:

$$5x + 2 - 4x = 4$$

$$x = 2$$

We finally set $x = 2$ in $y = 1 - 2x$ and obtain $y = 1 - 2(2) = 1 - 4 = -3$.

Thus, the unique point of intersection is $(2, -3)$. The graph of these two lines is shown in Figure 10.

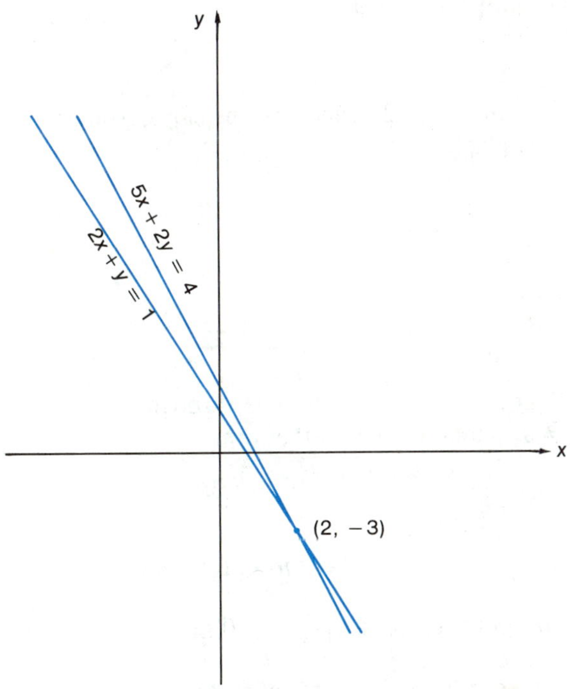

Figure 10

EXERCISES 1.3

In Problems 1 through 10, find an equation of the line through the two points and sketch its graph.

1. $(2, 3)$, $(4, 7)$

2. $(3, 5)$, $(-2, 4)$

3. $(4, -1)$, $(3, 6)$

4. $(2, -3)$, $(4, -2)$

5. $(-1, 0)$, $(-4, -2)$

6. $(1, 2)$, $(-3, -4)$

7. $(-4, 0)$ $(0, -4)$

8. $\left(\frac{1}{2}, 2\right), \left(3, -\frac{3}{2}\right)$

9. $(\sqrt{2}, 1)$, $(3 + \sqrt{2}, 5)$

10. $(1 - \sqrt{3}, 0)$, $(1 + \sqrt{3}, 3\sqrt{3})$

In Problems 11 through 16, find an equation of the line through the point with slope m. Sketch the graph.

11. $(3, -4)$; $m = 2$

12. $\left(-2, \frac{1}{2}\right)$; $m = -3$

13. $(1, -2)$; $m = 0$

14. $\left(-\frac{1}{3}, 3\right)$; $m = \frac{1}{2}$

15. $(2, -1)$ and there is no slope

16. $(-1, \sqrt{2})$; $m = \frac{1}{\sqrt{2}}$

In Problems 17 through 22, determine the slope, y-intercept, and x-intercept of the line. Sketch the graph.

17. $x + 2y - 4 = 0$

18. $x = 3y - 10$

19. $3x + 7 = 2y$

20. $2x = -4y + 5$

21. $x = 5$

22. $y = -5$

23. Show that an equation of the line through points $(a, 0)$ and $(0, b)$, with $a \neq 0$ and $b \neq 0$, can be written in the form

$$\frac{x}{a} + \frac{y}{b} = 1$$

This is called the *intercept form* for the equation of a line.

24. Write the equation $3x + 4y - 5 = 0$ in intercept form.

25. The graph of a linear function has slope $m = 3$. If $P(-2, 3)$ and $Q(1, c)$ are points on the graph, find c.

26. Repeat Problem 25 with $m = -2$ and the points $P(c, -1)$ and $Q(3, c)$.

27. Find a real number c such that the point $P(2, -3)$ is on the line $2x + cy - 4 = 0$.

28. Find a real number c such that the line $3x - cy - 2 = 0$ has y-intercept $= -4$.

29. (a) Write an equation whose graph is the x-axis.
(b) Write an equation whose graph is the y-axis.

30. Write an equation whose graph consists of all points with an ordinate of -4.

31. Write an equation whose graph consists of all points with an abscissa of 5.

In Problems 32 through 37, find the slope of each line that is: (a) parallel to the line; (b) perpendicular to the line.

32. $2x - 7y + 10 = 0$

33. $x = 4y - 7$

34. $2x = -4y + 1$

35. $-y = x$

36. $x = 3$

37. $y = 2$

38. Find an equation of the line passing through the point $(1, 1)$ and parallel to
(a) The line in problem 32.
(b) The line in problem 34.
(c) The line in problem 36.
(d) The line in problem 37.

39. Find an equation of the line passing through the point $(1, 1)$ and perpendicular to
(a) The line in problem 32.
(b) The line in problem 34.
(c) The line in problem 36.
(d) The line in problem 37.

40. A line passes through the points $(2, c)$ and $(1, -3)$. Find the number c so that the line: (a) has slope 3; (b) has slope zero; (c) is parallel to the line $3x + y + 6 = 0$; (d) is perpendicular to the line $x - y + 1 = 0$.

In Problems 41 through 45, prove that the three points are collinear; use (a) slopes; (b) the distance formula.

41. $(0, 1)$, $(1, 3)$, and $(-1, -1)$

42. $(1, 2)$, $(-1, 4)$, and $(2, 1)$

43. $(1, 2)$, $(0, -3)$, and $(-1, -8)$

44. $\left(1, \frac{1}{2}\right)$, $(2, 0)$, and $\left(\frac{1}{2}, \frac{3}{4}\right)$

45. $\left(0, -\frac{1}{2}\right)$, $(1, -2)$, and $\left(2, -\frac{7}{2}\right)$

46. Prove that the triangle with vertices at $A(1, 1)$, $B(-1, 4)$, and $C(5, 8)$ is a right triangle; use (a) the converse of the theorem of Pythagoras; (b) slopes.

47. Prove that the four points $A(-4, -1)$, $B(1, 2)$, $C(3, 1)$, and $D(-2, -2)$ are the vertices of a parallelogram. [*Hint*: Opposite sides are parallel.]

48. Prove that the four points $A(-10, 9)$, $B(-2, 24)$, $C(5, 1)$, and $D(13, 16)$ are the vertices of a square.

49. Find an equation of the line that is the perpendicular bisector of the line segment with endpoints $A(2, 3)$ and $B(-1, 5)$.

50. Find a relationship between x and y such that the point (x, y) is equidistant from the points $A(2, 3)$ and $B(-1, 5)$. (Compare with Problem 49 above.)

In problems 51 through 60, determine the point of intersection of the given lines.

51. $x - 2y = 3, \quad x + y + 4 = 0$

52. $2x + y - 5 = 0, \quad x - y + 16 = 0$

53. $3x - y + 2 = 0, \quad 2x - 3y + 1 = 0$

54. $4x - 2y + 7 = 0, \quad x + y + 6 = 0$

55. $-x + 5y + 2 = 0, \quad 3x + y + 1 = 0$

56. $y = 5x - 6, \quad y = \frac{2}{5}x + 4$

57. $y = 16x + 2, \quad 3y = 4x - 2$

58. $4x - 6y = 3, \quad x + 3y = 6$

59. $x = 4y - 3, \quad x + 4y = 7$

60. $2x + 7y = 8, \quad 3x - 2y = 4$

1.4 FUNCTIONS

In the seventeenth century, Leibniz[1], one of the developers of the calculus, introduced the word *function* into the mathematical vocabulary. The concept of function is undoubtedly one of the most basic in all of mathematics.

Intuitively, we call a quantity y a function of another quantity x if there is a rule (method, procedure) by which a unique value of y is associated with a corresponding value of x. For example, the equation

$$y = 3x - 1$$

determines a function, because if we replace x with various numbers, we get a corresponding unique value of y. For instance,

$$\text{if } x = 0, \text{ then } y = 3(0) - 1 = -1$$

$$\text{if } x = 2, \text{ then } y = 3(2) - 1 = 5$$

We observe that for each x that "goes into" (input) in this equation, only one value of y "comes out" (output). Other examples of functions are:

$$y = 7x^2 - 2x + 5$$

$$y = \sqrt{x^2 + 25}$$

$$y = -(x^2 + 2) + \sqrt{1 - x^2}$$

In each of these examples, we have a rule (in fact, formula) which tells us how to find the number y for a given x.

In general, it is not necessary that a function be defined by a formula relating x and y. For example, federal income tax (y) is a function of income (x), and the amount (y) of postage required to mail a letter is a function of its weight (x). In these examples

---

[1] GOTTFRIED WILHELM LEIBNIZ (1646-1716), philosopher and metaphysician, was born in Leipzig, Germany, the son of a professor of philosophy. Leibniz is considered one of the greatest thinkers of modern times, proficient in law, theology, geology, and history, as well as serving as a diplomat. Leibniz wrote with ease in five different languages: Greek, Latin, French, English, and his native German. He received his bachelor's degree in philosophy at age seventeen and had a doctorate in law when he was twenty. During his lifetime, he invented one of the first calculating machines and founded the Academy of Sciences in Berlin.

In mathematics, Leibniz contributed immensely to the development of the calculus and symbolic logic. In spite of his accomplishments, his life ended sadly; he was troubled in his final years by ill health and an ongoing bitter dispute with Newton over priority for the discovery of the calculus.

there is no equation relating x and y, but the relationships are given by tables.

We now have a formal definition of function.

Definition. A *function* f from a set X to a set Y, denoted $f: X \to Y$, is a correspondence that associates with each $x \in X$ a unique $y \in Y$. The element y is called the *image* of x under f and is denoted by $f(x)$. The set X is called the *domain* of f. The set of all images of elements of X is called the *range* of f.

We usually write $y = f(x)$, read "y equals f of x," to denote that y is a function of x. This notation is due to the Swiss mathematician Leonhard Euler.[2]

We call x the *independent variable* because x can be selected arbitrarily from the domain set X. We call y the *dependent variable* because the value of y depends upon the value of x selected. We sometimes represent the correspondence pictorially as shown in Figure 1.

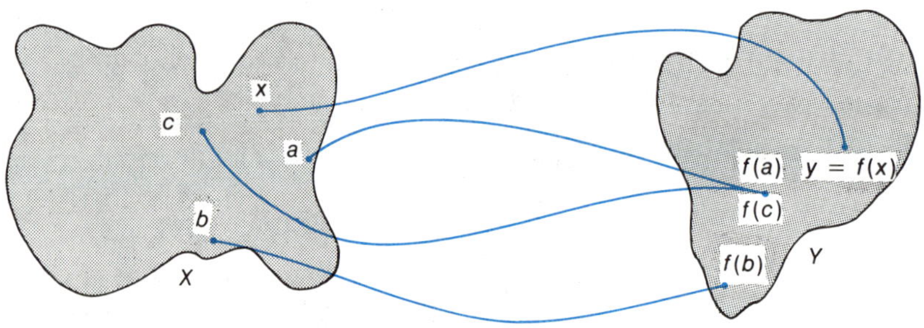

Figure 1

The sets X and Y need not be distinct. In fact, we often have $X = Y$. In particular, we shall deal with the sets X and Y where X and Y are subsets of the real numbers. The letters f, g, h, and F, G, H are usually used to denote functions.

Let us repeat for emphasis: To each $x \in X$ there corresponds precisely one element $f(x) = y \in Y$; however, different elements in X, such as a and c in Figure 1, may have the same image in Y.

[2] LEONHARD EULER (1701-1783) was a Swiss mathematician, receiving much of his early training from his father, who had studied mathematics with Jacob Bernoulli. Euler received his university education at Basel, majoring in theology. He received his Master's degree at age seventeen and wrote his first research paper two years later. Euler became one of the most prolific mathematicians of all times, writing more than 700 research papers in science.

During his lifetime, Euler was appointed to the Chair of Mathematics at the Academy of Sciences in St. Petersburg, Russia, and served as the Director of the Department of Mathematics in Berlin. In 1735, he lost the sight of one eye and became totally blind some years before his death. The handicap notwithstanding, Euler continued to work to the end, calculating the orbit for the planet Uranus on the day he died.

Two functions f and g which have the same domain are said to be *equal*, denoted by $f = g$, if $f(x) = g(x)$ for all x in their domain.

DOMAIN AGREEMENT

Before we give examples of *numerical functions*, that is, functions whose domain and range are each subsets of real numbers, we agree that if a function is given by a formula, and the domain is not specified, then we shall assume that the domain consists of all real numbers for which the formual is meaningful.

EXAMPLE 1. Find the domain and range of the function defined by

$$y = f(x) = \frac{x}{x-3} \tag{1}$$

SOLUTION. Since the expression $\dfrac{x}{x-3}$ is meaningful for every real number x, except $x = 3$ (why?), by our agreement the domain D of f is the set

$$D = \{x \mid x \in R \quad \text{and} \quad x \neq 3\}.$$

By direct inspection of (1) it is not easy to find the range of f. One way to find the range of f is to solve Equation (1) for x in terms of y.

$$\begin{aligned} y &= \frac{x}{x-3} \\ y(x-3) &= x \\ xy - 3y &= x \\ xy - x &= 3y \\ x(y-1) &= 3y \end{aligned}$$

or

$$x = \frac{3y}{y-1} \tag{2}$$

From (2) we see that y can take on any real number except 1. Therefore, the range of f defined by (1) is given by

$$\text{range of } f = \{y \mid y \text{ is a real number and } y \neq 1\}$$

The domain of the function f defined by

$$f(x) = \sqrt{2+x} \tag{3}$$

is understood to be

$$D = \{x \mid 2 + x \geq 0\} = \{x \mid x \geq -2\}$$

since for a real number b, $\sqrt{b}$ is real if and only if $b \geq 0$. However, if a certain set of values for the independent variable x is specified, such as

$$g(x) = \sqrt{2+x} \qquad 3 \leq x \leq 9 \tag{4}$$

then the domain of the function g is the closed interval [3, 9]. We note that the functions f and g defined by (3) and (4) are not equal, since they have different domains.

GRAPH OF A FUNCTION

The *graph* of a function f is the graph of the equation $y = f(x)$. In other words, the graph of f is the set of all points $(x, f(x))$ in the plane as x varies over the domain of f.

Let us consider the function f defined by

$$f(x) = x^2 - 2x + 1$$

The domain of f is assumed to be (by agreement stated earlier) the set of all real numbers. So any real number can be substituted for x. For example,

$$\begin{aligned} f(2) &= (2)^2 - 2(2) + 1 = 4 - 4 + 1 = 1 \\ f(-3) &= (-3)^2 - 2(-3) + 1 = 9 + 6 + 1 = 16 \\ f(0) &= (0)^2 - 2(0) + 1 = 0 - 0 + 1 = 1 \\ f(a) &= a^2 - 2a + 1 \\ f(a+h) &= (a+h)^2 - 2(a+h) + 1 \\ f(3c) &= (3c)^2 - 2(3c) + 1 \end{aligned}$$

and so on

EXAMPLE 2. Let f be a function defined by

$$f(x) = x^2 - 2, \qquad x \in R \tag{5}$$

If $a, h \in R$, where $h \neq 0$, find

(a) $\dfrac{f(a+h) - f(a)}{h}$

(b) Sketch the graph of f.

SOLUTION.

(a) We have

$$f(a+h) = (a+h)^2 - 2 \quad \text{and} \quad f(a) = a^2 - 2$$

Hence

$$\begin{aligned} \frac{f(a+h) - f(a)}{h} &= \frac{(a+h)^2 - 2 - (a^2 - 2)}{h} \\ &= \frac{a^2 + 2ah + h^2 - 2 - a^2 + 2}{h} \\ &= \frac{2ah + h^2}{h} = 2a + h \end{aligned}$$

(b) To sketch the graph of f, we list coordinates $(x, f(x))$ of some points on the graph of f in tabular form as follows.

x	-3	-2	-1	0	1	2	3
$f(x)$	7	2	-1	-2	-1	2	7

We then draw a smooth curve through these points, obtaining the sketch shown in Figure 2.

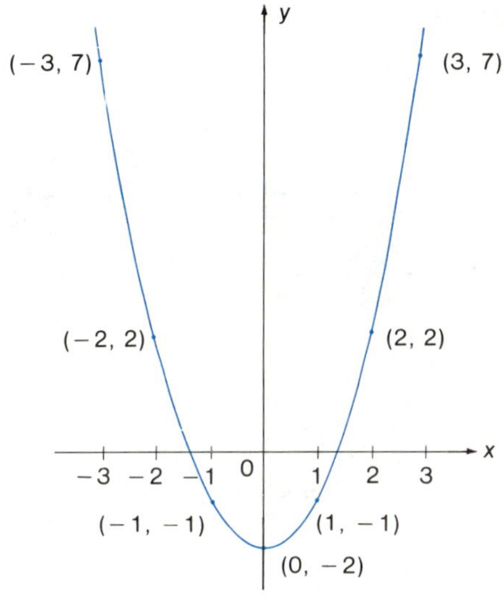

Figure 2

EXAMPLE 3. For the *identity function* defined by $f(x) = x$, find $f(-2), f(-1), f(0), f(1), f(2)$. Describe the domain and range of f. Sketch the graph of f.

SOLUTION. $f(-2) = -2, f(-1) = -1, f(0) = 0, f(1) = 1$, and $f(2) = 2$

$$\text{domain } f = \{x \mid x \in R\} \qquad \text{range } f = \{y \mid y \in R\}$$

The graph of f is a straight line shown in Figure 3.

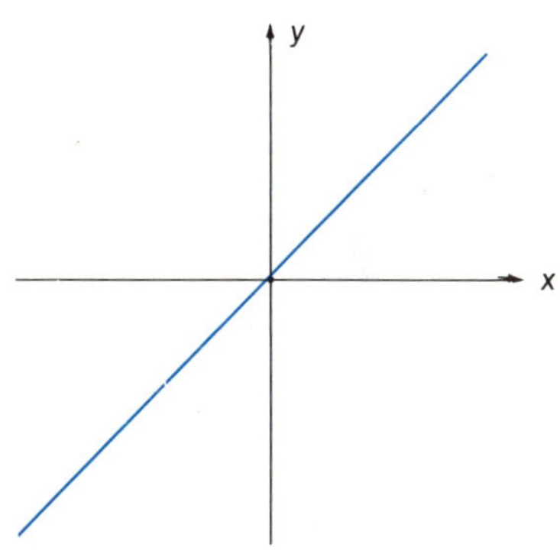

Figure 3

EXAMPLE 4. The graphs of the *constant functions* $f(x) = 1$, $f(x) = 2$, and $f(x) = -1$ are shown in Figure 4. The domain of a constant function $f(x) = c$, where c is a constant, is the set R of real numbers and its range is $\{c\}$. For example, if $f(x) = c$, then $f(3) = c, f(-2) = c, f(0) = c, f(-108) = c, f(139) = c$, etc.

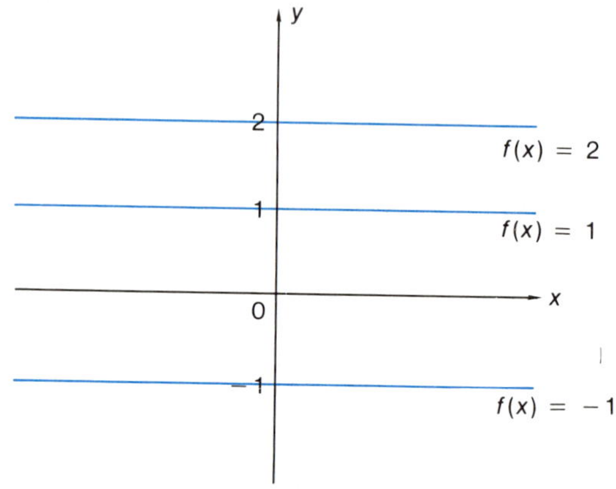

Figure 4

EXAMPLE 5. For the *absolute value function*, $f(x) = |x|$, find $f(-2)$, $f(-1)$, $f(0)$, $f(1)$, and $f(2)$. Describe the domain and range of f. Sketch its graph.

SOLUTION. By defintion of the absolute value, we have

$$f(x) = \begin{cases} x & \text{if } x \geq 0 \\ -x & \text{if } x < 0 \end{cases}$$

so that $f(-2) = -(-2) = 2, f(-1) = 1, f(0) = 0, f(1) = 1$, and $f(2) = 2$. The domain of f is R; the range of f is all non-negative real numbers. The graph of f is shown in Figure 5.

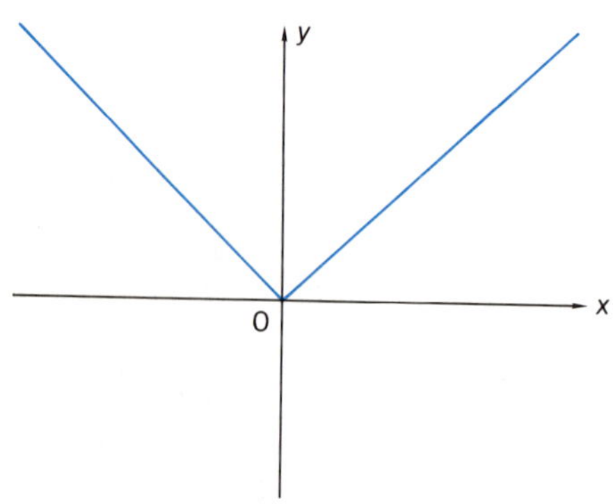

Figure 5

EXAMPLE 6. For the *greatest integer function*

$$f(x) = [\![x]\!]$$

where $[\![x]\!]$ denotes the greatest integer less than or equal to x, find $f(3.5), f(3.4), f(3)$, $f(-2), f(-2.5), f(0.5)$, and $f(-0.5)$. Describe the domain and range of f. Sketch the graph of f.

SOLUTION. $f(3.5) = [\![3.5]\!] = 3, f(3.4) = 3, f(3) = 3, f(-2) = -2, f(-2.5) = -3$, $f(0.5) = 0$, and $f(-0.5) = -1$. The domain of f is R and its range is the set I of integers. The graph of f is shown in Figure 6.

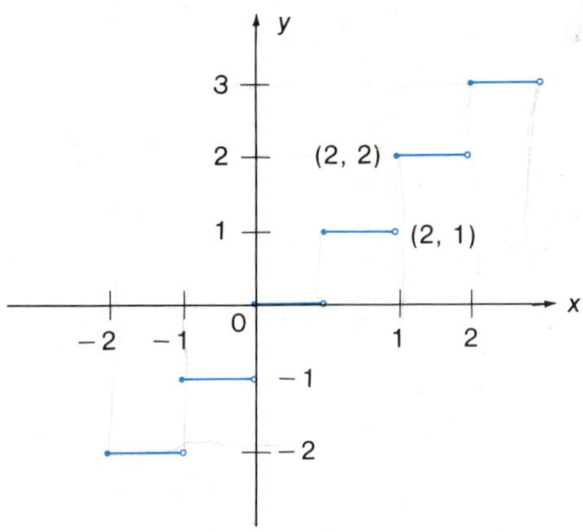

Figure 6

In the figure, the point (2, 1) with empty circle means that this point is not included in the graph, whereas the point (2,2) with full circle means that this point is included in the graph.

In business applications a function often has to be given by using more than one formula.

EXAMPLE 7. Annette is allowed to work at CDB Pizza for a maximum of 60 hours per week. For the first 40 hours she is paid \$4.00 per hour, and for the next 20 hours she is paid overtime wages at the rate of \$6.00 per hour. What are her weekly wages, $W(x)$, as a function of the number of hours x she works per week?

SOLUTION. If Annette works for x hours at the rate of k dollars per hour, then her wages will be kx dollars. Hence, $W(x)$ is given by

$$W(x) = \begin{cases} 4x & \text{if} \quad 0 \le x \le 40 \\ 160 + 6(x - 40) & \text{if} \quad 40 < x \le 60 \end{cases}$$

If she works for, say, 30 hours, then the first formula describes her wages: $4(30) = 120$ dollars. If she works for 50 hours, then the second formula describes her wages as: $160 + 6(50 - 40) = 160 + 6(10) = 160 + 60 = 220$ dollars.

The domain of this function is the closed interval $[0, 60]$, while the range is the closed interval $[0, 280]$.

We shall now describe some of the frequently used functions in economics.

Demand. It is generally true that during a fixed period of time the quantity of a good or service that the customers will buy depends upon the price charged. Usually, as the price of an item increases, the quantity demanded decreases. Thus, we see that the quantity demanded of a commodity is a function of the selling price. If p is the selling price per unit, and x is the numbers of units which can be sold at price p, we can write the *demand schedule* or *demand function* as

$$x = f(p) \tag{6}$$

Of course, the demand for a commodity depends upon many variables besides the selling price, for instance, the general status of the economy, the weather conditions (for seasonal items), supply of the item, consumer tastes, and the price of related goods and services. We shall assume that all such factors remain constant. Economists refer to holding other factors fixed as *ceteris paribus* assumptions. So, a *demand function* is any function that relates p, the price per unit of a commodity, with x, the number of units demanded by the consumer (that is, the number of units that the consumer is willing to buy).

When we solve equation (6) for p in terms of x, we may obtain

$$p = g(x) \tag{7}$$

In this equation, we have the price p (per unit) as a function of the quantity demanded x. Hence, both equations (6) and (7) are referred to as the *demand functions.*

Total Revenue. The total revenue is the total amount of money a company makes by selling x units of a commodity. So if $p(x)$ is the selling price per unit when x units are sold, then the total revenue is a function of x. Hence, if $R(x)$ denotes this total revenue, then

$$\begin{aligned} R(x) &= \text{(quantity sold)(price per unit)} \\ &= x \cdot p(x) \end{aligned}$$

Supply Function. The amount of goods that a company (producer) is willing to supply to a market depends upon the price that the market assigns to the goods. The higher the price per unit of the goods, the more goods the producer is willing to supply to the market. Thus, the supply is a function of the price p per unit. We denote the supply function by $S(p)$.

Equilibrium. When demand is equal to supply, we say that the market is in equilibrium. So, if $S(p)$ is the supply function, and $x(p)$ is the demand function, then the market is in equilbrium when

$$S(p) = x(p)$$

since at that price p (per unit) every item supplied is purchased.

Profit. Suppose a company produces x items of a product. We let $R(x)$ denote the total revenue received when x units are sold and $C(x)$ denote the total cost incurred in the production of x units ($C(x)$ is called the *cost function*). Then the profit P is given by

$$P(x) = R(x) - C(x), \qquad x \geq 0$$

and is called the *profit function*.

Break-Even Point. Typically, when no units are produced ($x = 0$), then $R(x) = xp(x) = 0p(0) = 0$. However, $C(0)$, the total cost of producing zero units, may still be positive. This may be due to the presence of *fixed* costs such as rents, insurance, development, etc. So, $P(0)$, the profit where zero units are produced, may be 0 or negative (loss). We say that a break-even point is reached when $P(x) = 0$; that is, the number of units x necessary to produce and sell, such that the set up costs are recovered and the product becomes profitable.

Polynomial Functions. The simplest types of functions that we shall consider extensively in calculus will be the polynomial functions. If $a_0, a_1, a_2, \ldots, a_n$ are fixed real numbers and $a_0 \neq 0$, then the function f given by

$$f(x) = a_0x^n + a_1x^{n-1} + \cdots a_{n-1}x + a_n$$

is called a *polynomial function in x of degree n*. Some of these polynomial functions are worth singling out because they occur so frequently.

Linear Functions. Let m and b be real numbers; then the function f defined by

$$f(x) = mx + b$$

is called a *linear function*. The graph of $f(x) = mx + b$ is a straight line with slope m and y-intercept b. If $m = 0$, then the function $f(x) = b$ is called the *constant function*. The graph of $f(x) = b$ is a horizontal line at height b.

When we say that in a particular situation *a linear model applies*, we mean that the function defined is a linear function.

Economics is one area in which linear functions have useful applications. Let us consider a manufacturer producing x units of an item. Suppose that the total cost $y = C(x)$ is a linear function of the number of units produced. We shall assume that the graph of the function relating y to x is given by Figure 7.

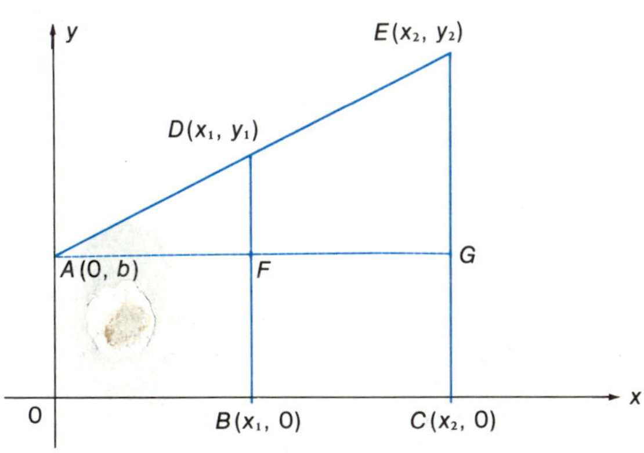

Figure 7

We interpret the graph as follows. If x_1 units are manufactured, then the total cost is y_1, which is the length of the line segment *BD*. Similarly, y_2, the length of the line segment *CE*, is the total cost of producing x_2 units of the item. If zero units of the item are produced, the cost is *b*; this is called the *fixed cost*.

The fixed cost of producing x_2 units is given by the line segment *CG* whose length is *b*, and the line segment *GE* represents the *variable cost*. Variable cost is the change in cost as the number of units produced changes. Thus, the variable cost of making x_1 units is given by the line segment *FD*.

We now observe that the slope of the line segment *AE* is given by

$$m = \frac{y_2 - b}{x_2 - 0} = \text{change in total cost when one more unit is produced.}$$

Economists refer to this as the *marginal cost*. When the total cost is linearly related to output, the marginal cost is constant.

EXAMPLE 8. In Figure 8 the line *L* is the graph of a function relating the total manufacturing cost *y* and the number of units *x* of a product manufactured. Find the following:

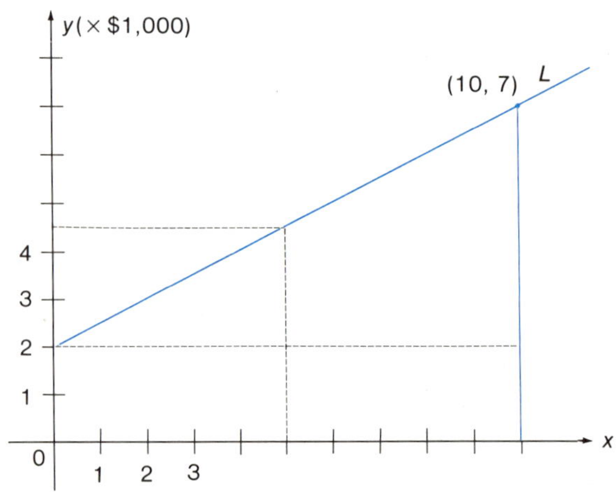

Figure 8

(a) The fixed cost

(b) The total cost when 10 units are produced

(c) The variable cost when 10 units are produced

(d) The marginal cost

(e) The variable cost when 5 units are produced

SOLUTION.

(a) Fixed cost = \$2000

(b) Total cost when 10 units are produced is \$7000

(c) Variable cost when 10 units are produced is total cost minus fixed cost:

$$\$7000 - \$2000 = \$5000$$

(d) Marginal cost $=$ slope of $L = \dfrac{7-2}{10-0} = \dfrac{1}{2}$

(e) First we find the total cost for any x; that is, we find the equation of the line L. Using the slope-intercept form, we have

$$y = mx + b = \tfrac{1}{2}x + 2$$

Setting $x = 5$, we obtain

$$y = \tfrac{1}{2}(5) + 2 = 4.5$$

In other words, the total cost is $4.5 \times \$1000 = \$4500$. Therefore, the variable cost when 5 units are produced is

$$(\text{total cost}) - (\text{fixed cost}) = \$4500 - \$2000 = \$2500$$

EXAMPLE 9. A company expects the variable cost to be \$25,000 at a revenue of \$40,000 with a fixed cost of \$12,000.

Assuming that the linear model applies, find the following.

(a) An equation relating total cost and revenue

(b) The *break-even point*, that is, the volume at which the company experiences neither loss nor profit

(c) The net profit if sales should turn out to be \$50,000

SOLUTION.

(a) First we determine the marginal cost:

$$\text{Marginal cost} = \frac{\text{variable cost}}{\text{sales volume}} = \frac{25{,}000}{40{,}000} = \frac{5}{8}$$

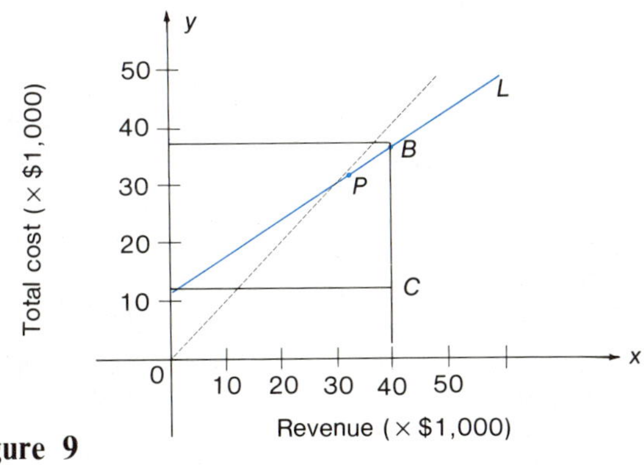

Figure 9

This result represents the slope of the line segment AB (see Figure 9). Using the slope-intercept form of the equation of the line L_1, we have the total cost

$$y = \tfrac{5}{8}x + b \tag{8}$$

where b represents the fixed cost (y-intercept). Hence

$$y = \tfrac{5}{8}x + 12{,}000 \tag{9}$$

is the equation relating total cost and sales.

(b) To determine the break-even point, we are asked to find a point on L where total cost equals revenue, that is, where $y = x$. In other words, we are asked to solve simultaneously the equations

$$y = \tfrac{5}{8}x + 12{,}000 \tag{10}$$

$$y = x \tag{11}$$

Subtracting (10) from (11), we obtain

$$\tfrac{3}{8}x - 12{,}000 = 0$$

from which we can calculate that the break-even level of revenue is

$$x = 32{,}000$$

so

$$y = 32{,}000$$

This result tells us that, when the revenue is at \$32,000, the total cost to the company is also \$32,000. The company's revenue must be above \$32,000 to bring in a profit. If the revenue is less than \$32,000, the company will have a loss.

(c) Setting $x = 50{,}000$ in Equation (9), we obtain the total cost:

$$y = \tfrac{5}{8}(50{,}000) + 12{,}000 = 31{,}250 + 12{,}000 = 43{,}250$$

The net profit (before taxes) is

$$\text{net profit} = (\text{revenue}) - (\text{cost}) = \$50{,}000 - \$43{,}250 = \$6750$$

Quadratic Functions. Let a, b, and c be real numbers with $a \neq 0$; then the function f defined by

$$f(x) = ax^2 + bx + c$$

is called a *quadratic function*. The graph of this function is a parabola which opens upward if $a > 0$ and opens downward if $a < 0$.

EXERCISES 1.4

In Problems 1 through 10, find $f(0), f(1), f(-3), f(5), f(a), f(a^2), f(\frac{1}{a}), f(a+h)$.

1. $f(x) = 6$

2. $f(x) = x + 3$

3. $f(x) = 2x - 1$

4. $f(x) = 1/x$

5. $f(x) = x - x^2$

6. $f(x) = \sqrt{x}$

7. $f(x) = \sqrt{1-x}$

8. $f(t) = \dfrac{1}{t+2}$

9. $f(x) = \dfrac{1}{\sqrt{x}}$

10. $f(x) = \dfrac{x-1}{x^2+1}$

11. For $f(x) = \begin{cases} 2x+1, & \text{if } x < 0 \\ x^2, & \text{if } x \geq 0 \end{cases}$

find $f(-2), f(0)$, and $f(4)$

12. For $f(x) = \begin{cases} \dfrac{1}{x+4}, & \text{if } -3 \leq x < 5 \\ x^2+3, & \text{if } 5 \leq x \leq 16 \end{cases}$

find $f(-1), f(0), f(6)$, and $f(17)$

In Problems 13 through 22, find the domain of each function.

13. $f(x) = 3x + 100$

14. $f(x) = \dfrac{1}{x}$

15. $f(x) = \dfrac{1}{x^2+9}$

16. $f(x) = \dfrac{1}{(x-1)(x-2)}$

17. $f(x) = \sqrt{x-1}$

18. $f(x) = \sqrt[3]{x-1}$

19. $f(x) = \sqrt{\dfrac{x}{x+2}}$

20. $f(x) = \dfrac{\sqrt{1+x}}{\sqrt{1-x^2}}$

21. $f(x) = \sqrt{x^2 - 6x + 8}$

22. $f(x) = \dfrac{3}{\sqrt{x^2+3x+2}}$

23. For the function f defined by $f(x) = x - |x|$, find $f(-3). f(-1), f(0), f(1)$, and $f(3)$. What is the domain and range of f? Sketch the graph of f.

24. If $g(x) = 3x + 8$ defines a function g, find the number x in the domain of g such that

(a) $g(x) = 2$ (b) $g(x) = -3$ (c) $g(x) = 0$

25. Repeat Problem 24 for $g(x) = x^2 - 1$.

In Problems 26 through 31, find the domain and range of the given function and sketch its graph.

26. $g(x) = [\![2x]\!]$

27. $f(x) = x - [\![x]\!]$

28. $g(x) = 2[\![x]\!]$

29. $f(x) = |x| + |x+1|$

30. $f(x) = \begin{cases} x & \text{if } x \le 2 \\ 2 & \text{if } x > 2 \end{cases}$

31. $f(x) = \begin{cases} -x+1 & \text{if } x \le -3 \\ -2 & \text{if } -3 < x < -1 \\ x^2 & \text{if } x > -1 \end{cases}$

32. In Figure 10 the line L is the graph of a function that relates the total manufacturing cost y and the number of units x of a product manufactured. Find the following:

Figure 10

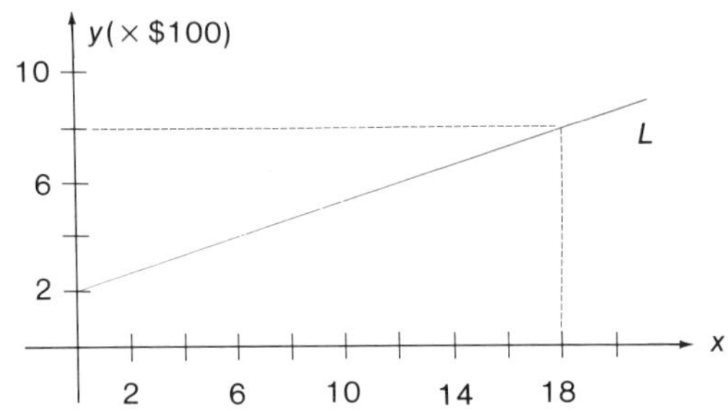

(a) The total cost when 20 units are produced
(b) The fixed cost
(c) The variable cost when 20 units are produced
(d) The marginal cost
(e) The variable cost when 7 units are produced

33. The total cost changes from \$500 to \$800 as the number of units manufactured increases from 300 to 900. Assuming that the total cost y is a linear function of x (the total number of units manufactured), determine

(a) The fixed cost
(b) The marginal cost
(c) The total cost when 500 units are manufactured
(d) The variable cost when 500 units are manufactured

34. On an average weekday in the bustling city of Garbville, the pollution index p starts to rise at 8:00 A.M. At 10:00 A.M., $p = 30$ and at 2:00 P.M., $p = 50$. Assume that p is a linear function of time between 8:00 A.M. and 6:00 P.M.

(a) Find an equation relating p and t.
(b) Sketch the graph of the line.
(c) Find p at 8:00 A.M., 12:00 NOON, and 6:00 P.M.
(d) What is the rate of increase of p with respect to time (that is, what is the slope of the line)?

35. Suppose that every night the pollution index at Garbville remains constant between 6:00 P.M. and 8:00 P.M. and then drops linearly to a fixed level at 8:00 A.M.

(a) Find the equation of the line relating p and t between 8:00 P.M. and 8:00 A.M.
(b) Calculate p at 12:00 midnight, at 2:00 A.M., and at 4:00 A.M.
(c) What is the rate of decrease (slope) of p with respect to t?

36. A demand function is given by $p(x) = 16 - (0.5)x$ dollars. What is the total revenue from the sale of (a) 2 units (b) 4 units.

37. A demand function is known to be linear. The price for 2 units is \$14.00 a unit, and the price for 6 units is \$6.00 a unit. What is the total revenue when 8 units are sold?

38. Suppose that the equation

$$y = \tfrac{1}{3}x + 24{,}000$$

relates the total cost y and the sales volume x. Find the following.

(a) The fixed cost
(b) The variable cost on sales of \$50,000
(c) The break-even point
(d) Net profit (before taxes) on sales of \$40,000

39. Given that the freezing point of water is measured at 0°C (Celcius) and 32°F (Fahrenheit), and the boiling point of water is measured at 100°C and 212°F,

(a) Find C as a linear function of F.
(b) Find F as a linear function of C.
(c) At what temperatures will °C = °F?
(d) Sketch a graph of each function and indicate the point of intersection.

1.5 OPERATIONS ON FUNCTIONS

We now consider the operations of addition, subtraction, multiplication, and division on functions.

Let f and g be two functions, and let D_f and D_g denote the domains of f and g, respectively.

The sum of f and g, denoted by $f+g$, is defined by

$$(f+g)(x) = f(x) + g(x) \quad \text{and} \quad x \in D_f \cap D_g$$

The difference of f and g, denoted by $f-g$, is defined by

$$(f-g)(x) = f(x) - g(x) \quad \text{and} \quad x \in D_f \cap D_g$$

The product of f and g, denote by fg, is defined by

$$(fg)(x) = f(x)g(x) \quad \text{and} \quad x \in D_f \cap D_g$$

The quotient of f by g, denoted by $\dfrac{f}{g}$, is defined by

$$\left(\frac{f}{g}\right)(x) = \frac{f(x)}{g(x)};\ g(x) \neq 0 \quad \text{and} \quad x \in D_f \cap D_g$$

EXAMPLE 1. Let $f(x) = 3x + 4$ and $g(x) = 3x^2 - 12$. Find $f+g$, $f-g$, fg, $\dfrac{f}{g}$, and $\dfrac{g}{f}$. Also find the domain of each of these functions.

SOLUTION. The domain of f and g is the set R of real numbers. Thus the domains of $f+g$, $f-g$, and fg is also R.

$$(f+g)(x) = f(x) + g(x) = (3x+4) + (3x^2 - 12) = 3x^2 + 3x - 8$$
$$(f-g)(x) = f(x) - g(x) = (3x+4) - (3x^2 - 12) = -3x^2 + 3x + 16$$

$$(fg)(x) = f(x)g(x) = (3x+4)(3x^2-12) = 9x^3+12x^2-36x-48.$$

Now $g(x) = 0$ if $3x^2 - 12 = 0$ or $3(x^2-4)=0$ or $3(x-2)(x+2)=0$ or $x=2$ or -2.

Hence the domain of $\dfrac{f}{g}$ is the set R execpt 2 and -2.

$$\left(\frac{f}{g}\right)(x) = \frac{f(x)}{g(x)} = \frac{3x+4}{3x^2-12}$$

Similarly, $f(x) = 0$ if $3x+4 = 0$ or $x = -\dfrac{4}{3}$. Hence the domain of $\dfrac{g}{f}$ is the set R except $-\dfrac{4}{3}$.

$$\left(\frac{g}{f}\right)(x) = \frac{g(x)}{f(x)} = \frac{3x^2-12}{3x+4}$$

Another useful operation on functions is the operation of composition. If f and g are two functions, the *composite functions* $f \circ g$ and $g \circ f$ have values given by (see Figures 1 and 2)

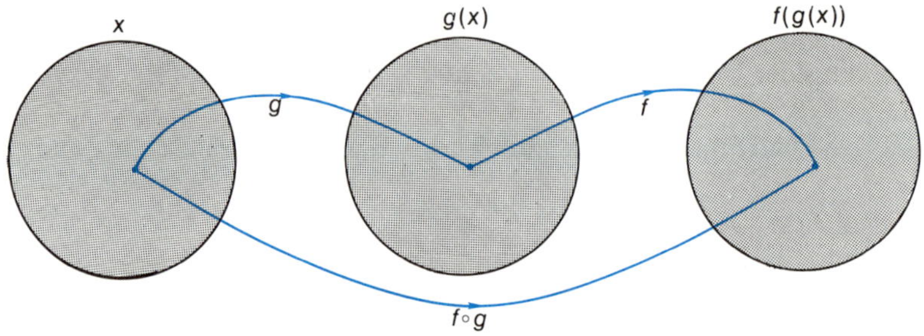

Figure 1

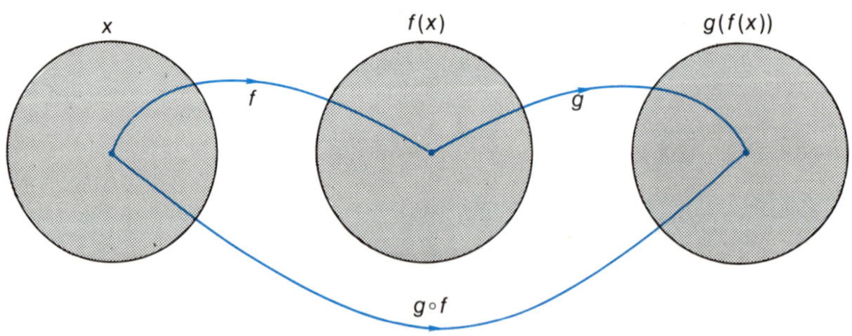

Figure 2

$$(f \circ g)(x) = f(g(x)); \qquad \text{domain of } f \circ g = \{x \mid x \in D_g,\ g(x) \in D_f\} \tag{1}$$

and

$$(g \circ f)(x) = g(f(x)); \qquad \text{domain of } g \circ f = \{x \mid x \in D_f,\ f(x) \in D_g\} \tag{2}$$

EXAMPLE 2. Let $f(x) = 2x^2 + 4$, $g(x) = \sqrt{x-1}$. Determine $f \circ g$ and $g \circ f$.

SOLUTION. Here $D_f = R$ and $D_g = \{x \mid x \geq 1\}$. Using (1), we have, for $f \circ g$,

$$(f \circ g)(x) = f(g(x)) = f(\sqrt{x-1}) = 2(\sqrt{x-1})^2 + 4$$
$$= 2(x-1) + 4 = 2x + 2$$

and

$$D_{f \circ g} = \{x \mid x \in D_g \text{ and } g(x) \in D_f\}$$
$$= \{x \mid x \geq 1, \sqrt{x-1} \in R\} = \{x \mid x \geq 1\}.$$

For $g \circ f$, we have

$$(g \circ f)(x) = g(f(x)) = g(2x^2 + 4) = \sqrt{2x^2 + 4 - 1} = \sqrt{2x^2 + 3}$$

and

$$D_{g \circ f} = \{x \mid x \in D_f, f(x) \in D_g\}$$
$$= \{x \mid x \in R \text{ and } 2x^2 + 4 \geq 1\} = \{x \mid x \in R\}.$$

Note that $f \circ g \neq g \circ f$. For example, we find that

$$(f \circ g)(3) = f(g(3)) = f(\sqrt{2}) = 2(\sqrt{2})^2 + 4 = 8$$

and

$$(g \circ f)(3) = g(f(3)) = g(22) = \sqrt{22 - 1} = \sqrt{21}.$$

EXERCISES 1.5

In Problems 1 through 6, given the values of the functions f and g, describe the functions $f + g$, $f - g$, fg, and f/g.

1. $f(x) = 2x + 5,\ g(x) = x^2$

2. $f(x) = \sqrt{x},\ g(x) = x^2 + 1$

3. $f(x) = \sqrt{x^2 - 4},\ g(x) = x^2 + 2x - 3$

4. $f(x) = \dfrac{1}{x+1},\ g(x) = \dfrac{x}{2x-1}$

5. $f(x) = \dfrac{1}{\sqrt{x}},\ g(x) = \dfrac{1}{x^2}$

6. $f(x) = \sqrt[3]{x},\ g(x) = \sqrt{x^2 - 1}$

In Problems 7 through 14, describe $f \circ g$ and $g \circ f$ for the given functions f and g.

7. $f(x) = 2 + x,\ g(x) = 3x$

8. $f(x) = 3x^2 - 4,\ g(x) = 2 - x$

9. $f(x) = x^2 + 4,\ g(x) = \sqrt{x}$

10. $f(x) = \sqrt[3]{x + 1},\ g(x) = 2x^3 + 1$

11. $f(x) = \dfrac{1}{x+1},\ g(x) = x - 1$

12. $f(x) = \dfrac{x}{x^2 + 2},\ g(x) = \dfrac{1}{x}$

13. $f(x) = \sqrt{2x^2 + 3},\ g(x) = \dfrac{2}{x^2}$

14. $f(x) = x^{-1/2},\ g(x) = x - 2$

15. If $f(x) = 3x + 2$ and $g(x) = 2x^2$, find

(a) $(f + g)(2)$ (b) $(f - g)(6)$ (c) $(fg)(-1)$

(d) $(f/g)(\frac{1}{2})$ (e) $(f \circ g)(3)$ (f) $(g \circ f)(3)$

16. Repeat Problem 15 for $f(x) = \sqrt{x^2 - 16}$, $g(x) = \sqrt{x^2 + 4}$.

17. For each function f defined, determine $f \circ f$.

(a) $f(x) = \dfrac{1}{x}$ (b) $f(x) = 2 + 3x$

(c) $f(x) = \sqrt{4 + x}$ (d) $f(x) = \dfrac{1}{2 + x}$

(e) $f(x) = 2x + \dfrac{1}{x}$

18. If $f(x) = 3x - 1$, show that $f(3x) - 3f(x) = 2$.

19. If $f(x) = 2x^2$, show that $f(a + b) - f(a - b) = 8ab$.

20. If $f(x) = 2x - 1$ and $g(x) = 3x - 2$, show that $f \circ g = g \circ f$.

In Problems 21 through 24, let $f(x) = 2x + 1$, $g(x) = x + 4$ and $h(x) = x^2$; find the given functions.

21. $f\big(g(h(x))\big)$ **22.** $g\big(f(h(x))\big)$

23. $h\big(g(f(x))\big)$ **24.** $h\big(f(g(x))\big)$

1.6 LIMITS

The concept of the limit of a function is fundamental in the development of calculus. Roughly speaking, limits examine how a function $f(x)$ behaves as the values x in the domain get closer and closer to (but do not take on) a certain value c.

Both Newton and Leibniz considered the idea of limit, but their notion of the concept was vague and lacking in mathematical precision. It was the great mathematician Weierstrass[1] who finally gave the notion of limit a rigorous mathematical foundation.

[1] KARL W. T. WEIERSTRASS (1815-1897) was born in Ostenfelde, Germany, the eldest in a family of two brothers and two sisters. At 19 he entered the University of Bonn, where he indulged in fencing and beer drinking; he returned home after four years without a degree. At the age of 26, he took his teacher's certificate examinations and then taught in secondary schools until 1853.

Weierstrass is called the father of modern analysis. His early publications in an unknown high school paper went unnoticed. In 1854 he published a memoir on Abelian functions which created a sensation and won him an honorary doctorate from the University of Königsberg. He was elected to the Berlin Academy and was given a professorship at the University of Berlin, where he stayed until his death. His fame as an outstanding mathematician and an excellent teacher spread all over Europe and to America. He emphasized rigor and logic.

Consider the following examples:

EXAMPLE 1. Let the function f be defined by

$$f(x) = x + 2$$

The reader can determine the graph of f, which is given in Figure 1. Let us see what happens to $f(x)$ as x gets closer and closer to 2, as shown in the following tables.

x	1	1.5	1.75	1.80	1.90	1.97	1.99	1.9999
$f(x) = x + 2$	3	3.5	3.75	3.80	3.90	3.97	3.99	3.9999

Graph of $f(x) = x + 2$

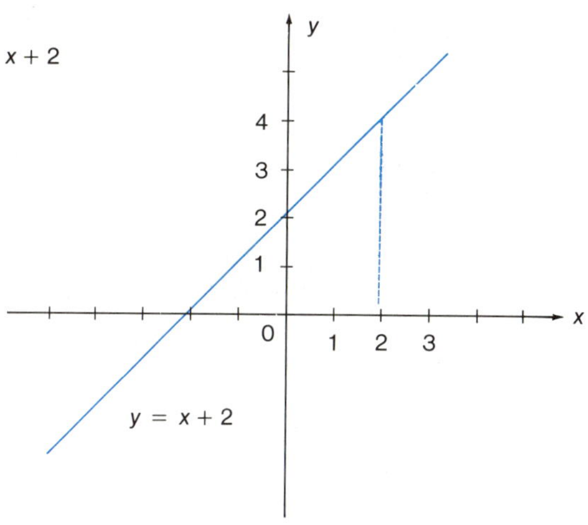

Figure 1

Obviously, as x approaches 2, $f(x)$ approaches 4. Of course, x could approach 2 through values greater than 2. Thus

x	3	2.50	2.40	2.30	2.20	2.10	2.05	2.001	2.0001
$f(x) = x + 2$	5	4.50	4.40	4.30	4.20	4.10	4.05	4.001	4.0001

Again $f(x)$ approaches 4 as x approaches 2 through values greater than 2. In the first case we say that $f(x)$ approaches 4 as x approaches 2 from the left (*Left-hand limit*), and we write

$$f(x) \to 4 \quad \text{as} \quad x \to 2^-$$

In the second case we say that $f(x)$ approaches 4 as x approaches 2 from the right (*Right-hand limit*), and we write

$$f(x) \to 4 \quad \text{as} \quad x \to 2^+$$

Therefore, the function $f(x)$ approaches 4 as x approaches 2 from the left $(x \to 2^-)$ or from the right $(x \to 2^+)$. In such a situation we say that the limit of $f(x)$ as x approaches 2 is the number 4, and write

$$\lim_{x \to 2} f(x) = 4$$

EXAMPLE 2. Consider the function f defined by Figure 2.

$$f(x) = x^2, \qquad x \neq -1$$

To illustrate a point, we have removed the number -1 from the domain of f; $f(-1)$ is not defined, but we have

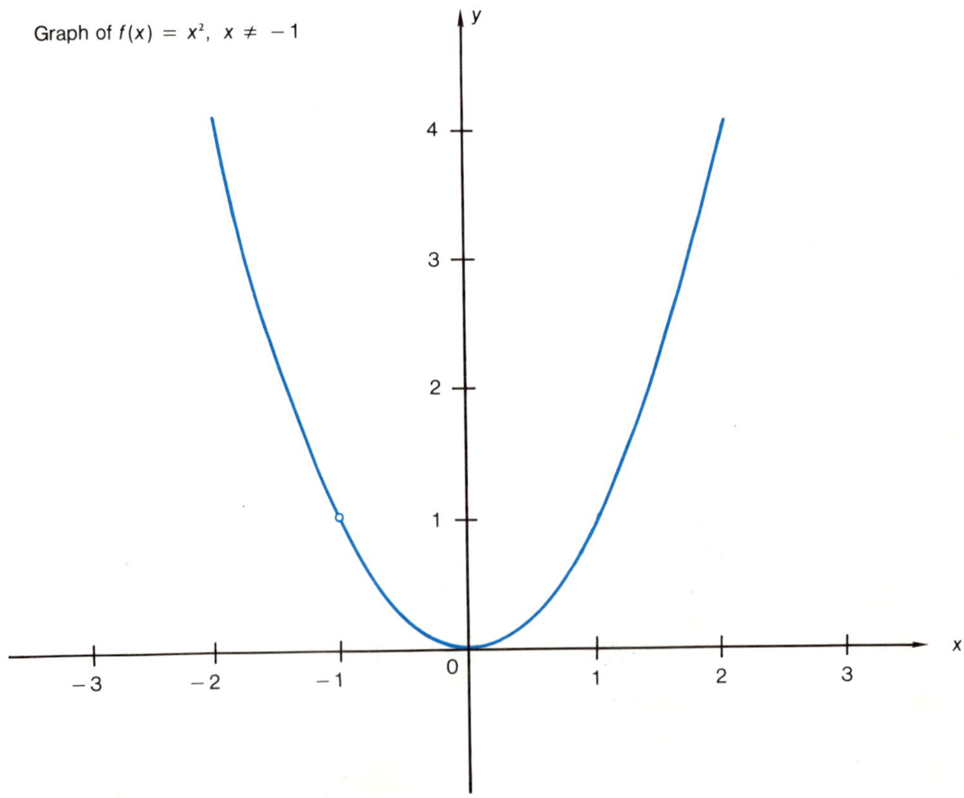

Figure 2

x	-2	-1.50	-1.40	-1.30	-1.20	-1.10	-1.01	-1.001
$f(x) = x^2$	4	2.25	1.96	1.69	1.44	1.21	1.0201	1.002001

In other words,

$$f(x) \to 1 \text{ as } x \to -1^-$$

Also, we have the following:

x	0	-0.50	-0.60	-0.70	-0.80	-0.90	-0.99	-0.999
$f(x) = x^2$	0.00	0.25	0.36	0.49	0.64	0.81	0.9801	0.998001

or

$$f(x) \to 1 \text{ as } x \to -1^+$$

Thus we may write

$$\lim_{x \to -1} f(x) = 1$$

We note that even though the function is not defined at -1, the limit of $f(x)$ as x approaches -1 still exists and is equal to 1.

EXAMPLE 3. Let f be the function defined by

$$f(x) = \begin{cases} x - 1, & x > 3 \\ x^2, & x \leq 3 \end{cases}$$

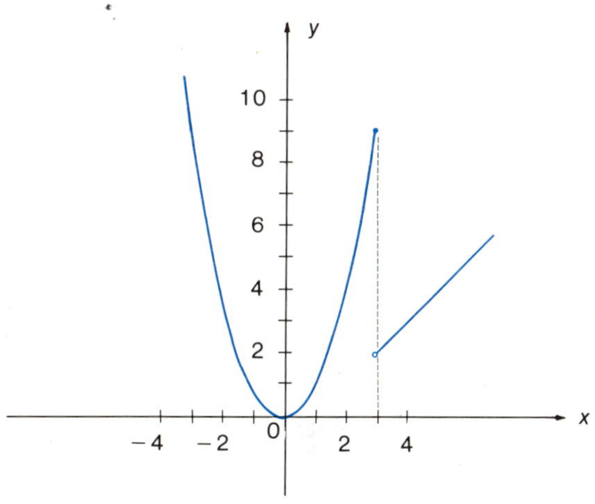

Figure 3

A graph of f is shown in Figure 3. The reader who moves a pencil on the curve just to the left of the line $x = 3$ will find that $f(x)$ approaches 9 as the pencil gets closer and closer to the line $x = 3$. Similarly, if the pencil is moved along the curve just to the right of the line $x = 3$, the reader will find that $f(x)$ approaches 2 as the pencil approaches the line $x = 3$. Thus

$$f(x) \to 9 \quad \text{as} \quad x \to -3^-$$

whereas

$$f(x) \to 2 \quad \text{as} \quad x \to -3^+$$

Since $f(x)$ approaches two different numbers depending upon whether x approaches 3 from the left or the right, we say that

$$\lim_{x \to 3} f(x) \text{ does not exist.}$$

EXAMPLE 4. Consider the function f defined by

$$f(x) = \sqrt{x+4}$$

Here the domain of f is $\{x \mid x \geq -4\}$. The graph of f is shown in Figure 4. We find that

$$f(x) \to 0 \quad \text{as} \quad x \to -4^+$$

However, when $x < -4$, $f(x)$ is not defined. Consequently, $f(x)$ does not approach a real number as $x \to -4^-$. We say that

$$\lim_{x \to -4} f(x) \quad \text{does not exist}$$

Graph of $f(x) = \sqrt{x+4}$

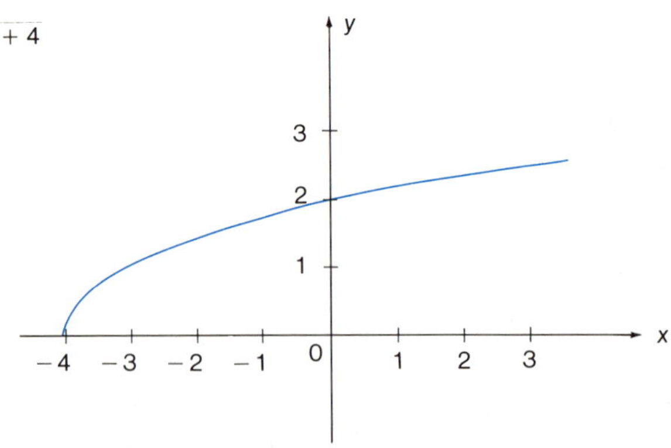

Figure 4

In view of the preceding examples, for the expression

$$\lim_{x \to c} f(x) = L$$

(read "the limit of $f(x)$ as x approaches c is L"), we make the following observations:

(1) The value of $f(x)$ approaches (gets closer to) L as x approaches c.

(2) $f(x)$ approaches L whether x approaches c from the left or from the right.

(3) The function need not even be defined at $x = c$.

In general, the expression

$$\lim_{x \to c} f(x) = L$$

intuitively means that the value of $f(x)$ can be made as close to L as we wish by making x close enough to c.

In Examples 1 and 2, you may have noted that instead of making tables for the values of $f(x)$ as x gets closer and closer to c (from the left and from the right), we could have computed limits by substituting c for x in $f(x)$. For example, in Example 1,

$$\lim_{x \to 2} f(x) = \lim_{x \to 2} (x + 2) = 2 + 2 = 4$$

and in Example 2,

$$\lim_{x \to -1} f(x) = \lim_{x \to -1} x^2 = (-1)^2 = 1$$

The technique is quite valid and will work for this type of problem. Unfortunately, this is not always the case. Let us consider another example.

EXAMPLE 5. Evaluate $\lim_{x \to 1} \dfrac{x^2 - 1}{x - 1}$

SOLUTION. The function

$$f(x) = \frac{x^2 - 1}{x - 1}$$

is not defined when $x = 1$, because it causes division by zero. So we cannot take the shortcut of substituting for x. However, we note that

$$\frac{x^2 - 1}{x - 1} = \frac{(x - 1)(x + 1)}{x - 1} \tag{1}$$

Since x is only approaching 1 and is not equal to 1, in Equation (1), we have $x - 1 \neq 0$. Consequently, we can divide the numerator and the denominator of the right-hand side of (1) be $x - 1$. So when $x - 1 \neq 0$, we have

$$\frac{x^2 - 1}{x - 1} = \frac{(x - 1)(x + 1)}{x - 1} = x + 1$$

and, therefore,

$$\lim_{x \to 1} \frac{x^2 - 1}{x - 1} = \lim_{x \to 1} \frac{(x - 1)(x + 1)}{x - 1} = \lim_{x \to 1} (x + 1) = 1 + 1 = 2$$

We now state (without proof) several theorems which can be used as an aid in evaluating limits.

THEOREM 1. If c and k are constants, then

$$\lim_{x \to c} k = k$$

Theorem 1 states that the limit of a constant is that constant. For example,

$$\lim_{x \to 3} 6 = 6 \qquad \text{and} \qquad \lim_{x \to -5} 6 = 6$$

THEOREM 2. If a, b, and c are constants, then

$$\lim_{x \to c} (ax + b) = ac + b$$

In other words, if x approaches c, then the limit of $ax + b$ is obtained by substituting c for x. For example,

$$\lim_{x \to 3} (2x - 5) = 2(3) - 5 = 6 - 5 = 1$$

$$\lim_{x \to 4} (7 - 3x) = 7 - 3(4) = 7 - 12 = -5$$

THEOREM 3. If f and g are functions and c is a constant, then

$$\lim_{x \to c} [f(x) \pm g(x)] = \lim_{x \to c} f(x) \pm \lim_{x \to c} g(x)$$

provided that each limit exist.

In words, the limit of a sum or difference is equal to the sum or difference of the limits.

THEOREM 4. If f and g are functions and c is a constant, then

$$\lim_{x \to c} [f(x) \cdot g(x)] = \lim_{x \to c} f(x) \cdot \lim_{x \to c} g(x)$$

provided that each limit exists.

In words, the limit of the product is equal to the product of the limits.

EXAMPLE 6.

(a) $$\lim_{x \to 5} (2 - x)(x + 3) = \lim_{x \to 5} (2 - x) \cdot \lim_{x \to 5} (x + 3)$$
$$= (2 - 5)(5 + 3) = -3 \cdot 8 = -24$$

(b) $$\lim_{x \to 4} (3x^2 + 7) = \lim_{x \to 4} 3 \cdot \lim_{x \to 4} x^2 + \lim_{x \to 4} 7 \qquad \text{(Theorems 3 and 4)}$$
$$= 3 \lim_{x \to 4} (x \cdot x) + 7 \qquad \text{(Theorem 1)}$$
$$= 3 \lim_{x \to 4} x \cdot \lim_{x \to 4} x + 7 \qquad \text{(Theorem 4)}$$
$$= 3 \cdot 4 \cdot 4 + 7 \qquad \text{(Theorem 2)}$$
$$= 48 + 7 = 55$$

As a special case of Theorem 4, we observe that if $f(x) = k$, where k is a constant, then

$$\lim_{x\to c} kg(x) = \lim_{x\to c} k \cdot \lim_{x\to c} g(x) = k\lim_{x\to c} g(x)$$

For example,

$$\lim_{x\to 4} 7(3x-2) = 7\lim_{x\to 4}(3x-2) = 7(3\cdot 4 - 2) = 70$$

THEOREM 5. If f and g are functions and c is a constant, then,

$$\lim_{x\to c}\left[\frac{f(x)}{g(x)}\right] = \frac{\lim_{x\to c} f(x)}{\lim_{x\to c} g(x)}$$

provided that each limit exists and $\lim_{x\to c} g(x) \neq 0$.

In words, the limit of a quotient is the quotient of the limits.

EXAMPLE 7. $\lim_{x\to 5}\frac{3x+4}{2x-3} = \frac{\lim_{x\to 5}(3x+4)}{\lim_{x\to 5}(2x-3)}$ (Theorem 5)

$= \frac{3\cdot 5 + 4}{2\cdot 5 - 3}$ (Theorem 2)

$= \frac{19}{7}$

THEOREM 6. If f is a function and c and $n \in I$ are constants,

$$\lim_{x\to c}\left[f(x)\right]^{1/n} = \left[\lim_{x\to c} f(x)\right]^{1/n} \tag{2}$$

provided the right side of (2) exists, and (1) $\lim_{x\to c} f(x) > 0$ if n is even; (2) $\lim_{x\to c} f(x) \neq 0$ if $n < 0$.

EXAMPLE 8.

(a) $\lim_{x\to 2} x^{1/3} = \left(\lim_{x\to 2} x\right)^{1/3}$ (Theorem 6)

$= (2)^{1/3}$ (Theorem 2)

(b) $\lim_{x\to 2}\sqrt{x^2+5} = \lim_{x\to 2}(x^2+5)^{1/2} = \left[\lim_{x\to 2}(x^2+5)\right]^{1/2}$

$= (9)^{1/2} = 3$

Using the previous theorems, it is easy to see that the limits of polynomial functions can be obtained by substitution. For example, if

$$f(x) = a_0x^n + a_1x^{n-1} + \cdots + a_{n-1}x + a_n$$

where $a_0, a_1, \ldots, a_n$ are constants, then

$$\lim_{x \to c} f(x) = a_0c^n + a_1c^{n-1} + \cdots + a_{n-1}c + a_n$$

However, in calculus some important limits cannot be obtained by direct substitution. Let us consider some more examples.

EXAMPLE 9. Evaluate $\lim_{x \to 2} \dfrac{x^2 - 5x + 6}{x - 2}$

SOLUTION. If 2 is substituted for x, we obtain

$$\frac{x^2 - 5x + 6}{x - 2} = \frac{(2)^2 - 5(2) + 6}{2 - 2} = \frac{4 - 10 + 6}{2 - 2} = \frac{0}{0}$$

Before we jump to the conclusion that the limit does not exist, we try factoring.

$$\frac{x^2 - 5x + 6}{x - 2} = \frac{(x - 2)(x - 3)}{x - 2}$$

Now it is permissible to cancel the common factor $x - 2$, since $x \neq 2$. Therefore

$$\begin{aligned} \lim_{x \to 2} \frac{x^2 - 5x + 6}{x - 2} &= \lim_{x \to 2} \frac{(x - 2)(x - 3)}{x - 2} = \lim_{x \to 2} (x - 3) \\ &= 2 - 3 \qquad \text{(Theorem 2)} \\ &= -1 \end{aligned}$$

EXAMPLE 10. Evaluate $\lim_{x \to 0} \dfrac{\sqrt{3 + h} - \sqrt{3}}{h}$

SOLUTION. If we substitute 0 for h, we obtain

$$\frac{\sqrt{3 + h} - \sqrt{3}}{0} = \frac{\sqrt{3} - \sqrt{3}}{0} = \frac{0}{0}$$

Again we have the form 0/0. In this example the numerator cannot be factored as in the last example. So we try rationalizing the numerator by multiplying and dividing the given expression by $\sqrt{3 + h} + \sqrt{3}$.

$$\begin{aligned} \frac{\sqrt{3 + h} - \sqrt{3}}{h} &= \frac{\sqrt{3 + h} - \sqrt{3}}{h} \cdot \frac{\sqrt{3 + h} + \sqrt{3}}{\sqrt{3 + h} + \sqrt{3}} = \frac{(3 + h) - 3}{h(\sqrt{3 + h} + \sqrt{3})} \\ &= \frac{h}{h(\sqrt{3 + h} + \sqrt{3})} = \frac{1}{\sqrt{3 + h} + \sqrt{3}} \qquad \text{(since } h \neq 0) \end{aligned}$$

Therefore

$$\lim_{h \to 0} \frac{\sqrt{3 + h} - \sqrt{3}}{h} = \lim_{h \to 0} \frac{1}{\sqrt{3 + h} + \sqrt{3}}$$

$$= \frac{1}{\sqrt{3+0}+\sqrt{3}} = \frac{1}{2\sqrt{3}}$$

EXAMPLE 11. Evaluate $\lim_{x \to 2} \sqrt[3]{\dfrac{x^3-8}{x^2-4}}$

SOLUTION. $\lim_{x \to 2} \sqrt[3]{\dfrac{x^3-8}{x^2-4}} = \lim_{x \to 2} \left(\dfrac{x^3-8}{x^2-4}\right)^{1/3} = \left(\lim_{x \to 2} \dfrac{x^3-8}{x^2-4}\right)^{1/3}$

$$= \left[\lim_{x \to 2} \frac{(x-2)(x^2+2x+4)}{(x-2)(x+2)}\right]^{1/3}$$

$$= \left(\lim_{x \to 2} \frac{x^2+2x+4}{x+2}\right)^{1/3} = \left(\frac{12}{4}\right)^{1/3}$$

$$= (3)^{1/3} = \sqrt[3]{3}$$

In the next example, we consider the illustration where the limit does not exist.

EXAMPLE 12. Evaluate (a) $\lim_{x \to 2} \dfrac{5}{x-2}$ and (b) $\lim_{x \to -3} \sqrt{x+1}$

SOLUTION.

(a) The limit does not exist, since no factoring process or rationalization can change the division by zero.

(b) The limit does not exist, since
$\lim_{x \to -3} \sqrt{x+1} = [\lim_{x \to -3} (x+1)]^{1/2} = \sqrt{-3+1} = \sqrt{-2}$
but $\sqrt{-2}$ is not a real number.

EXAMPLE 13. Evaluate $\lim_{h \to 0} \dfrac{(x+h)^2 - x^2}{h}$

SOLUTION. Here the variable x is to be treated as a constant, since we are interested in $h \to 0$. Consequently,

$$\lim_{h \to 0} \frac{(x+h)^2 - x^2}{h} = \lim_{h \to 0} \frac{x^2 + 2xh + h^2 - x^2}{h} = \lim_{h \to 0} \frac{2xh + h^2}{h}$$

$$= \lim_{h \to 0} \frac{(2x+h)h}{h} = \lim_{h \to 0} (2x+h) \qquad (h \neq 0)$$

$$= 2x + 0 = 2x$$

EXERCISES 1.6

In Problems 1 through 24, evaluate the indicated limits (if the limit exists).

1. $\lim_{x \to 2} (3 + 5x)$

2. $\lim_{x \to 3} (-7x)$

3. $\lim_{x \to -2} (2x - 3)$

4. $\lim_{x \to -4} (3 - 8x)$

5. $\lim_{x \to 2} (5x^2 + 4)$

6. $\lim_{x \to 1} (3x^2 - 8x + 2)$

7. $\lim_{x \to 2} (x - 2)(x^2 + 3)$

8. $\lim_{x \to -3} (2x^3 - 3x^2 + 6x + 1)$

9. $\lim_{x \to 4} \dfrac{x^2}{2x - 1}$

10. $\lim_{x \to 2} \dfrac{x^3 + 5}{x^2 + 3x}$

11. $\lim_{x \to 2} \dfrac{x^2 - 6}{3x + 1}$

12. $\lim_{x \to 4} \dfrac{2x^2 - 7}{x^3 + 3}$

13. $\lim_{x \to 0} \dfrac{-x^2 + 4x - 8}{x^3 + 3x + 2}$

14. $\lim_{x \to 2} \dfrac{3x + 1}{2x - 2}$

15. $\lim_{h \to 0} \dfrac{h + h^2}{h}$

16. $\lim_{h \to 0} \dfrac{x^2 + 2xh + h^2}{x}$

17. $\lim_{x \to 2} \dfrac{x^2 - 4}{x^2 - 3x + 2}$

18. $\lim_{x \to -2} \dfrac{x^3 + 8}{2x^2 + x - 1}$

19. $\lim_{x \to 1} \dfrac{x^3 - 1}{x^2 - 1}$

20. $\lim_{x \to -1} \sqrt[3]{\dfrac{x + 1}{x^3 + 1}}$

21. $\lim_{h \to 0} \dfrac{\sqrt{1 + h} - \sqrt{h}}{h}$

22. $\lim_{h \to 0} \dfrac{\sqrt{x + h} - \sqrt{x}}{h}$

23. $\lim_{x \to 0} \dfrac{\sqrt{x + 16} - 4}{x}$

24. $\lim_{y \to 0} \dfrac{\sqrt{y + 36} - 6}{y}$

In Problems 25 through 38, form the new function F, where

$$F(h) = \frac{f(x + h) - f(x)}{h}, \qquad h \neq 0$$

and evaluate $\lim_{h \to 0} F(h)$.

25. $f(x) = x$

26. $f(x) = x + 1$

27. $f(x) = 1 - 3x$

28. $f(x) = x^2$

29. $f(x) = \sqrt{x}$

30. $f(x) = \sqrt{x + 1}$

31. $f(x) = \sqrt{2x + 3}$

32. $f(x) = \sqrt{3 - 4x}$

33. $f(x) = x^3$

34. $f(x) = 3\sqrt{x}$

35. $f(x) = x^2 - 2x + 5$

36. $f(x) = x^3 + 2x^2 - 3x + 7$

37. $f(x) = 2$

38. $f(x) = -4$

1.7 CONTINUITY

Let us consider the functions whose graphs are illustrated in Figures 1 through 4. In each case we are interested in what happens to the graph at the point $x = c$. In the first three graphs we note that there is a break in the curve at c. In other words, if you move your pencil along the curve, you will have to lift the pencil at $x = c$. We say that the functions involved are *discontinuous* at $x = c$. On the other hand, you could move your pencil along the curve in Figure 4 near c without lifting the pencil at c. We say that the function whose graph is shown in Figure 4 is *continuous* at $x = c$.

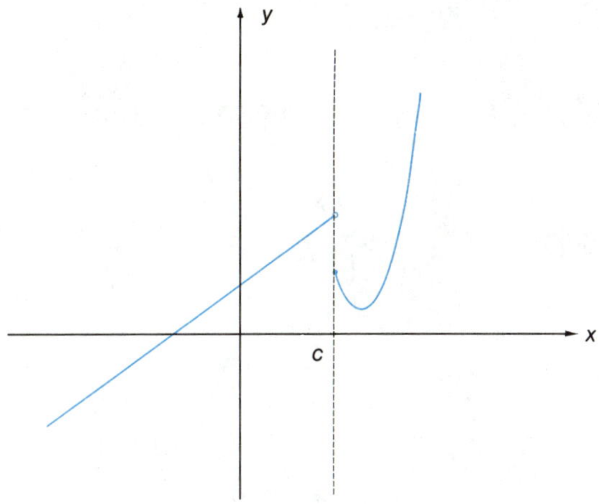

Figure 1

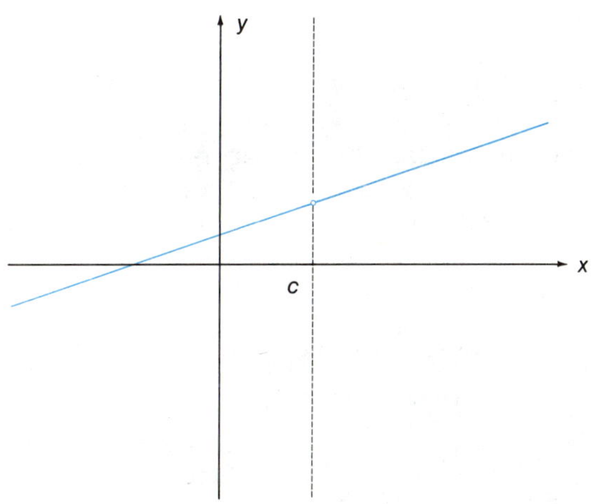

Figure 2

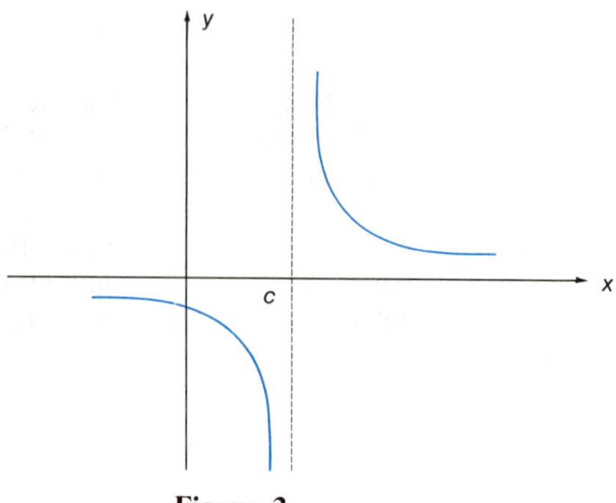

Figure 3

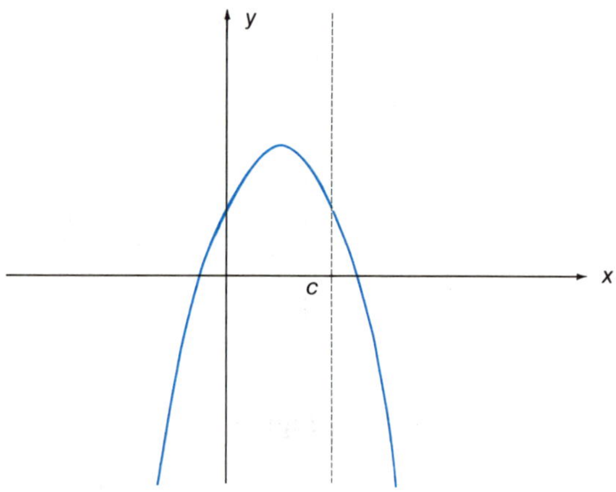

Figure 4

The mathematicians Bolzano[1] and Weierstrass offered definitions of the concept of continuity, which, as we have seen, is connected with the notion of unbroken curves. Bolzano's definition was imprecise. Weierstrass offered the more rigorous definition, which is essentially the one that is used today.

We now see that continuity of a function at c is closely related to the limit of a function at $x = c$.

Definition. A function f is said to be *continuous* at the number c if and only if

(1) $f(c)$ exists; that is, f is defined at c

(2) $\lim_{x \to c} f(x)$ exists, and

---

[1] BERNHARD BOLZANO (1781-1848), an Austrian mathematician and theologian, was a professor of religious philosophy at the University of Prague. In 1820 he was dismissed from his post by the Austrian government because his sermons were considered subversive. His mathematical work *Paradoxes of Infinity* was published after his death.

(3) $\lim_{x \to c} f(x) = f(c)$

When a function f is not continuous at c, we say that f is *discontinuous* at c. If a function is continuous at every point in an open interval I, it is said to be *continuous* on I.

EXAMPLE 1. Let f be the function defined by

$$f(x) = \begin{cases} 1, & x \geq 2 \\ -1, & x < 2 \end{cases}$$

Is f continuous at $x = 2$?

SOLUTION. Figure 5 illustrates the graph of f. Here $f(2) = 1$ and so condition (1) is satisfied. Now

$$\lim_{x \to 2^+} f(x) = 1 \quad \text{and} \quad \lim_{x \to 2^-} f(x) = -1$$

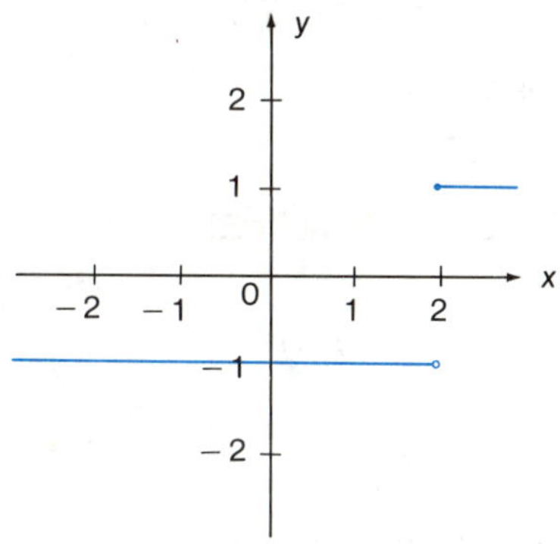

Figure 5

Hence $\lim_{x \to 2} f(x)$ does not exist and so condition (2) is not satisfied.
Therefore f is discontinuous at $x = 2$.

EXAMPLE 2. Let f be the function defined by

$$f(x) = \begin{cases} \dfrac{(x+2)(x-1)}{x-1}, & x \neq 1 \\ 2, & x = 1 \end{cases}$$

Is f continuous at $x = 1$?

SOLUTION. A graph of f is shown in Figure 6. Here $f(1) = 2$. Furthermore,

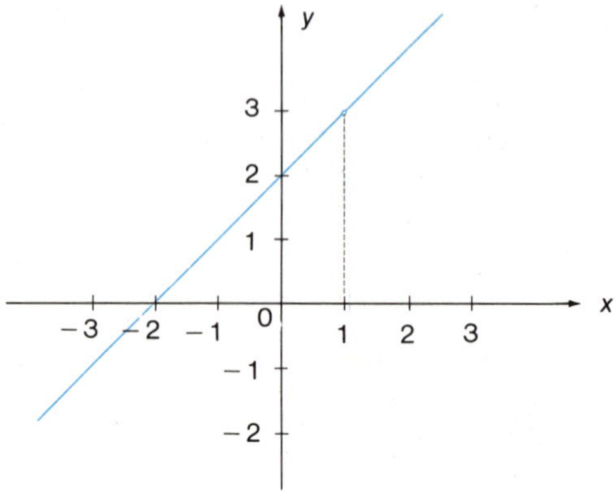

Figure 6

$$\lim_{x \to 1^+} f(x) = 3 \quad \text{and} \quad \lim_{x \to 1^-} f(x) = 3$$

Hence

$$\lim_{x \to 1} f(x) = 3$$

Thus conditions (1) and (2) are satisfied. However,

$$\lim_{x \to 1} f(x) \neq f(1)$$

Therefore f is discontinuous at $x = 1$.

EXAMPLE 3. Let the function f be defined by

$$f(x) = \frac{1}{x}$$

Is f continuous at $x = 0$?

SOLUTION. Since $f(0)$ is not defined, condition (1) is not satisfied and so f is discontinuous at $x = 0$ (see Figure 7).

Graph of $f(x) = 1/x$

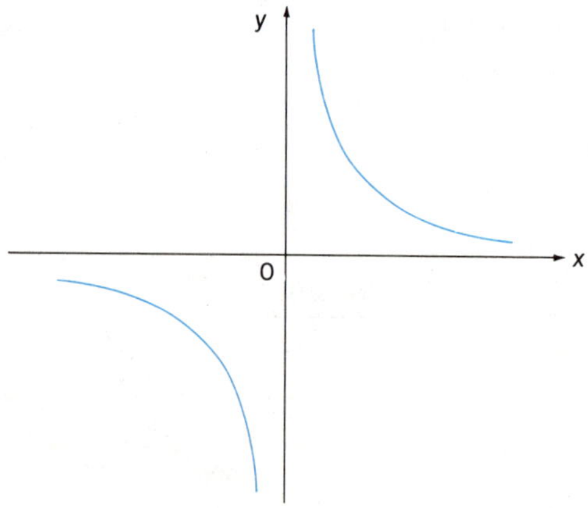

Figure 7

EXAMPLE 4. Let f be defined by

$$f(x) = \begin{cases} \dfrac{1}{x}, & x \neq 0 \\ 1, & x = 0 \end{cases}$$

Is f continuous at $x = 0$?

SOLUTION. Here $f(0) = 1$ and so condition (1) is satisfied. Now[2]

$$\lim_{x \to 0^+} f(x) = +\infty \qquad \text{and} \qquad \lim_{x \to 0^-} f(x) = -\infty$$

Therefore, condition (2) is not satisfied and so f is discontinuous at $x = 0$ (see Figure 8).

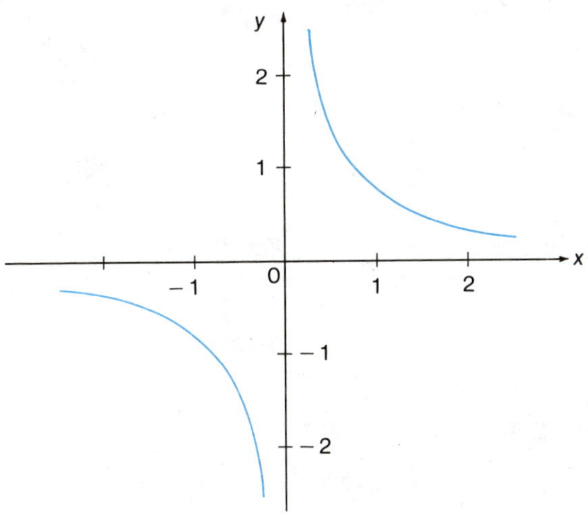

Figure 8

EXAMPLE 5. Let f be the function defined by

$$f(x) = \sqrt{x+1}$$

Is f continuous at $x = -1$?

SOLUTION. A graph of f is shown in Figure 9. We find that $f(-1) = 0$. Now

---

[2] The symbol $\lim\limits_{x \to a} f(x) = +\infty$ $(-\infty)$ means that $f(x)$ increases (decreases) without bound as x approaches a.

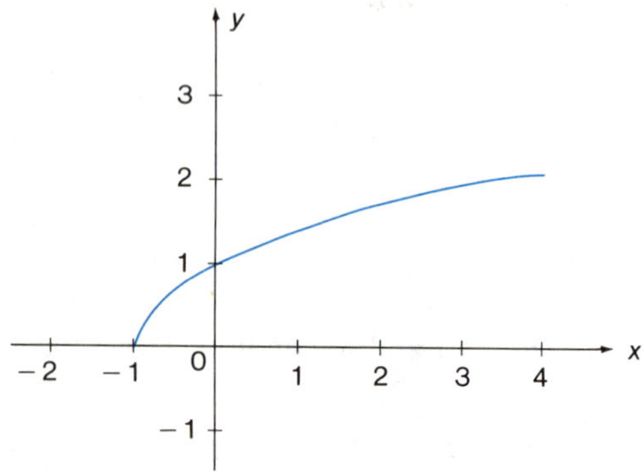

Figure 9

$$\lim_{x \to -1^+} f(x) = 0 \text{ but } \lim_{x \to -1^-} f(x) \text{ does not exist.}$$

Hence $\lim_{x \to -1} f(x)$ does not exist, and so condition (2) is not satisfies. Therefore, f is discontinuous at $x = -1$.

EXAMPLE 6. Let f be the function defined by

$$f(x) = \frac{1}{x+1}$$

(a) Is f continuous on $I_1 = \{x \mid 0 < x < 4\}$?

(b) Is f continuous on $I_2 = \{x \mid -3 < x < 4\}$?

SOLUTION.

(a) Let c be any number in I_1. Then $f(c) = \dfrac{1}{(c+1)}$. Also

$$\lim_{x \to c} f(x) = \lim_{x \to c} \frac{1}{x+1} = \frac{1}{c+1}, \qquad c \neq -1$$

Hence

$$\lim_{x \to c} f(x) = f(c)$$

Conditions (1), (2), and (3) are satisfied for every c in I_1. Therefore, f is continuous on I_1.

(b) Here -1 is in I_2, and $f(-1)$ is not defined. Thus f is not continuous at -1. Therefore, f is not continuous on I_2.

We have the following example:

EXAMPLE 7. Define $f(2)$ so that the function

$$f(x) = \frac{x^2 - x - 2}{x - 2}, \qquad x \neq 2$$

is continuous at $x = 2$.

SOLUTION. We have

$$f(x) = \frac{x^2 - x - 2}{x - 2} = \frac{(x-2)(x+1)}{x-2} = x + 1, \qquad x \neq 2$$

Now,

$$\lim_{x \to 2} f(x) = \lim_{x \to 2} (x + 1) = 3$$

Therefore, if we define $f(2) = 3$, then the three conditions of the definition will be satisfied. Thus, the function

$$f(x) = \begin{cases} \dfrac{x^2 - x - 2}{x - 2}, & x \neq 2 \\ 3, & x = 2 \end{cases}$$

is continuous at $x = 2$.

EXERCISES 1.7

In Problems 1 through 14, sketch the graphs of each of the functions defined and determine if each is continuous at the given value c.

1. $f(x) = \dfrac{3}{x-2}$ at $c = 2$

2. $f(x) = \begin{cases} \dfrac{1}{x+4}, & x \neq -4 \\ 0, & x = -4 \end{cases}$ at $c = -4$

3. $f(x) = \begin{cases} \dfrac{x^2 - 3x + 2}{x - 1}, & x \neq 1 \\ -5, & x = 1 \end{cases}$ at $c = 1$

4. $f(x) = \sqrt{3x - 1}$ at $c = \frac{1}{3}$

5. $f(x) = \begin{cases} 2x + 1, & x \leq 1 \\ -x + 4, & x > 1 \end{cases}$ at $c = 1$

6. $f(x) = \begin{cases} 2x + 1, & x \leq 2 \\ -x + 4, & x > 2 \end{cases}$ at $c = 2$

7. $f(x) = |x - 2|$ at $c = 2$

8. $f(x) = \begin{cases} |x-2|, & x \neq 2 \\ 1, & x = 2 \end{cases}$ at $c = 2$

9. $f(x) = \begin{cases} \dfrac{x^2-9}{x+3}, & x \neq -3 \\ 3, & x = -3 \end{cases}$ at $c = -3$

10. $f(x) = \begin{cases} \dfrac{x^2-4}{x-2}, & x \neq 2 \\ 4, & x = 2 \end{cases}$ at $c = 2$

11. $f(x) = \sqrt[3]{x}$ at $c = -1$

12. $f(x) = \sqrt[3]{x-1}$ at $c = 1$

13. $f(x) = \begin{cases} \dfrac{x-4}{x^2-16}, & x \neq \pm 4 \\ \frac{1}{8}, & x = 4 \\ 0, & x = -4 \end{cases}$ at $c = 4, -4$

14. $f(x) = \begin{cases} x+1, & x \leq -1 \\ x^2, & -1 < x < 1 \\ 3x+1, & x \geq 1 \end{cases}$ at $c = -1, 1$

In Problems 15 through 22, a function f is defined on an interval I. Determine whether f is continuous on I.

15. $f(x) = \dfrac{1}{x^2+1}$, $I = \{x \mid -1 < x < 5\}$

16. $f(x) = \dfrac{x}{x-1}$, $I = \{x \mid -3 < x < 1\}$

17. $f(x) = \dfrac{x}{x-1}$, $I = \{x \mid -1 < x < 4\}$

18. $f(x) = \dfrac{x^2+3x-1}{2x^2+3}$, $I = \{x \mid -2 < x < 2\}$

19. $f(x) = \begin{cases} \dfrac{x^2+x-2}{x+2}, & x \neq -2 \\ -3, & x = -2 \end{cases}$ $I = \{x \mid -3 < x < 3\}$

20. $f(x) = \begin{cases} \dfrac{x^3-1}{x-1}, & x \neq 1 \\ 2, & x = 1 \end{cases}$ $I = \{x \mid 0 < x < \frac{3}{2}\}$

21 $f(x) = \begin{cases} \dfrac{4x^2 - 1}{2x + 1}, & x \neq -\frac{1}{2} \\ 2, & x = -\frac{1}{2} \end{cases}$ $\qquad I = \{x \mid -1 < x < 1\}$

22. $f(x) = \dfrac{x + 2}{x^2 - 4}, \quad I = \{x \mid 0 < x < 2\}$

In Problems 23 through 26, define $f(c)$ so that the function f is continuous at c.

23. $f(x) = \dfrac{(x - 1)(x + 2)}{x - 1}, \quad x \neq 1, \quad c = 1$

24. $f(x) = \dfrac{x^2 - 3x + 2}{x^2 - 4}, \quad x \neq \pm 2, \quad c = 2$

25. $f(x) = \dfrac{x^2 - 1}{x + 1}, \quad x \neq -1, \quad c = -1$

26. $f(x) = \dfrac{x^3 + 8}{x + 2}, \quad x \neq -2, \quad c = -2$

CHAPTER 1 REVIEW

Part 1 (ORAL)

1. If $P(x_1, y_1)$ and $Q(x_2, y_2)$ are in R^2, then the distance between P and Q is $d(P, Q) =$ ______________________.
2. The coordinates of the midpoint M of a line segment joining $P(x_1, y_1)$ and $Q(x_2, y_2)$ are (____, ____).
3. Define a function.
4. Define the graph of a function.
5. Explain the meaning of $[\![x]\!]$.
6. The point-slope form of an equation of a line is ______________________.
7. What is a linear function?
8. What is a quadratic function?
9. Explain composition of two functions.
10. What is the definition of continuity of $f(x)$ at $x = c$?

Part 2 (WRITTEN)

1. Find the distance between $P(3, 5)$ and $Q(7, 9)$. Also find the coordinates of the midpoint.
2. Do Problem 1 for $P(-4, 0)$ and $Q(1, -3)$.

3. Find the equation of a circle satisfying the given conditions:
   (a) Center at $(-1, 2)$ and radius 5
   (b) Center at $(-1, 3)$ and touching the line $y = -1$
4. Find the equation of a line passing through the points $(2, 3)$ and $(5, -7)$.
5. Find the slope, x-intercept, and y-intercept of the line $3x - 4y = 12$.
6. Find an equation of the line passing through the point $(1, -1)$ and
   (a) perpendicular to the line $3x + 4y - 7 = 0$
   (b) parallel to the line $2x - 3y + 5 = 0$
7. Sketch the graph of $y = f(x) = 2x^2 + 3x + 1$.
8. Sketch the graph of $y = f(x) = -x^2 - 6x + 16$.
9. Find a linear cost function from the following data: A cost of \$425.00 occurs when 50 units are produced and a cost of \$525.00 occurs when 250 units are produced. Also find the following.
   (a) The total cost when 100 units are produced
   (b) The fixed cost
   (c) The variable cost when 80 units are produced
   (d) The marginal cost

In Problems 10 through 16, let $f(x) = 1 - 2x^2$ and $g(x) = \dfrac{1}{x}$.

10. $f(-\frac{1}{2}) =$ ____________________
11. $g(a + h) =$ ____________________
12. $(f + g)(1) =$ ____________________
13. $(f \cdot g)(2) =$ ____________________
14. $(\dfrac{f}{g})(5) =$ ____________________
15. $(f \circ g)(3) =$ ____________________
16. $(g \circ f)(3) =$ ____________________

In Problems 17 through 24, find the limit (if it exists).

17. $\lim_{x \to -1} \dfrac{x + 3}{2x + 4}$
18. $\lim_{x \to -1} \dfrac{x + 1}{x^2 - 1}$
19. $\lim_{x \to 9} \dfrac{\sqrt{x} - 3}{x - 9}$
20. $\lim_{x \to 3} \sqrt{x - 3}$
21. $\lim_{x \to 3} 57$
22. $\lim_{x \to 0} |x|$
23. $\lim_{x \to 0} \dfrac{x}{|x|}$
24. $\lim_{x \to 3} (x - [\![x]\!])$

25. Let $f(x) = 2x + 5$. Find $\lim_{h \to 0} \frac{f(3+h) - f(3)}{h}$.

26. Let $f(x) = x^2$. Find $\lim_{h \to 0} \frac{f(1+h) - f(1)}{h}$.

In Problems 27 through 30, find the values of x at which the given function is discontinuous.

27. $f(x) = \frac{1}{x^2 - 3x + 2}$

28. $f(x) = \frac{3}{x^2 + 5}$

29. $f(x) = \begin{cases} 1, & \text{if } x \geq 3 \\ -1, & \text{if } x < 3 \end{cases}$

30. $f(x) = \begin{cases} x, & \text{if } x < 0 \\ [\![x]\!], & \text{if } x \geq 0 \end{cases}$

Chapter Two

The Derivative

In this chapter we begin our study of calculus. In Chapter 1, we discussed functions, slope, limits of functions and continuous functions. Now we shall attempt to blend these topics together to introduce one of the most basic concepts in the development of calculus: the derivative of a function. Invented by Newton[1]*, the derivative has become a very valuable tool in physical sciences, business, and social sciences. The objective of this chapter is to understand what the so-called "derivative" of a function is and what it means. After we define the derivative of a function, we shall discover that it is often difficult to compute the derivative from scratch, that is from the definition. Consequently, we shall learn some simple rules for computing derivatives.*

---

[1] ISAAC NEWTON (1642-1727) was born on Christmas Day at Woolsthorpe, England. Newton's father died before the frail child's birth. Newton received his A.B. in 1664 at Trinity College in Cambridge. The Great Plague (1664-1665) forced Newton to return to Woolsthrope to meditate for two years, during which time he invented calculus, proved that light is composed of different colors, and discovered the universal law of gravitation. In 1669 he became professor of mathematics at Trinty. He constructed a telescope and observed the satellites of Jupiter, and in 1686 he published (at Halley's expense) his masterpiece, *The Mathematical Principles of Natural Philosophy*.

Newton had a strong interest in theology and in alchemy. In addition, he spent some time in Parliament and was made Master of the Mint in 1699. Four years later he was elected President of the Royal Society and was knighted by Queen Anne.

2.1 DEFINITION OF THE DERIVATIVE

The concept of the derivative arose from the problem of finding the tangent line to an arbitrary curve. This problem was successfully solved by the French mathematician Fermat[2].

In geometry, we learned that a tangent line to a circle is a line that touches the circle at one and only one point. However, we have to be careful when we extend this notion of a tangent line to an arbitrary curve. Intuitively, we know that in Figure 1 the lines L_1 and L_2 are tangent lines to the curve $y = f(x)$ at the points P_1 and P_2, respectively. The line L_3 is not a tangent line (although it meets the curve in only one point). We would not expect to find a tangent line at A, where there is a break or discontinuity. At B, the curve has a cusp, and there is no tangent line there, either.

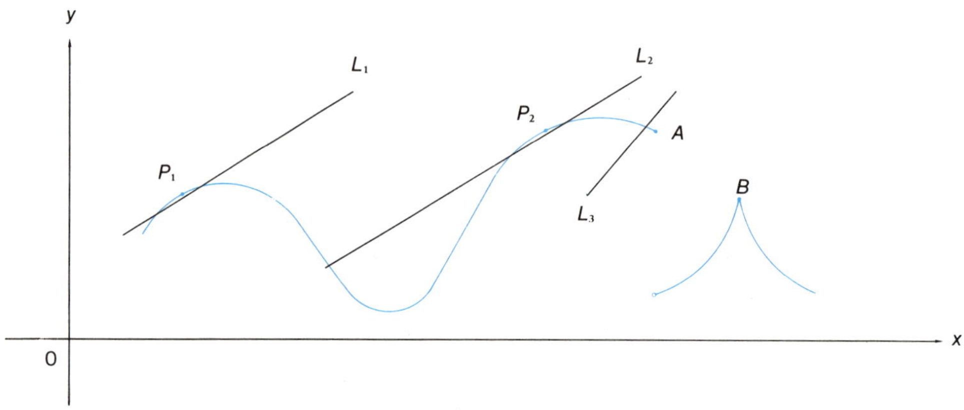

Figure 1

[2] PIERRE DE FERMAT (1610-1665) was born at Beaumont-de-Lomagne, France. His father, Dominique Fermat, was a leather merchant. Fermat led an uneventful life. He received his education in his home town and later at Toulouse. At the age of 30 he married, and he fathered three children, a boy and two girls. The girls became nuns. In 1648 he became Counselor of the Parliament of Toulouse, a position which gave him ample time to do mathematics.

Fermat is considered as one of the creators of calculus (along with Newton and Leibniz). He solved the problem of finding the extrema of the graph of a function and deduced the laws of reflection and refraction. He extended Decartes's analytic geometry to three dimensions and was a cofounder (with Pascal) of the theory of probability. His greatest work, however, was done in the theory of numbers. He stated and claimed to have proved a famous problem (Fermat's Last Theorem): If n is an integer, $n > 2$, then there exist no rational numbers x, y, and z $(xyz \neq 0)$ such that $x^n + y^n = z^n$. This remains an unproved theorem today.

It is natural to ask for a suitable definition of a tangent line to the graph of $y = f(x)$ at any point on the graph. We shall use the limit concept to arrive at such a definition.

Consider a function f whose graph is shown in Figure 2. Let P be a point on the graph of f with coordinates $(c, f(c))$.

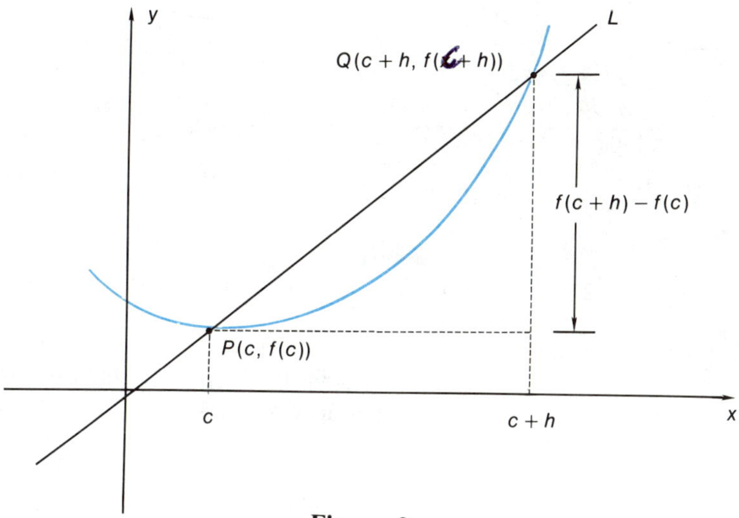

Figure 2

Let h be a number ($h \neq 0$) such that $c + h$ is in the domain of f. Then there is a point Q on the graph of f with coordinates $(c + h, f(c + h))$. The line L containing points P and Q is called a *secant line* (in geometry, a line which crosses a circle at two points is called a secant line). Now the slope of L (see Chapter 1) is given by

$$\text{slope of } PQ = \frac{f(c+h) - f(c)}{h}, \qquad h \neq 0 \tag{1}$$

We shall assume that the function f is continuous on an interval containing the points c and $c + h$. Now let the point Q approach P along the curve. Clearly, as Q approaches P, h approaches zero. Taking the limit in Equation (1), we have

$$\lim_{Q \to P} (\text{slope of } PQ) = \lim_{h \to 0} \frac{f(c+h) - f(c)}{h} \tag{2}$$

If the limit on the right side of (2) exists and is equal to, say, m, then we define the *tangent line* to the graph of f at P as the line L_{PT}, having slope m and containing the point P (Figure 3). Thus, using the point slope form of the equation of a line, we obtain the equation of this line as

$$y - f(c) = m(x - c) \tag{3}$$

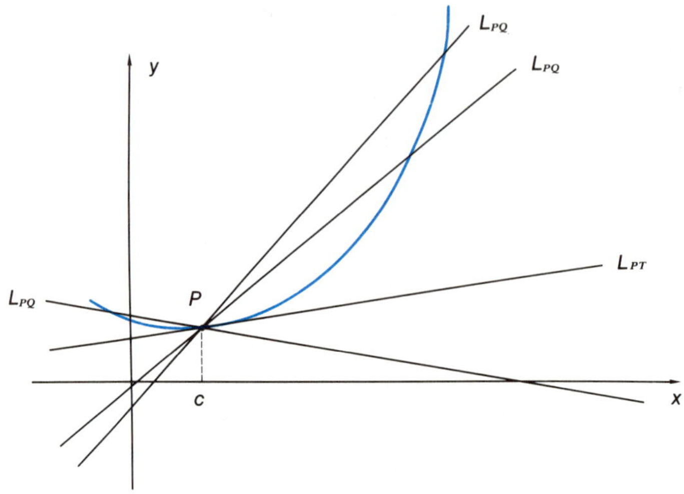

Figure 3

We can generalize the result in (2) to any point $(x, f(x))$ on the curve by replacing c with x. We thus have the following definition, which forms the basis of differential calculus.

Definition 1. Let $y = f(x)$ be a function. The limit

$$\lim_{h \to 0} \frac{f(x+h) - f(x)}{h} \tag{4}$$

if it exists, is called the *derivative* of f at x and is denoted by $f'(x)$ (read "f prime of x"). If the limit in (4) exists, the function f is said to be *differentiable* at x. The process of finding the derivative is called differentiation.

From our definition of the derivative, we have

$$f'(x) = \lim_{h \to 0} \frac{f(x+h) - f(x)}{h} \tag{5}$$

provided that the limit exists. So, we observe that f' is a new function obtained from f by the process of differentiation.

In (5), the fraction

$$\frac{f(x+h) - f(x)}{h} \tag{6}$$

is frequently called the *difference quotient*, because if $P(x, f(x))$ and $Q(x+h, f(x+h))$ are two points on $y = f(x)$, then the numerator in (6) is the difference of the y coordinates of P and Q, and the denominator in (6) is the difference of the corresponding x coordinates of P and Q.

Notation The derivative notation f' was introduced by Lagrange[3]. For a function f defined by the equation $y = f(x)$, Leibniz denoted the derivative of f by

$$\frac{dy}{dx} \quad \text{and} \quad \frac{df}{dx}$$

read "the derivative of y (or f) with respect to x." Other notations used are

$$Dy, \quad Df, \quad y'$$

When we wish to evaluate the derivative at $x = c$, we write

$$f'(c) = \left.\frac{dy}{dx}\right|_{x=c} = \left.\frac{df}{dx}\right|_{x=c} = y'(c) = Dy|_{x=c} = Df|_{x=c}$$

To find the derivative $f'(x)$ of a function $f(x)$, we perform the following steps:

Step 1. Find $f(x+h)$.

Step 2. Find the difference: $f(x+h) - f(x)$.

Step 3. Find the difference quotient: $\dfrac{f(x+h) - f(x)}{h}$.

Step 4. If this limit exists, $f'(x) = \lim\limits_{h \to 0} \dfrac{f(x+h) - f(x)}{h}$.

EXAMPLE 1. *Let f be the function defined by*

$$f(x) = x^3$$

Find $f'(x), f'(2), f'(-4), f'(c)$.

SOLUTION.

Step 1. Find $f(x+h)$. In this example,

$$f(x+h) = (x+h)^3 = x^3 + 3x^2h + 3xh^2 + h^3$$

---

[3] JOSEPH LOUIS LAGRANGE (1736-1813) was born in Turin, Italy. He did important mathematical work in partial differential equations, and he contributed immensely to the development of the calculus of variations.

At the age of 19, Lagrange became professor of mathematics at the Royal Artillery School of Turin. At 28 he won the coveted prize of mathematics of the Academy of Sciences of Paris, a prize he was to win several times in the succeeding years. In 1766 at the invitation of Frederick II of Prussia, he replaced Euler as director of the mathematics department of the Academy of Berlin. In 1787 at the invitation of Louis XVI, Lagrange went to Paris to work at the French Academy. Napoleon treated Lagrange as a friend, giving him the title of count and making him a senator and a high officer of the Legion of Honor.

Step 2. Find the difference $f(x+h)-f(x)$. Here,

$$\begin{aligned} f(x+h)-f(x) &= (x^3+3x^2h+3xh^2+h^3)-x^3 \\ &= 3x^2h+3xh^2+h^3 \end{aligned}$$

Step 3. Find the difference quotient.

$$\frac{f(x+h)-f(x)}{h} = \frac{3x^2h+3xh^2+h^3}{h} = 3x^2+3xh+h^2$$

Step 4. See if $\lim_{h\to 0} \frac{f(x+h)-f(x)}{h}$ exists. Here,

$$f'(x) = \lim_{h\to 0} \frac{f(x+h)-f(x)}{h} = \lim_{h\to 0} (3x^2+3xh+h^2) = 3x^2$$

Hence, if $f(x) = x^3$, then the derivative $f'(x) = 3x^2$. Therefore, for $x = 2$, we have

$$f'(2) = 3(2)^2 = 12$$

Similarly,

$$f'(-4) = 3(-4)^2 = 48 \quad \text{and} \quad f'(c) = 3(c)^2 = 3c^2.$$

To avoid excessive writing, we may, in fact, go directly to Step 4. We do this in the remaining examples.

EXAMPLE 2. Let

$$f(x) = \frac{1}{x} \qquad x > 0.$$

Find $f'(x)$.

SOLUTION. Using Definition 1, we have for $h \neq 0$,

$$\begin{aligned} f'(x) &= \lim_{h\to 0} \frac{f(x+h)-f(x)}{h} \\ &= \lim_{h\to 0} \frac{1}{h}\left(\frac{1}{x+h}-\frac{1}{x}\right) = \lim_{h\to 0} \frac{-1}{x(x+h)} = -\frac{1}{x^2} \end{aligned}$$

EXAMPLE 3. Let $f(x) = \sqrt{x}$. Find $f'(x)$.

SOLUTION. We have

$$f'(x) = \lim_{h\to 0} \frac{f(x+h)-f(x)}{h} = \lim_{h\to 0} \frac{1}{h}(\sqrt{x+h} - \sqrt{x})$$

We observe that if we substitute $h = 0$ in the right-hand side, we obtain the indeterminate form 0/0. A useful trick at this point is to rationalize the numerator. Thus,

$$\sqrt{x+h} - \sqrt{x} = (\sqrt{x+h} - \sqrt{x})\left(\frac{\sqrt{x+h} + \sqrt{x}}{\sqrt{x+h} + \sqrt{x}}\right)$$

$$= \frac{x+h-x}{\sqrt{x+h} + \sqrt{x}}$$

$$= \frac{h}{\sqrt{x+h} + \sqrt{x}}$$

Hence,

$$f'(x) = \lim_{h\to 0} \frac{1}{h}\left(\frac{h}{\sqrt{x+h} + \sqrt{x}}\right) = \lim_{h\to 0} \frac{1}{\sqrt{x+h} + \sqrt{x}} = \frac{1}{\sqrt{x} + \sqrt{x}} = \frac{1}{2\sqrt{x}}$$

EXAMPLE 4. Find $f'(x)$ if $f(x) = 3x^2 + 4x + 8$.

SOLUTION. We have

$$f'(x) = \lim_{h\to 0} \frac{f(x+h) - f(x)}{h}, \qquad h \neq 0$$

$$= \lim_{h\to 0} \frac{1}{h}\{[3(x+h)^2 + 4(x+h) + 8] - [3x^2 + 4x + 8]\}$$

$$= \lim_{h\to 0} \frac{1}{h}[3x^2 + 6xh + 3h^2 + 4x + 4h + 8 - 3x^2 - 4x - 8]$$

$$= \lim_{h\to 0} \frac{1}{h}(6xh + 3h^2 + 4h) = \lim_{h\to 0} (6x + 3h + 4) = 6x + 4$$

EXAMPLE 5. Let $f(x) = |x|$. Show that $f'(0)$ does not exist.

SOLUTION. We point out that the function f is defined for all values of x. Now for $h \neq 0$, we have

$$f'(0) = \lim_{h\to 0} \frac{f(0+h) - f(0)}{h} = \lim_{h\to 0} \frac{|h|}{h}$$

We observe from the definition of the absolute value that

$$\lim_{h\to 0^+} \frac{|h|}{h} = \lim_{h\to 0^+} \frac{h}{h} = 1 \quad \text{and} \quad \lim_{h\to 0^-} \frac{|h|}{h} = \lim_{h\to 0^-} \frac{-h}{h} = -1$$

(See Figure 4.) Hence, the $\lim_{h\to 0} \frac{|h|}{h}$ does not exist. Consequently, $f'(0)$ does not exist.

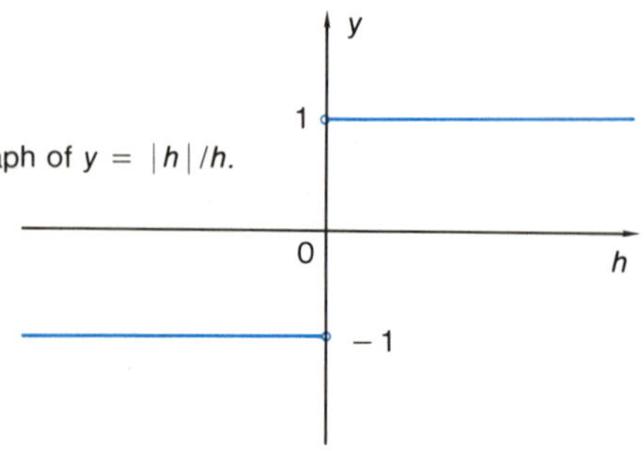

Graph of $y = |h|/h$.

Figure 4

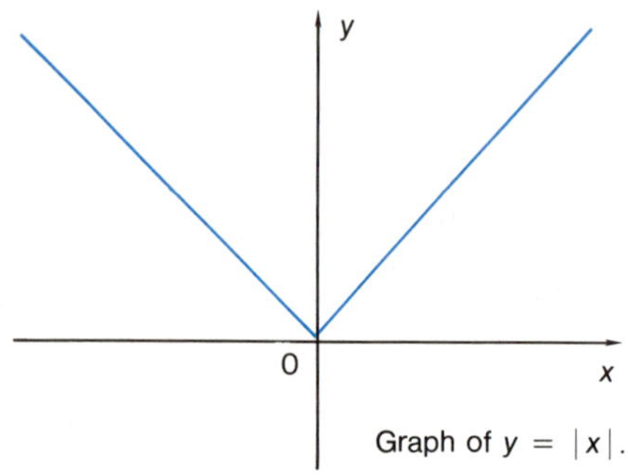

Graph of $y = |x|$.

Figure 5

In Example 5, we note that the function $f(x) = |x|$ is continuous at $x = 0$ (see Figure 5) but is not differentiable at $x = 0$. However, the converse of this statement is always true. We have the following theorem:

THEOREM 1. If a function f is differentiable at $x = a$, then it is continuous at $x = a$.

From the preceding examples and Theorem 1, we have the following information:

(1) If $f'(a)$ exists, then f' is continuous at $x = a$.

(2) If f is discontinuous at $x = a$, then $f'(a)$ does not exist.

(3) If f is continuous at $x = a$, then $f'(a)$ may exist (Example 1) or may not exist (Example 5).

EXAMPLE 6. Find an equation of the tangent line to the graph of $y = 3x^2 + 4x + 8$ at the point (1, 15).

SOLUTION. We know that $f'(x)$ or $\dfrac{dy}{dx}$ represents the slope of the tangent line of the curve at the point $(x, f(x))$. From Example 4, we have

$$\frac{dy}{dx} = f'(x) = 6x + 4$$

When $x = 1$, we have the slope at (1, 15) given by

$$m = \left.\frac{dy}{dx}\right|_{x=1} = f'(1) = 6(1) + 4 = 10$$

Using the point slope form, we have

$$y - 15 = 10(x - 1) \quad \text{or} \quad y = 10x + 5$$

as the required equation of the tangent line.

EXERCISES 2.1

In Problems 1 through 14, find $f'(x)$ and indicate the domain in which the function f is differentiable.

1. $f(x) = x$

2. $f(x) = x^2$

3. $f(x) = 5x^2$

4. $f(x) = 7$

5. $f(x) = 3x^2 + 2$

6. $f(x) = x^2 + 3x + 6$

7. $f(x) = \dfrac{2}{x}$

8. $f(x) = \dfrac{1}{x + 3}$

9. $f(x) = x + \dfrac{1}{x}$

10. $f(x) = \dfrac{x}{x^2 + 1}$

11. $f(x) = \dfrac{1}{\sqrt{x}}$

12. $f(x) = x - \sqrt{x}$

***13.** $f(x) = \sqrt{x^2 + 1}$

***14.** $f(x) = x^3 + \dfrac{1}{\sqrt{x}}$

In Problems 15 through 20, find the values of x for which $f'(x) = 0$, if any. Sketch the graph of f in each case.

15. $f(x) = 3x^2$

16. $f(x) = 2x^2 - 4x$

17. $f(x) = 2x^2 + 4x$

18. $f(x) = \sqrt{x + 1}$

19. $f(x) = x^3$

20. $f(x) = \dfrac{1}{x}$

***21.** Let $f(x) = |x - 2|$. Show that $f'(2)$ does not exist.

***22.** Let $f(x) = \sqrt[3]{x}$. Show that $f'(0)$ does not exist.

In Problems 23 through 26, find the equation of the tangent line to the graph of the given function at the given point.

23. $y = x^3$ at $(2, 8)$

24. $y = \sqrt{x}$ at $(4, 2)$

25. $y = \dfrac{1}{x}$ at $(2, \frac{1}{2})$ **26.** $y = x^2 + x$ at $(1, 2)$

2.2 RULES OF DIFFERENTIATION

In Section 2.1, much of our effort was devoted to the computation of the derivative of a function by using the definition. It would be foolish to continue computing derivatives using the somewhat cumbersome definition (the four-step procedure) each time we desire the derivative of a function, because formulas are available which give us the derivative much more quickly. We shall consider these rules of differentiation in the next two sections. Does this mean that our study of the difference quotient was a waste? Certainly not! We shall see in Section 2.4 that some of the most important concepts in business, physical sciences, biological sciences, and social sciences grow out of the notion of the difference quotient.

We shall state some of the important formulas as rules.

To begin with, we recall that the graph of a constant function $f(x) = c$ is a horizontal line. The slope of a horizontal line is zero everywhere. This means that $f'(x) = 0$, which is our first rule.

RULE 1. If $f(x) = c$, where c is a constant, then $f'(x) = 0$.

That is, the derivative of a constant function is zero.

$$\text{If } f(x) = c, \text{ then } f'(x) = 0 \tag{1}$$

EXAMPLE 1. (a) If $f(x) = 6$, then $f'(x) = 0$.
(b) If $g(x) = \sqrt{10}$, then $g'(x) = 0$. For example, $g'(3) = 0$.
(c) If $s(t) = \pi^2$, then $s'(t) = 0$, or in another notation: $\dfrac{ds}{dt} = 0$.

RULE 2. (*Power rule*). If $f(x) = x^n$, where n is any real number, then $f'(x) = nx^{n-1}$.

That is, the derivative of a constant power of x is the exponent times x raised to a power one less than the given power.

$$\text{If } f(x) = x^n, \quad \text{then} \quad f'(x) = nx^{n-1} \tag{2}$$

Note that if $n = 0$, Rule 1 applies.

EXAMPLE 2. (a) If $f(x) = x^{17}$, then $f'(x) = 17x^{17-1} = 17x^{16}$
(b) If $f(x) = x^{\frac{2}{3}}$, then $f'(x) = \frac{2}{3}x^{\frac{2}{3}-1} = \frac{2}{3}x^{-\frac{1}{3}}$
(c) If $f(x) = x^{\pi}$, then $f'(x) = \pi x^{\pi-1}$

(d) If $f(x) = x^{-7}$, then $f'(x) = (-7)x^{-7-1} = -7x^{-8}$

(e) If $f(x) = 5^3$, then $f'(x) = 0$. (Why?)

RULE 3. If $g(x) = cf(x)$, where c is a constant and $f'(x)$ exists, then

$$g'(x) = cf'(x)$$

That is, the derivative of a constant times a function is the constant times the derivative of the function.

We can combine Rules 2 and 3 and obtain:

$$\text{If } f(x) = cx^n, \quad \text{then} \quad f'(x) = cnx^{n-1} \tag{3}$$

EXAMPLE 3. If $f(x) = 3x^5$, find $f'(x)$ and $f'(2)$.

SOLUTION. By (3), $f'(x) = 3(5)x^{5-1} = 15x^4$. Thus,

$$f'(2) = 15(2)^4 = 15(16) = 240$$

EXAMPLE 4. Compute $\dfrac{dy}{dx}$ if (a) $y = \dfrac{9}{x^4}$; (b) $y = \dfrac{6}{\sqrt{x}}$.

SOLUTION. We first transform each equation into the form cx^n and then apply formula (3).

(a) $y = \dfrac{9}{x^4} = 9x^{-4}$, so that $\dfrac{dy}{dx} = 9(-4)x^{-4-1} = -36x^{-5}$

(b) $y = \dfrac{6}{\sqrt{x}} = 6x^{-1/2}$, so that $\dfrac{dy}{dx} = 6(-\frac{1}{2})x^{(-1/2)-1} = -3x^{-3/2}$

EXAMPLE 5. Find an equation of the tangent line to the graph of $f(x) = \sqrt[3]{x}$ at $x = 27$.

SOLUTION. When $x = 27$, we obtain the y-coordinate of the point on the graph by $y = f(27) = \sqrt[3]{27} = 3$. So the tangent line passes through the point (27, 3), and its slope is $m = f'(27)$. Now, since

$$f(x) = \sqrt[3]{x} = x^{1/3}$$

we have

$$f'(x) = \frac{1}{3}x^{-2/3}$$

Thus,

$$m = f'(27) = \frac{1}{3}(27)^{-2/3} = \frac{1}{3}\cdot\frac{1}{(27)^{2/3}} = \frac{1}{3}\cdot\frac{1}{(\sqrt[3]{27})^2} = \frac{1}{3}\cdot\frac{1}{9} = \frac{1}{27}$$

Using the point-slope form, we have

$$y - 3 = \frac{1}{27}(x - 27)$$

or

$$x - 27y + 54 = 0$$

which is the required equation of the tangent line.

Next, we state the rules for computing the derivative of the sum or difference of two differentiable functions in terms of their derivatives.

If u and v are differentiable functions of x, then

RULE 4. (*Sum Rule*)

$$\text{If } f(x) = u(x) + v(x), \quad \text{then } f'(x) = u'(x) + v'(x) \tag{4}$$

RULE 5. (*Difference Rule*)

$$\text{If } f(x) = u(x) - v(x), \quad \text{then } f'(x) = u'(x) - v'(x) \tag{5}$$

EXAMPLE 6. Find $\dfrac{dy}{dx}$ if

(a) $y = x^4 + 2x^3$

(b) $y = x^3 - x$

SOLUTION. By the sum rule we compute the derivative of x^4 and $2x^3$ and add them together. Thus,

(a) $$\frac{dy}{dx} = \frac{d}{dx}(x^4) + \frac{d}{dx}(2x^3) = 4x^3 + 2(3)x^2 = 4x^3 + 6x^2$$

(b) $$\frac{dy}{dx} = \frac{d}{dx}(x^3) - \frac{d}{dx}(x) = 3x^2 - 1x^0 = 3x^2 - 1$$

EXAMPLE 7. Let $f(x) = x^2 - 6x$. Find the point(s) on the graph of f where the tangent line is horizontal.

SOLUTION. By the difference rule,

$$f'(x) = 2x - 6$$

So we are seeking those point(s) on the graph of f where the slope $(f'(x))$ of the tangent line(s) is zero. Thus, setting

$$f'(x) = 2x - 6 = 0$$

we obtain $x = 3$. Substituting $x = 3$ in the original equation, we obtain

$$f(3) = (3)^2 - 6(3) = 9 - 18 = -9$$

Thus, at the point $(3, -9)$ the tangent line is horizontal (slope $= 0$).

THE PRODUCT RULE

In Rule 4, we saw that the derivative of the sum of two functions is equal to the sum of the derivatives. We might guess that the derivative of the product is equal to the product of the derivatives. That is, if $f(x) = u(x)v(x)$, then $f'(x) = u'(x)v'(x)$. To see that this guess is wrong, Let $u(x) = x^3$ and $v(x) = x^4$. Then

$$f(x) = x^3 \cdot x^4 = x^7$$

so that

$$f'(x) = 7x^6$$

On the other hand, $u'(x) = 3x^2$ and $v'(x) = 4x^3$. Thus, $u'(x) \cdot v'(x) = 3x^3 \cdot 4x^3 = 12x^5$, which is not the same as $f'(x)$. So our guess is really false. Thus, in general,

$$\text{if} \quad f(x) = u(x)v(x), \quad \text{then} \quad f'(x) \neq u'(x)v'(x)$$

The correct formula for the derivative of the product is given in

RULE 6. (*Product Rule*).

$$\text{If } f(x) = u(x)v(x), \quad \text{then} \quad f'(x) = u(x)v'(x) + v(x)u'(x) \tag{6}$$

In words: the derivative of the product of two differentiable functions is the first times the derivative of the second plus the second times the derivative of the first.

In our example above, we have $u(x) = x^3$ and $v(x) = x^4$. We can use (6) to write

$$\begin{aligned} f'(x) &= u(x)v'(x) + v(x)u'(x) \\ &= x^3 4x^3 + x^4 3x^2 \\ &= 4x^6 + 3x^6 \\ &= 7x^6 \end{aligned}$$

which is the correct answer.

EXAMPLE 8. If $f(x) = (3x^3 + 5x^2 + 7)(2x^5 - 8x^3 + 6x)$, find $f'(x)$.

SOLUTION. Let $u(x) = 3x^3 + 5x^2 + 7$ and $v(x) = 2x^5 - 8x^3 + 6x$. Then

$$u'(x) = 9x^2 + 10x, \qquad \text{and} \qquad v'(x) = 10x^4 - 24x^2 + 6$$

Now

$$f = uv$$

and, by (6), we have

$$f'(x) = u(x)v'(x) + v(x)u'(x)$$

Substitution then yields

$$f'(x) = (3x^3 + 5x^2 + 7)(10x^4 - 24x^2 + 6) + (2x^5 - 8x^3 + 6x)(9x^2 + 10x)$$

THE QUOTIENT RULE

Just as the derivative of the product is not equal to the product of the derivatives, the derivative of the quotient is not equal to the quotient of the derivatives. The correct formula for the derivative of the quotient is given below:

RULE 7. *(Quotient Rule)* If $f(x) = u(x)/v(x)$, and $v(x) \neq 0$, then

$$f'(x) = \frac{v(x)u'(x) - u(x)v'(x)}{[v(x)]^2} \tag{7}$$

In words: the derivative of the quotient of two differentiable functions is the denominator times the derivative of the numerator minus the numerator times the derivative of the denominator all divided by the square of the denominator if the denominator is not zero.

EXAMPLE 9. If $f(x) = \dfrac{(x^2 + 1)}{(2x + 1)}$, find $f'(x)$.

SOLUTION. Let $u(x) = x^2 + 1$ and $v(x) = 2x + 1$. Then

$$u'(x) = 2x \qquad \text{and} \qquad v'(x) = 2$$

and for $f = u/v$, we have

$$\begin{aligned} f'(x) &= \frac{u'(x)v(x) - u(x)v'(x)}{[v(x)]^2} \\ &= \frac{(2x)(2x + 1) - (x^2 + 1)(2)}{(2x + 1)^2} \\ &= \frac{4x^2 + 2x - 2x^2 - 2}{(2x + 1)^2} \\ &= \frac{2x^2 + 2x - 2}{(2x + 1)^2} \end{aligned}$$

$$= \frac{2(x^2 + x - 1)}{(2x + 1)^2}$$

EXAMPLE 10. If $f(x) = \dfrac{(x^2 + 2)(x^3 + 1)}{(x^2 - 2)}$ and $x^2 \neq 2$, find $f'(x)$.

SOLUTION. Set $f = u/v$, where

$$u(x) = (x^2 + 2)(x^3 + 1), \qquad \text{and} \qquad v(x) = x^2 - 2$$

We find that $v'(x) = 2x$, and application of Rule 6 yields

$$u'(x) = (x^2 + 2)(3x^2) + (x^3 + 1)(2x) = 5x^4 + 6x^2 + 2x$$

Now

$$f'(x) = \frac{u'(x)v(x) - u(x)v'(x)}{[v(x)]^2}$$

Hence

$$f'(x) = \frac{(5x^4 + 6x^2 + 2x)(x^2 - 2) - (x^2 + 2)(x^3 + 1)(2x)}{(x^2 - 2)^2}$$

$$= \frac{3x^6 - 8x^4 - 12x^2 - 8x}{(x^2 - 2)^2}$$

EXERCISES 2.2

In Problems 1 through 40, find the derivative of the given function.

1. $f(x) = 7x$

2. $f(x) = 3x^4$

3. $f(x) = -5x^2$

4. $f(x) = 2x^2 - 3x + 7$

5. $f(x) = \frac{1}{3}x^3 + \frac{3}{2}x^2$

6. $f(x) = x^3 + 7x^2 - 9x$

7. $f(x) = -4x^4 - 3x^3 + 1$

8. $f(x) = 5x^4 - 3x^3 - 10x^2 + 8x + 4$

9. $f(x) = 2x^{3/2}$

10. $f(x) = -\frac{3}{2}x^{1/2}$

11. $f(x) = \sqrt{x}$

12. $f(x) = 2\sqrt[3]{x}$

13. $f(x) = 2x^{-4}$

14. $f(x) = -3x^{-2}$

15. $f(x) = -3x^{-1} + 1000$

16. $f(x) = -2\sqrt{x} + x^{3/2}$

17. $f(x) = 5x^{101} + 17x - 8$

18. $f(x) = (x + 3)^2$

19. $f(x) = \dfrac{1}{x}$

20. $f(x) = \dfrac{4}{x^3}$

21. $f(x) = \dfrac{1}{x} + \sqrt{x}$

22. $f(x) = \dfrac{1}{\sqrt{x}} - x^3$

23. $f(x) = \dfrac{1}{x^2} - 2$

24. $f(x) = \dfrac{x + 1}{x}$

25. $f(t) = 3t^{1/2} - 9t^3$

26. $s(t) = \dfrac{t^2 - 10t - 3}{5}$

27. $w(s) = \left(s + \dfrac{1}{s}\right)^2$

28. $h(u) = \left(u - \dfrac{1}{u}\right)^2$

29. $f(x) = \left(x + \dfrac{1}{\sqrt{x}}\right)^2$

30. $w(s) = \left(s - \dfrac{1}{\sqrt{s}}\right)^3$

31. $f(x) = (2x^3 - x^2)^2$

32. $f(x) = \sqrt{x}(2x - 2)$

33. $f(x) = 2\sqrt{x}(x^3 - 2x^2)$

34. $f(x) = (3x^{1/2} - 4)(2x^{-1/2})$

35. $f(x) = \dfrac{x + 2}{x - 1}$

36. $f(x) = \dfrac{1}{x + 1}$

37. $f(x) = \dfrac{3x}{4 + x}$

38. $f(x) = \dfrac{2 + 3x}{1 - 4x}$

39. $\dfrac{2 - 3x}{2 + 5x}$

40. $\dfrac{2x - 1}{3 - 2x}$

41. If u, v, and w are differentiable functions of x, and if $f(x) = u(x)v(x)w(x)$, show that

$$f'(x) = u(x)v(x)w'(x) + u(x)v'(x)w(x) + u'(x)v(x)w(x).$$

In Problems 42 through 44, find $f'(x)$.

42. $f(x) = (2x - 1)(x^2 + 1)(4 - x)$

43. $f(x) = (x^5 - 3x^2)(4 - x^3)(5x^2 - x)$

44. $f(x) = (\sqrt{x} + 4x^{-1})(2x^{-1} - 3x^2)(4x^2 - x^{-2})$

In Problems 45 through 49, find the equation of the tangent line to the graph of the given function at the given point.

45. $f(x) = x^3$ at $(2, 8)$

46. $f(x) = \sqrt{x}$ at $(4, 2)$

47. $f(x) = \dfrac{1}{x}$ at $(2, \frac{1}{2})$

48. $f(x) = \frac{1}{4}x^2 + \frac{3}{4}x + \frac{1}{2}$ at $(2, 3)$

49. $f(x) = x^2 + 2x - 3$ at points where the x-coordinates are -1 and 1.

50. At what point on the curve $y = 2x - x^2$ is the slope of the tangent line 5?

51. If $f(x) = 3x^2$, show that the slope of the tangent line to the graph of f at $(4, 48)$ is twice the slope at $(2, 12)$.

52. At what point on the curve $y = x + (4/x)$ is the tangent line parallel to the x-axis?

53. Let $y = 6x - x^2$. Find the point at which the slope is -4.

54. Find the coordinates of the point on the curve $y = 5x^2 + (80/x)$ at which the tangent line is parallel to the x-axis.

2.3 THE CHAIN RULE

Suppose we wish to find the derivative of f, where

$$y = f(x) = (3x^3 + 7x^2 + 8)^{70} \tag{1}$$

From what we have learned so far, the only way to find $f'(x)$ is to compute the seventieth power of $3x^3 + 7x^2 + 8$ and then differentiate the resulting polynomial. To avoid this unpleasant task, we can fortunately use what is called the *chain rule*. By using the chain rule, we shall see that from (1) we obtain

$$y' = f'(x) = 70(3x^3 + 7x^2 + 8)^{69}(9x^2 + 14x) \tag{2}$$

Before studying the chain rule, the reader should review Section 1.5 on the composition of functions. We recall that if g and u are functions, then the composition of g with u is a function f defined by

$$f(x) = g(u(x)).$$

Let us examine the structure of the function defined by (1). It is a composite function, consisting of two pieces: $y = u^{70}$ and $u = 3x^3 + 7x^2 + 8$. The function is constructed by composing the two pieces

$$y = u^{70} \quad \text{and} \quad u = 3x^3 + 7x^2 + 8 \quad \text{to give} \quad y = f(x) = (3x^3 + 7x^2 + 8)^{70}$$

To find the derivatives of these functions we use the *chain rule*:

RULE 8. *(Chain Rule).*

If the function u is differentiable at x, and the function g is differentiable at $u(x)$, then $f(x) = g(u(x))$ is differentiable at x, and

$$f'(x) = g'(u(x)) \cdot u'(x) \tag{3}$$

Using Leibniz' notation, if $f(x) = g(u(x))$, then

$$\frac{df}{dx} = \frac{dg}{du} \cdot \frac{du}{dx} \tag{4}$$

EXAMPLE 1. Let us return to the function f defined by

$$f(x) = (3x^3 + 7x^2 + 8)^{70}$$

Find $f'(x)$.

SOLUTION. We can write $f(x)$ as

$$f(x) = g(u(x))$$

where

$$u(x) = 3x^3 + 7x^2 + 8 \quad \text{and} \quad g(u) = u^{70}$$

Now

$$g'(u) = 70u^{69} \quad \text{and} \quad u'(x) = 9x^2 + 14x$$

Therefore, application of Rule 8 yields

$$\begin{aligned} f'(x) &= g'(u(x)) \cdot u'(x) \\ &= 70[u(x)]^{69}(9x^2 + 14x) \\ &= 70(3x^3 + 7x^2 + 8)^{69}(9x^2 + 14x) \end{aligned}$$

EXAMPLE 2. For each of the following functions, find $f'(x)$

(a) $f(x) = (x^3 + 5x^2 + 7)^{9/2}$ (b) $f(x) = \dfrac{1}{\sqrt[3]{4x^3 + 6x + 7}}$

SOLUTION.

(a) Let

$$u(x) = x^3 + 5x^2 + 7 \quad \text{and} \quad g(u) = u^{9/2}$$

Then

$$g'(u(x)) = \tfrac{9}{2}u(x)^{9/2-1}\ [u(x)]^{(9/2)-1} = \tfrac{9}{2}[u(x)]^{7/2} = \tfrac{9}{2}(x^3 + 5x^2 + 7)^{7/2}$$

and

$$u'(x) = 3x^2 + 10x$$

Hence, by the chain rule,

$$f'(x) = g'(u(x)) \cdot u'(x) = \tfrac{9}{2}(x^3 + 5x^2 + 7)^{7/2}(3x^2 + 10x)$$

(b) Let

$$u(x) = 4x^3 + 6x + 7 \quad \text{and} \quad g(u) = u^{-1/3}$$

Then

$$g'(u(x)) = -\tfrac{1}{3}[u(x)]^{(-1/3)-1} = -\tfrac{1}{3}u^{-4/3} = -\tfrac{1}{3}(4x^3 + 6x + 7)^{-4/3}$$

and

$$u'(x) = 12x^2 + 6$$

Hence, by the chain rule, since $f(x) = g(u(x))$,

$$\begin{aligned} f'(x) &= g'(u(x)) \cdot u'(x) = -\tfrac{1}{3}(4x^3 + 6x + 7)^{-4/3}(12x^2 + 6) \\ &= -2(4x^3 + 6x + 7)^{-4/3}(2x^2 + 1) \end{aligned}$$

EXAMPLE 3. If $y = x^2\sqrt{x^2 - 1}$, find dy/dx.

SOLUTION. Applying Rule 6, we have

$$\frac{dy}{dx} = x^2 \frac{d}{dx}[(x^2 - 1)^{1/2}] + (x^2 - 1)^{1/2} \frac{d}{dx}(x^2)$$

and by the chain rule

$$\begin{aligned} \frac{d}{dx}[(x^2 - 1)^{1/2}] &= \tfrac{1}{2}(x^2 - 1)^{(1/2)-1} \frac{d}{dx}(x^2 - 1) \\ &= \tfrac{1}{2}(x^2 - 1)^{-1/2}(2x) = x(x^2 - 1)^{-1/2} \end{aligned}$$

Hence

$$\begin{aligned} \frac{dy}{dx} &= x^2[x(x^2 - 1)^{-1/2}] + (x^2 - 1)^{1/2}(2x) \\ &= \frac{x^3}{\sqrt{x^2 - 1}} + 2x\sqrt{x^2 - 1} = \frac{x^3 + 2x(x^2 - 1)}{\sqrt{x^2 - 1}} = \frac{3x^3 - 2x}{\sqrt{x^2 - 1}} \end{aligned}$$

EXERCISES 2.3

In Problems 1 through 4, use two methods to find $f'(x)$ and compare your results.

1. $f(x) = (3x + 1)^2$

2. $f(x) = \left(x - \dfrac{1}{x}\right)^2$

3. $f(x) = (x + 1)^{-1}$

4. $f(x) = (x + 1)^3$

In Problems 5 through 20, find $f'(x)$.

5. $f(x) = (x - 5)^4$

6. $f(x) = (2 - 3x)^4$

7. $f(x) = (4x^2 + x)^6$

8. $f(x) = (2x^2 + 1)^{1/2}$

9. $f(x) = (5x^3 + 6x^2 + 9)^{12}$

10. $f(x) = (4x^2 + 5)^{3/2}$

11. $f(x) = 2x(4x - 1)^5$

12. $f(x) = -3x(1 - 4x^2)^3$

13. $f(x) = 5x\sqrt{3x + 1}$

14. $f(x) = -3x^2\sqrt{1 - 2x^2}$

15. $f(x) = (2x + 1)\sqrt{x^2 + 5}$

16. $f(x) = \dfrac{3x}{(x^2 + 1)^5}$

17. $f(x) = \dfrac{-2x}{(2x^2 - x)^{3/2}}$

18. $f(x) = \dfrac{1}{\sqrt{2x^2 + x - 1}}$

19. $f(x) = x^2 + 3 + \dfrac{1}{\sqrt{x^2 + 3}}$

20. $f(x) = \sqrt{\dfrac{3x + 4}{5x + 4}}$

2.4 THE DERIVATIVE AS A RATE OF CHANGE

Although the concept of the derivative arose from the problem of tangents, the notion of derivative leads to the definition of the instantaneous rate of change of a function. There are numerous applications of this notion in a wide variety of disciplines, including the biological, social, and physical sciences, as well as economic theory.

THE DERIVATIVE IN PHYSICAL SCIENCES

Suppose an object is moving in such a way that at any time t (seconds) the distance s (feet) of the object from a fixed starting point is $f(t)$. The equation $s = f(t)$ is called the *equation of motion* for the object. The question we are interested in is: What is the *velocity* of the object at some given time t? The clue as to how velocity should be defined is found in the notion of *average velocity*. If an object travels 80 miles in 2 hours, then the average velocity of the object is 80/2 = 40 miles per hour. That is,

$$\text{Average velocity} = \frac{\text{Change in distance}}{\text{Change in time}}$$

Now, at time t the object is $f(t)$ units from a fixed reference point, and at time $t + h$, the object is $f(t + h)$ units from the same fixed reference point. Thus, the

$$\text{Change in distance} = f(t + h) - f(t)$$

and the

$$\text{Change in time} = (t + h) - t = h.$$

Hence,

$$\text{Average velocity from time } t \text{ to time } t+h = \frac{f(t+h) - f(t)}{h} \tag{1}$$

It seems reasonable to expect that, when h is close to zero, the difference quotient in (1) should approximate the velocity at time t. We therefore define the *instantaneous velocity* at time t, denoted by $v(t)$, as the limit of the difference quotient (1) as h (change in time) tends to zero. That is,

$$v(t) = \lim_{h \to 0} \frac{f(t+h) - f(t)}{h} \tag{2}$$

We observe that the right side of (2) is, in fact, the derivative of $f(t)$. Hence

$$v(t) = f'(t) \tag{3}$$

EXAMPLE 1. The height s of a ball thrown upward is given by the equation

$$s = f(t) = 144t - 16t^2$$

where s is in feet and t is in seconds measured from the instant the ball is thrown. Find:

(a) the average velocity of the ball over the first 2 seconds;

(b) the instantaneous velocity of the ball at $t = 0$ and $t = 2$

SOLUTION. (a) If $f(t) = 144t - 16t^2$, then the average velocity over the first 2 seconds is given by

$$\frac{f(2) - f(0)}{2 - 0} = \frac{[(144)(2) - 16(2)^2] - 0}{2} = \frac{224}{2} = 112 \text{ feet per second}$$

(b) The instantaneous velocity $v(t)$ at time t is given by

$$v(t) = f'(t) = 144 - 32t$$

so that $v(0) = 144$ and $v(2) = 144 - 32(2) = 80$.

We observe that the instantaneous velocities, 144 feet per second and 80 feet per second, are both different from the average velocity of 112 feet per second.

The concepts involved in the average rate of change and instantaneous rate of change can be applied to any pair of related quantities. Suppose y is a function of x. We let Δx denote the change in x. Here Δx (Δ is the Greek capital letter delta) should be thought of as a single entity, pronounced "delta x," and not as the product of the symbols Δ and x. For example, if x changes from 1 to 4 then $\Delta x = 4 - 1 = 3$. We let Δy (read "delta y") denote the corresponding change in the dependent variable y. That is, for the function $y = f(x)$,

$$\Delta y = f(x + \Delta x) - f(x)$$

Then the ratio

$$\frac{\Delta y}{\Delta x} = \frac{f(x + \Delta x) - f(x)}{\Delta x} \tag{4}$$

is the average rate of change of y per unit change in x. If we take the limit in (4), we obtain the derivative

$$\frac{dy}{dx} = f'(x) = \lim_{\Delta x \to 0} \frac{\Delta y}{\Delta x} \tag{5}$$

and this derivative is called the *instantaneous rate of change* of y per unit change in x.

EXAMPLE 2. Find the (instantaneous) rate of change of $y = x^4$ with respect to x. Interpret your result when $x = 3$ and $x = -2$.

SOLUTION. The rate of change is given by

$$\frac{dy}{dx} = 4x^3$$

Thus, the rate at which x^4 changes with respect to x is $4x^3$. When $x = 3$, then

$$\left.\frac{dy}{dx}\right|_{x=3} = 4(3)^3 = 4(27) = 108$$

This means that y is increasing 108 times as fast as x. When $x = -2$, then

$$\left.\frac{dy}{dx}\right|_{x=-2} = 4(-2)^3 = 4(-8) = -32$$

This means that y is decreasing (because of the minus sign) 32 times as fast as x is increasing.

THE DERIVATIVE IN ECONOMICS

If $C(x)$ is the total cost of producing x units of a product, then the *average cost* per unit, denoted by $\overline{C}(x)$, is given by (total cost) ÷ (number of units produced). That is,

$$\overline{C}(x) = \frac{C(x)}{x} \tag{6}$$

By multiplying both sides of equation (6) by x, we obtain

$$C(x) = x\overline{C}(x) \tag{7}$$

Equation (7) tells us that the total cost is the product of the number of units produced and the average cost per unit.

For example, if the total cost of producing 15 units is \$90.00, then the average cost per unit is $\bar{C} = \dfrac{90}{15} = 6$ dollars, and $90 = 6(15)$.

Suppose a manufacturer produces x units at a total cost of $C(x)$. Then the cost of producing the 99th item is $C(99) - C(98)$. So, in general, if the production level is increased from x units to $x + \Delta x$ units, then the total cost increases from $C(x)$ to $C(x + \Delta x)$. If we let ΔC denote the change in cost, then

$$\Delta C = C(x + \Delta x) - C(x)$$

The ratio

$$\frac{\Delta C}{\Delta x} = \frac{C(x + \Delta x) - C(x)}{\Delta x} \tag{8}$$

represents the average cost for a change in production level from x to $x + \Delta x$.

Now, suppose the production level is increased by one unit. Then $\Delta x = 1$, and

$$\frac{\Delta C}{\Delta x} = \frac{C(x + 1) - C(x)}{1} \tag{9}$$

represents the average change in cost corresponding to a one-unit increase in production. That is, equation (9) gives the cost of producing the $(x + 1)$st item.

In equation (8), if we let $\Delta x \to 0$, then the limit

$$\lim_{\Delta x \to 0} \frac{\Delta C}{\Delta x} = \lim_{\Delta x \to 0} \frac{C(x + \Delta x) - C(x)}{\Delta x} = \frac{dC}{dx} \tag{10}$$

represents the instantaneous rate of change of $C(x)$ with respect to x. Economists refer to this derivative $\dfrac{dC}{dx}$ as the marginal cost.

That is,

$$\text{marginal cost} = \frac{dC}{dx}$$

EXAMPLE 3. If a manufacturer's cost function is given by

$$C(x) = 30 + 2\sqrt{x}$$

find:

(a) The marginal cost when $x = 49$

(b) The cost of producing the fiftieth item

SOLUTION.

(a) The marginal cost $= \dfrac{dC}{dx} = 0 + 2 \cdot \dfrac{1}{2\sqrt{x}} = \dfrac{1}{\sqrt{x}}$

when $x = 49$, we have

$$\left.\frac{dC}{dx}\right|_{x=49} = \frac{1}{\sqrt{49}} = \frac{1}{7} = 0.1428 \tag{11}$$

(b) The cost of producing the fiftieth item

$$\begin{aligned} &= C(50) - C(49) \\ &= (30 + 2\sqrt{50}\,) - (30 + 2\sqrt{49}\,) \\ &= 2(\sqrt{50} - 7) \\ &= 0.1421 \end{aligned} \tag{12}$$

In (11) and (12), we observe that the two numbers are close to each other, and this is generally true. For this reason we interpret marginal cost as the approximate change in cost resulting from one additional unit of output.

What we have said about the cost function also applies to the revenue and profit functions. For example, if $R(x)$ and $P(x)$ represent the revenue and profit functions, respectively, then

$$\text{Average revenue} = \frac{R(x)}{x}$$

$$\text{Average profit} = \frac{P(x)}{x}$$

$$\text{Marginal revenue} = \frac{dR}{dx} = R'(x)$$

$$\text{Marginal profit} = \frac{dP}{dx} = P'(x)$$

We also have the similar interpretation for the marginal revenue and marginal profit. For example, if $R(x) = -20 + 6x + 2x^2$ (in dollars), then $R'(x) = 6 + 4x$. $R'(5) = 6 + 4(5) = 6 + 20 = 26$. Thus, the approximate additional revenue received from selling 6 units instead of 5 is 20 dollars.

EXAMPLE 4. The Loud Video Corporation estimates that the weekly profit from the sale of x videos is given by

$$P(x) = 50x + x^2 - 7000 \text{ dollars}$$

Find:

(a) The average profit per video if 200 videos are sold

(b) The average rate of change of profit if the weekly sales increase from 200 to 250

(c) The profit on the 200th video

(d) The marginal profit

(e) The marginal profit when $x = 199$

SOLUTION.

(a) Average profit when $x = 200$

$$= \frac{P(200)}{200} = \frac{(50)(200) + (200)^2 - 7000}{200}$$

$$= 215 \text{ dollars/video.}$$

(b) Average profit from 200 to 250 is given by

$$\frac{\Delta P}{\Delta x} = \frac{P(250) - P(200)}{250 - 200}$$

$$= \frac{[(50)(250) + (250)^2 - 7000] - [(50)(200) + (200)^2 - 7000]}{50}$$

$$= 480 \text{ dollars/video.}$$

(c) The profit on the 200th video is given by

$$P(200) - P(199) = [(50)(200) + (200)^2 - 7000] - [(50)(199) + (199)^2 - 7000]$$
$$= 449 \text{ dollars}$$

(d) The marginal profit is

$$P'(x) = \frac{dP}{dx} = 50 + 2x$$

(e) The marginal profit when $x = 199$ is

$$P'(199) = 50 + 2(199) = 448 \text{ dollars.}$$

We observe from (c) and (e) that the marginal profit when $x = 199$ is approximately equal to the profit on the next item—that is, the profit on the 200th video.

ELASTICITY OF DEMAND

If $x(p)$ represents the demand of a product at price p, then the proportional change in demand divided by the proportional change in price can be symbolically written as

$$\frac{x(p+h) - x(p)}{x(p)} \div \frac{(p+h) - p}{p} \tag{13}$$

We can rewrite (13) as

$$\frac{x(p+h) - x(p)}{h} \cdot \frac{p}{x(p)} \tag{14}$$

If $h \to 0$, then (14) becomes

$$x'(p) \cdot \frac{p}{x(p)} = p \cdot \frac{x'(p)}{x(p)} \tag{15}$$

The expression (15) denoted by $E(p)$ is called the *elasticity of demand* at price p. Thus,

$$E(p) = p \cdot \frac{x'(p)}{x(p)} \tag{16}$$

The elasticity of demand gives us an approximate percentage change in demand corresponding to a 1% increase in price.

The economists usually distinguish three ranges of values $E = E(a)$ at a fixed price $p = a$:

(1) If $E < -1$, we say that the demand is *elastic*. In this case the number of units demanded decreases (increases) relatively more than an incremental increase (decrease) in price.

(2) If $E > -1$, we say that the demand is *inelastic*. In this case the number of units demanded decreases (increases) relatively less than an incremental increase (decrease) in price.

(3) If $E = -1$, we say that the demand has *unit elasticity*. In this case an increase (decrease) in price results in a proportional decrease (increase) in the number of units demanded.

EXAMPLE 5. Find the elasticity of demand if the demand function is

$$x(p) = 40 - p^2 \qquad \text{and} \qquad p = 2$$

SOLUTION. We have

$$x'(p) = \frac{dx}{dp} = -2p$$

Substituting $p = 2$ in $x(p)$ and $x'(p)$, we have

$$x(2) = 40 - (2)^2 = 36$$

$$x'(p) = (-2)(2) = -4$$

Thus,

$$E(2) = x'(2) \cdot \frac{2}{x(2)} = -4 \cdot \frac{2}{36} = -\frac{2}{9}$$

This tells us that there is approximately a 2/9% decrease in demand if the price is increased by 1%.

Since $-2/9 > -1$ the demand is inelastic.

MARGINAL PROPENSITY TO CONSUME

A function that plays an important role in economic analysis is the *consumption function*. If c denotes the national consumption, and i denotes the total national income, then the relationship between c and i expressed by

$$c = f(i)$$

is called the *consumption function*.

The *marginal propensity to consume* (MPC) is defined as the rate of change of consumption with respect to income. This is merely the derivative of c with respect to i. Thus,

$$\text{MPC} = \frac{dc}{di} \tag{17}$$

If we assume that the difference between the total income i and the total consumption c is the savings s, then

$$s = i - c \tag{18}$$

Differentiating both sides of (18) with respect to i, we obtain

$$\frac{ds}{di} = 1 - \frac{dc}{di} \tag{19}$$

We define $\frac{ds}{di}$ *as the marginal propensity to save* (MPS).

So the MPC tells us at what rate the consumption changes with respect to income, and the MPS indicates at what rate savings change with respect to income.

EXAMPLE 6. Suppose the consumption function is given by:

$$c = 0.0005i^2 + 0.2i$$

where c and i are in billions of dollars.

Find MPC, and find MPS when $i = 400$.

SOLUTION. We have

$$\frac{dc}{di} = (0.0005)(2i) + 0.2$$

$$= (0.001)i + 0.2$$

$$\left.\frac{dc}{di}\right|_{i=400} = (0.001)(400) + 0.2$$

$$= 0.4 + 0.2$$

$$= 0.6$$

So the marginal propensity to consume when $i = 400$ is 0.6.

$$\text{Since } \frac{ds}{di} = 1 - \frac{dc}{di}, \text{ we have}$$

$$\left.\frac{ds}{di}\right|_{i=400} = 1 - 0.6 = 0.4$$

This tells us that if the current income of 400 billion dollars is increased by one billion dollars, then the nation will consume approximately 60% and save approximately 40% of that increase.

EXERCISES 2.4

1. The position of a particle moving on the x axis is given by $x(t) = 2t^2 - 4t + 5$, where $t \geq 0$ is in seconds. Find: (a) the initial position of the particle, (b) the average velocity of the particle between $t = 2$ and $t = 4$ seconds, (c) the instantaneous velocity at $t = 2$ and $t = 4$.

2. Do problem 1 if $x(t) = t^3 - t^2 + 2t$.

3. The height of a ball thrown upward is given by $s = -16t^2 + 128t$, where s is in feet, and t is in seconds measured from the moment the ball is thrown. Find: (a) the average velocity over the first two seconds, (b) the instantaneous velocity at $t = 0$ and $t = 2$, (c) the maximum height the ball will attain. [*Hint:* The maximum height is attained when $s'(t) = 0$.]

4. Do problem 3, if the height of a ball thrown upward from the roof of a building is given by

$$s = -16t^2 + 144t + 96$$

5. Given that the total cost C (in dollars) of producing x units of a product is $C(x) = 2x + 160$:
   (a) What is the average cost per unit if 80 units are made?
   (b) What is the average rate of change of cost?
   (c) What is the marginal cost of the 50th unit? 80th unit?

6. Repeat problem 5 with $C = 2x^2$.

7. Given that the total cost C (in dollars) of making x gallons of oil is $C(x) = 20\sqrt{x} + 5$:
   (a) Find the marginal cost of 160 gallons output.

(b) At what level of output is the marginal cost \$0.20? \$0.40? [*Hint:* Solve for x such that $C'(x) = \$0.20$.]

8. A manufacturer sells x units of product at a price of p dollars per unit, where $p(x) = 100 + (x/50)$.

(a) Find the total revenue if the total output is 2500 units.

(b) Find the marginal revenue when the ouput is 500 units, 1000 units, 2500 units.

(c) At what level of output is the marginal revenue \$150 per unit? [*Hint:* The revenue $R(x) = xp(x) = 100x + (x^2/50.]$

9. The profit function for Pepperoni Pizza Parlor is given by the equation $P(x) = 2x - 10 + (1/x)$, where x is the number of pizzas sold per day, and P is in dollars. Find the marginal profit at $x = 10$, and interpret your result.

10. Suppose that the total cost of making x units of a product is C dollars, where

$$C(x) = 20 + 18x^{2/3}$$

For what value of x will the marginal cost \$2?

11. If the total cost C of making x units of a product is

$$C(x) = 6x - 0.03x^2 + 0.00005x^3$$

construct a table showing the marginal cost at 50, 100, 150, 200, 250, and 300 units.

12. For what output will the marginal cost be \$22 if the cost of making x units is C dollars, where

$$C(x) = 12x - 0.045x^2 + 0.0001\frac{x^3}{3}?$$

13. A brush fire in a valley spreads along a straight line according to the law

$$s(t) = \tfrac{3}{2}t - \tfrac{1}{4}t^2$$

where the distance s is measured from a starting point, and t is the time measured in hours.

(a) What is the instantaneous rate of spreading (velocity) of the fire at $t = 1$ hr? $t = 1.5$ hr? $t = 2$ hr?

(b) At what distance from the starting point does the fire stop spreading?

(c) If the fire spreads radially at the given rate, find the area of the circular region devasted by the fire.

14. Suppose that W is the number of gallons of water in a pool being drained for cleaning, and t is the number of minutes after the draining starts

$$W(t) = 300(40 - t)^2$$

(a) Find the instantaneous rate at which the water is running out 5 min., 10 min., 20 min., and 30 min. after the pool has started to drain.

(b) In how many minutes after the pool has started to drain will the pool be empty?

15. An offshore oil well is leaking. The oil spreads on the surface of the water in a circle. If the radius r of the circle is increasing at the rate of 2 ft./sec., at what rate is the area A of the oil spill increasing when $r = 20$ ft.? 50 ft.? 200 ft.?

16. A spherical weather balloon is being inflated. At a certain instant the volume of the balloon is increasing at the rate of 500 cu. in./min. If at that instant the radius of the balloon is 40 in., how fast is the surface area increasing?

17. A spherical snowball is melting. At a certain instant the volume is decreasing ($V'(t) < 0$) at the rate of 10 cu. in./hr. At that instant the radius of the snowball is 25 in.

(a) How fast is the diameter decreasing?

(b) How fast is the surface area decreasing?

18. Suppose in a model the pollution index $I(t)$ is related to a dose $A(t)$ of an antipollution agent by the equation

$$I \cdot A^{3/2} = k \qquad \text{(a constant)}$$

At a given instant the index $I = 30\%$ and the dose $A = 9$ units. If at that moment the dose of the antipollution agent is decreasing at the rate of 3 units/min., how fast is the index I changing?

19. Smoke is being emitted into the atmosphere in the form of an inverted right circular cone (see Figure 1). At any instant, the diameter is equal to the height h. If

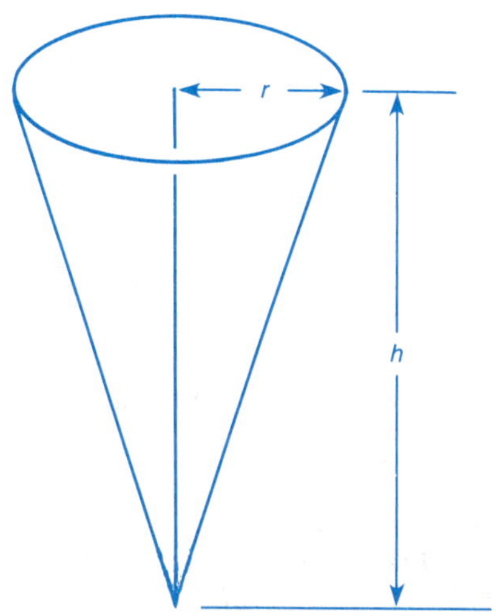

Figure 1

the rate of smoke emission is 200 cu. ft./min., how fast is the height increasing when $h = 50$ ft.? [*Hint*: Volume of cone $= \frac{1}{3}\pi r^2 h$.]

20. Midget Motors is introducting a new car called the Mongoose. Road tests show that $G(s)$, the number of miles traveled per gallon by this car, is given by

$$G(s) = 20 + \frac{s}{2} - \frac{s^2}{180}$$

where s is the speed of the in miles per hour. Find the number of miles per gallon that this car travels at the speed where $G'(s) = 0$.

In Problems 21 through 24, a demand function is given. Find the elasticity of demand at the indicated prices.

21. $x(p) = 30 - 2p^2$ at $p = 0.5$

22. $x(p) = 200 + \dfrac{1}{p}$ at $p = 3$

23. $x(p) = 25 + \dfrac{3}{p} - 2p$ at $p = 2$

24. $x(p) = \dfrac{p}{p^2 + 1} + 300 - p$ at $p = 5$

In Problems 25 through 28, let $C(x)$ represent the total cost function. The quantity $x \cdot \dfrac{C'(x)}{C(x)}$ is called the *elasticity of cost*; compute the elasticity of cost for the given value of x.

25. $C(x) = 2x + 500$ at $x = 4$

26. $C(x) = 5000 + 100x - 0.1x^2$ at $x = 10$

27. $C(x) = 3x + \dfrac{1}{x} + 50$ at $x = 2$

28. $C(x) = 5x + \dfrac{1}{\sqrt{x}} + 900$ at $x = 16$

In Problems 29 through 32, find the marginal propensity to consume and the marginal propensity to save for the given consumption function at the specified income level (i is in billions of dollars).

29. $C = \dfrac{5}{6}i + \dfrac{2}{i} + 50$ $\quad i = 200$

30. $C = 10\sqrt{i} + 30 \qquad i = 400$

31. $C = \dfrac{2}{3}i + \dfrac{5}{i^2} + 60 \qquad i = 10$

32. $C = \dfrac{i^2 + 1}{i + 1} \qquad i = 10$

2.5 IMPLICIT DIFFERENTIATION

Suppose that f is a function given by the equation $y = f(x)$. Then we say that y is defined *explicitly* as a function of x. The functions we have encountered so far have been of this form. However, a relationship between x and y expressed in the form

$$g(x, y) = 0$$

may also define one or more functions of x. In such a case the function or functions are said to be defined *implicitly*.

Thus, in the equations

$$3x^3 + y^2 - 8y = 0 \tag{1}$$

$$x^2 - 4 - 3xy = 0 \tag{2}$$

and

$$x^2 + y^2 - 1 = 0 \tag{3}$$

y is defined implicitly as a function of x. Sometimes it is possible to solve an equation for y as a function of x. For example, from Equation (2), we have

$$y = \frac{x^2 - 4}{3x}$$

Equation (3) can be solved for y, yielding two functions:

$$y = f_1(x) = \sqrt{1 - x^2} \qquad \text{and} \qquad y = f_2(x) = -\sqrt{1 - x^2}$$

Therefore, Equation (3) implicitly defines two functions of x (see Figure 1).

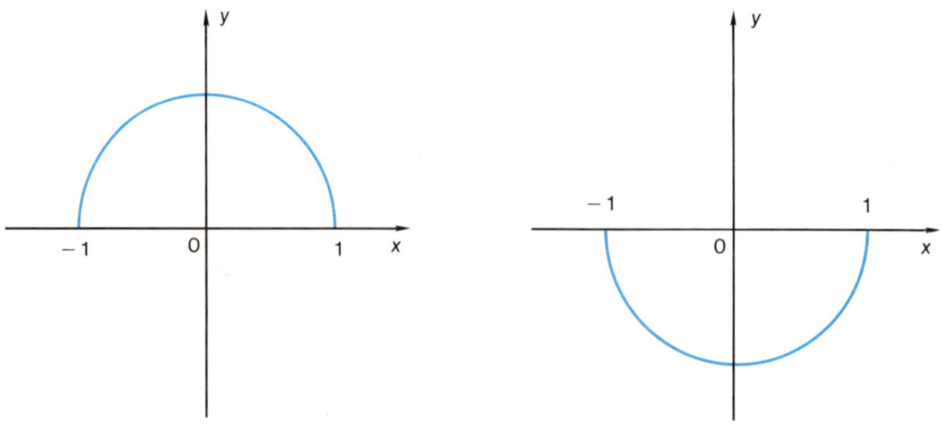

(a) $y = f_1(x) = \sqrt{1 - x^2}$ **Figure 1** (b) $y = f_2(x) = -\sqrt{1 - x^2}$

However, sometimes it may be difficult or even impossible to solve for y as a differentiable function of x. The technique of *implicit differentiation* allows us to compute the derivative of a function defined implicitly by an equation without first finding the formula that defines the function. We use the following two step process:

(1) We assume y is a differentiable function of x. We differentiate both sides of the equation with respect to x, and then

(2) We solve this equation for $\frac{dy}{dx}$.

EXAMPLE 1. Find $\frac{dy}{dx}$ for $x = y^3$.

SOLUTION. If we assume $y = f(x)$, then, by the chain rule,

$$\frac{d}{dx}(y^3) = \frac{d}{dx}[f(x)]^3 = 3[f(x)]^2\frac{df}{dx} = 3y^2\frac{dy}{dx}$$

Thus, differentiation of both sides of $x = y^3$ gives

$$\frac{d}{dx}(x) = \frac{d}{dx}(y^3)$$

or

$$1 = 3y^2\frac{dy}{dx}$$

Solving this equation for $\frac{dy}{dx}$, we obtain

$$\frac{dy}{dx} = \frac{1}{3y^2}$$

EXAMPLE 2. Find $\frac{dy}{dx}$ if

$$2x^2 + 3y^2 = 4 \tag{4}$$

SOLUTION 1. Differentiation both sides of Equation (4) with respect to x, we obtain

$$\frac{d}{dx}(2x^2 + 3y^2) = \frac{d}{dx}(4)$$

$$\frac{d}{dx}(2x^2) + \frac{d}{dx}(3y^2) = \frac{d}{dx}(4)$$

$$4x + 6y\frac{dy}{dx} = 0$$

and so

$$\frac{dy}{dx} = -\frac{4x}{6y} = -\frac{2x}{3y}$$

SOLUTION 2. Suppose that we solve Equation (4) for y. We have

$$y = \frac{1}{\sqrt{3}}\sqrt{4-2x^2} \qquad (2 - x^2 \geq 0) \tag{5}$$

or

$$y = -\frac{1}{\sqrt{3}}\sqrt{4-2x^2} \qquad (2 - x^2 \geq 0) \tag{6}$$

From (5) we obtain

$$\frac{dy}{dx} = \frac{1}{\sqrt{3}} \cdot \frac{1}{2}(4-2x^2)^{-1/2}\frac{d}{dx}(4-2x^2) = \frac{1}{2\sqrt{3}}(4-2x^2)^{-1/2}(-4x)$$

$$= -\frac{2x}{\sqrt{3}(4-2x^2)^{1/2}} = -\frac{\sqrt{3}(2x)}{\sqrt{3}\sqrt{3}(4-2x^2)^{1/2}} = -\frac{2x}{3y}$$

which agrees with the preceding result we obtained. The reader is encouraged to repeat the solution process starting with Equation (6) and show that the same result is obtained again.

EXAMPLE 3. Find the equation of the tangent line to the ellipse (see Figure 2)

$$9x^2 + 16y^2 = 144 \tag{7}$$

at the point $(\sqrt{7}, \frac{9}{4})$.

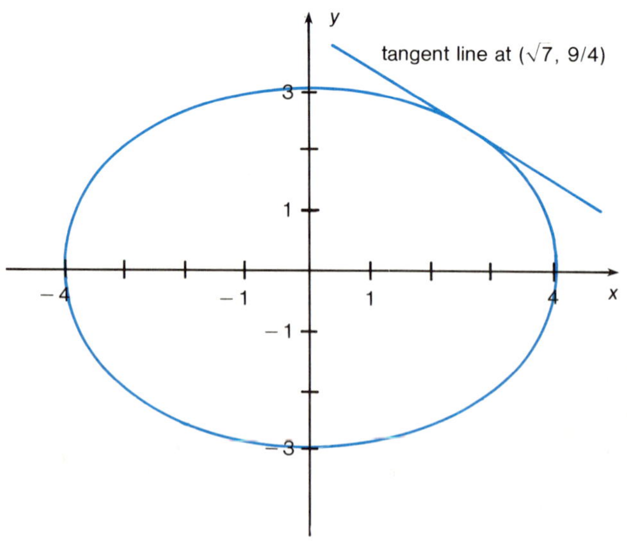

Figure 2

SOLUTION. Differentiating implicitly, we obtain

$$\frac{d}{dx}(9x^2 + 16y^2) = \frac{d}{dx}(144)$$

or

$$\frac{d}{dx}(9x^2) + \frac{d}{dx}(16y^2) = 0$$

$$18x + 32y\frac{dy}{dx} = 0$$

Hence

$$\frac{dy}{dx} = -\frac{9x}{16y}$$

and so the slope of the tangent line at the point $(\sqrt{7}, \frac{9}{4})$ is

$$\left.\frac{dy}{dx}\right|_{(\sqrt{7},9/4)} = -\frac{9(\sqrt{7})}{16(\frac{9}{4})} = -\frac{\sqrt{7}}{4}$$

Therefore, the equation of the tangent line at $(\sqrt{7}, \frac{9}{4})$ is given by

$$y - \frac{9}{4} = -\frac{\sqrt{7}}{4}(x - \sqrt{7})$$

Simplifying, we have

$$4y + \sqrt{7}x - 16 = 0$$

EXAMPLE 4. Two cars C_1 and C_2 are racing on two highways intersecting at right angles (see Figure 3). C_1 is travelling south at the rate of 100 ft./sec., and C_2 is travelling west at the rate of 120 ft./sec. At a certain instant, C_1 is 300 feet from point A, and C_2 is 400 feet from point A.

(a) How fast is the distance between the two cars changing at that instant?

(b) If the cars maintain the same speed and ignore stop signs, will they collide?

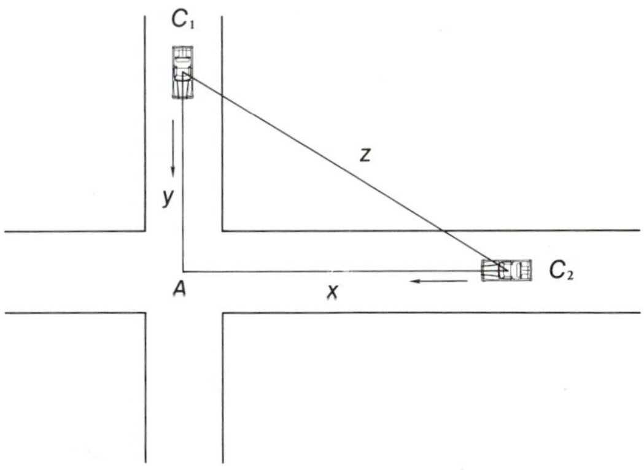

Figure 3

SOLUTION.

(a) Let x, y, and z be the distances from C_2 to A, C_1 to A, and C_1 to C_2, respectively. Here x, y, and z are functions of time t. Using the Pythagorean theorem, we have the relation

$$z^2 = x^2 + y^2 \tag{8}$$

Differentiating both sides of Equation (8) with respect to t, we obtain

$$\frac{d}{dt}(z^2) = \frac{d}{dt}(x^2 + y^2) \quad \text{and} \quad 2z\frac{dz}{dt} = 2x\frac{dx}{dt} + 2y\frac{dy}{dt}$$

Dividing by 2, we have

$$z\frac{dz}{dt} = x\frac{dx}{dt} + y\frac{dy}{dt} \tag{9}$$

At the instant in question, $x = 400$ ft. and $y = 300$ ft., and so

$$z = \sqrt{(400)^2 + (300)^2} = 500 \text{ ft.}$$

Therefore, from Equation (9),

$$\frac{dz}{dt} = \frac{1}{500}[(400)(120) + (300)(100)] = \frac{780}{5} = 156$$

(b) Since car C_2 is 400 ft. from point A and is traveling at 120 ft./sec., it will arrive at point A in $400/120 = 10/3$ sec. later, while car C_1 will arrive at point A in $300/100 = 3$ sec. In other words, the cars (we hope!) will miss each other by 1/3 of a second.

EXAMPLE 5. Find $\frac{dy}{dx}$, when

$$x^3 + y^3 - 3xy = 0 \tag{10}$$

SOLUTION.

$$\frac{d}{dx}(x^3 + y^3 - 3xy) = \frac{d}{dx}(0)$$

or

$$\frac{d}{dx}(x^3) + \frac{d}{dx}(y^3) + \frac{d}{dx}(-3xy) = 0$$

This implies that

$$3x^2 + 3y^2\frac{dy}{dx} - 3\left(x\frac{dy}{dx} + y\right) = 0$$

Collecting terms involving dy/dx and factoring, we obtain

$$(y^2 - x)\frac{dy}{dx} = y - x^2$$

Hence

$$\frac{dy}{dx} = \frac{y - x^2}{y^2 - x} \qquad \text{for} \quad y^2 \neq x$$

EXERCISES 2.5

In Problems 1 through 4, find dy/dx using two methods (assume that y is a differentiable function of x).

1. $x^2 + y^2 = 25$

2. $y^2 = 3x^3$

3. $y^2 = \dfrac{x+2}{x-2}$

4. $x^3 + y^3 = 1$

In Problems 5 through 14, find dy/dx.

5. $y^2 = 3x^3 + 5x^2 + 7$

6. $y^3 + 3xy = x$

7. $x^2y^2 = a^2(x^2 - y^2)$ (a is a constant)

8. $\dfrac{x^2}{a^2} + \dfrac{y^2}{b^2} = 1$ (a and b are constants)

9. $x^5 + y^5 = 5(x^3y^2 + x^2y^3)$

10. $x^{2/3} + y^{2/3} = a^{2/3}$ (a is a constant)

11. $\sqrt{1 - x^2} + \sqrt{1 - y^2} = 8(x - y)$

12. $(x + 4)^5 = 2y^3$

13. $2x^2 = \dfrac{x + y}{x - y}$

14. $y^2 = \dfrac{\sqrt[3]{x^2 + 7}}{3x}$

In Problems 15 through 18, find the equation of the tangent line to the given curve at the given point.

15. $2y^2 - 3x^2 = 6$ at $(2, 3)$

16. $x^2 + y^2 = 4$ at $(-\sqrt{3}, 1)$

17. $x - 1 - 2\sqrt{x} + \sqrt{y} = 0$ at $(4, 1)$

18. $\dfrac{y}{x} + \dfrac{x}{y} = 5xy$ at $(1, \frac{1}{2})$

19. Show that the two circles

$$(x - 6)^2 + (y - 3)^2 = 20 \quad \text{and} \quad x^2 + y^2 + 2x + y = 10$$

have a common tangent line at $(2, 1)$.

20. A painter is on top of a ladder 25 ft. long that is leaning against a vertical wall (Figure 4). If a prankster pulls the bottom of the ladder away from the wall at the rate of 6 in./sec., how fast is the painter sliding down when the bottom is 20 ft. from the wall?

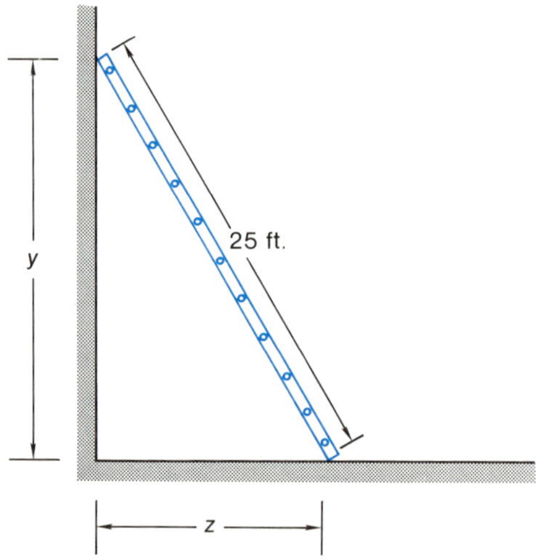

Figure 4

21. Ship A leaves port at 2:00 P.M. and travels due east at 15 knots. At 6:00 P.M. on the same day, ship B leaves the same port and travels south at 20 knots. How fast is the distance changing between A and B when B has traveled 80 nautical miles?

2.6 HIGHER DERIVATIVES

Consider the function f, where $f(x) = x^3$. The derivative of f is a new function f', where $f'(x) = 3x^2$. We can again differentiate the function f'. If we do so, we get a new function f'', where $f''(x) = 6x$. If we further differentiate the function f'', we obtain $f'''(x) = 6$, and so on.

In general, if a function f has a derivative f', and if f' itself is differentiable, we denote the derived function of f' by f'' and call f'' the *second derivative* of f.

Continuing in this manner, we obtain the functions $f''', f^{(4)}, \ldots, f^{(n)}$, each of which is the derivative of the preceding function. Of course, here we are assuming that each of the preceding functions is differentiable. There are various notations used for the higher derivatives. If $y = f(x)$, the derivatives can be written as follows:

first derivative $\qquad \dfrac{dy}{dx}, \dfrac{df}{dx}, y', f'$

second derivative $\qquad \dfrac{d^2y}{dx^2}, \dfrac{d^2f}{dx^2}, y'', f''$

third derivative $\dfrac{d^3y}{dx^3}, \dfrac{d^3f}{dx^3}, y''', f'''$

$\vdots$

nth derivative $\dfrac{d^ny}{dx^n}, \dfrac{d^nf}{dx^n}, y^{(n)}, f^{(n)}$

The reason for the peculiar location of the superscripts in the first two notations is that d/dx is a symbol of the operation of differentiation. Thus, we have

$$\frac{dy}{dx} = \frac{d}{dx}(y)$$

$$\frac{d^2y}{dx^2} = \frac{d}{dx}\left(\frac{dy}{dx}\right) = \frac{d}{dx}\left(\frac{d}{dx}(y)\right)$$

$$\frac{d^3y}{dx^3} = \frac{d}{dx}\left(\frac{d^2y}{dx^2}\right) = \frac{d}{dx}\left(\frac{d}{dx}\left[\frac{d}{dx}(y)\right]\right)$$

and so on.

EXAMPLE 1. Find the derivatives of all orders for the function defined by $y = x^5$.

SOLUTION. We have

$$\frac{dy}{dx} = 5x^4$$

$$\frac{d^2y}{dx^2} = \frac{d}{dx}\left(\frac{dy}{dx}\right) = \frac{d}{dx}(5x^4) = 20x^3$$

$$\frac{d^3y}{dx^3} = \frac{d}{dx}\left(\frac{d^2y}{dx^2}\right) = \frac{d}{dx}(20x^3) = 60x^2$$

$$\frac{d^4y}{dx^4} = \frac{d}{dx}\left(\frac{d^3y}{dx^3}\right) = \frac{d}{dx}(60x^2) = 120x$$

$$\frac{d^5y}{dx^5} = \frac{d}{dx}\left(\frac{d^4y}{dx^4}\right) = \frac{d}{dx}(120x) = 120$$

$$\frac{d^6y}{dx^6} = \frac{d}{dx}\left(\frac{d^5y}{dx^5}\right) = \frac{d}{dx}(120) = 0$$

and

$$\frac{d^ny}{dx^n} = 0 \qquad \text{for } n > 6.$$

If the motion of an object is described by distance as a function of time, say

$$s = f(t)$$

then, as we have indicated, the first derivative ds/dt gives the instantaneous velocity at time t. In this case the second derivative d^2s/dt^2 gives what is called the *instantaneous acceleration* of the object at time t. In other words, acceleration is the rate at which the velocity is changing with respect to time.

EXAMPLE 2. A toy rocket is launched vertically upward with an initial velocity of 120 ft./sec. The distance s traveled t seconds later is given by

$$s = 120t - 16t^2$$

Find the velocity and the acceleration. How high will the rocket rise?

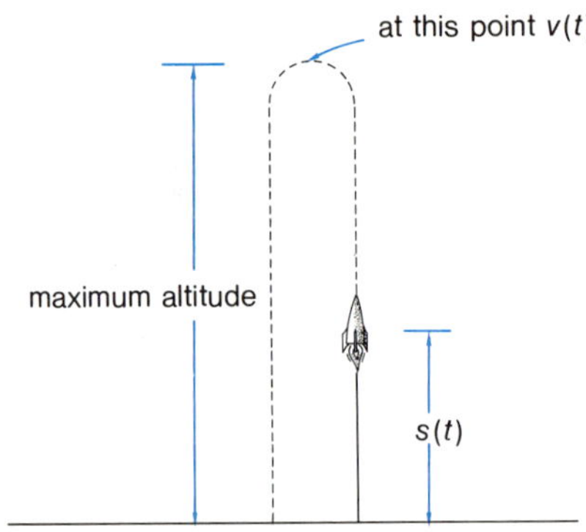

Figure 1

SOLUTION. Here we find that the velocity $v(t)$ is

$$v(t) = \frac{ds}{dt} = 120 - 32t$$

and the acceleration $a(t)$ is

$$a(t) = \left(\frac{d^2s}{dt^2}\right) = -32$$

This result tells us that the velocity of the rocket is changing at the rate of -32 ft./sec. every second. The negative sign indicates that the velocity is decreasing; it will become zero ($v = 0$) when

$$120 - 32t = 0 \quad \text{or when} \quad t = \tfrac{120}{32} = \tfrac{15}{4} \text{ sec.}$$

The distance travelled during this time interval is then

$$s(\tfrac{15}{4}) = 120(\tfrac{15}{4}) - 16(\tfrac{15}{4})^2 = 450 - 225 = 225 \text{ ft.}$$

EXERCISES 2.6

In Problems 1 through 12, find the first and second derivatives.

1. $f(x) = x^3 - 6$

2. $f(x) = 2x^5 - 3x^2 + 1$

3. $f(x) = x^3 + 2x^{3/2} - 4$

4. $f(x) = \sqrt{x}$

5. $f(x) = \dfrac{1}{x}$

6. $f(x) = (2x^3 + 1)^{3/2}$

7. $f(x) = \dfrac{x+1}{x-1}$

8. $f(x) = x^2\sqrt{x} - 3x$

9. $f(x) = \sqrt{2x} + \dfrac{1}{\sqrt{2x}}$

10. $f(x) = \dfrac{1}{\sqrt{3x^2+1}}$

11. $f(x) = \dfrac{x^2-1}{x^2+1}$

12. $f(x) = \dfrac{1-\sqrt{x}}{1+\sqrt{x}}$

For the equations of motion in Problems 13 through 16, find the velocity function $v(t)$ and the acceleration function $a(t)$.

13. $s(t) = 600 - 20t - 16t^2$

14. $s(t) = (4t + 7)^2$

15. $s(t) = 2t^3 - 12t^2 + 24t + 7$

16. $s(t) = t^3 + 6t^2 - 21t + 15$

17. For each of Problems 13 through 16, perform the following exercises.
(a) Find the initial velocity $v(0)$.
(b) Find the values of t^* for which $v(t^*) = 0$.
(c) Compute the corresponding values of $a(t^*)$.

18. Let the reaction R of a body as a function of the dosage x of a drug be given by the equation

$$R = 2x^2\left(\frac{1}{4} - \frac{x}{6}\right)$$

(a) Compute dR/dx and d^2R/dx^2.
(b) Find dR/dx at $x = \frac{1}{4}, \frac{1}{3}, \frac{1}{2}, \frac{2}{3}, \frac{3}{4}$.
(c) Find d^2R/dx^2 at $x = \frac{1}{4}, \frac{1}{3}, \frac{1}{2}, \frac{2}{3}, \frac{3}{4}$.

CHAPTER 2 REVIEW

Part 1 (ORAL)

1. Define the derivative of a function.

2. What is meant by saying that f is differentiable at the point c?

3. If a function f is differentiable at a point c, is it also continuous at c?

4. If f is continuous at c, is it also differentiable at c?

5. Define marginal cost and marginal revenue.

6. Define instantaneous velocity and instantaneous acceleration.

7. If f is a differentiable function and

(a) $f(x) = u(x) + v(x)$, then what is $f'(x)$?

(b) $f(x) = u(x)v(x)$, then what is $f'(x)$?

(c) $f(x) = \dfrac{u(x)}{v(x)}$, then what is $f'(x)$?

8. If f is a differentiable function at c, then what does $f'(c)$ represent geometrically?

9. What is the chain rule?

10. Explain:

(a) elasticity of demand

(b) marginal propensity to consume.

Part 2 (WRITTEN)

In Problems 1 through 4, use the definition of derivative to find $f'(x)$.

1. $f(x) = 2x^2 + 1$

2. $f(x) = 10$

3. $f(x) = \sqrt{2x}$

4. $f(x) = \frac{1}{2}x^{-1}$

In Problems 5 through 14, find $f'(x)$ using the rules in this chapter.

5. $f(x) = x^4 - 2x^{1/3} + 3$

6. $f(x) = 2\sqrt{x} + \dfrac{1}{2x}$

7. $f(x) = (3x^2 + x)(7x^2 + x^{-3} + 5)$

8. $f(x) = \dfrac{3x^2 + 1}{2x - 3}$

9. $f(x) = (3x^3 - 7x + 1)^{31}$

10. $f(x) = \sqrt{x}(4x^2 + 1)^{12}$

11. $f(x) = (x^2 + 1)\sqrt{2x + 1}$

12. $f(x) = \dfrac{x}{(x^2 + 2)^4}$

13. $f(x) = \dfrac{1}{x}(x^{-2} - x^2)^{-2}$

14. $f(x) = \left(\dfrac{x + 1}{1 - x}\right)^{3/2}$

15. Let $f(x) = \sqrt{x - 1}$. Show that $f'(1)$ does not exist.

16. Suppose the total cost C of producing x units of a product is given by $C(x) = (2/x) + 3x + 150$. Find the following:
    (a) The marginal cost function
    (b) The marginal cost when 40 units are produced
    (c) The cost of manufacturing 40 units

17. An object thrown vertically upward moves according to the law $s(t) = -16t^2 + 64t$, where s is the distance measured in feet from the ground, and t is the time measured in seconds. Find the following:
    (a) The initial velocity
    (b) The velocity at $t = 1$ sec.
    (c) How many seconds it will take the object to reach its highest point
    (d) How high the object will rise

18. An object is thrown vertically upward from the roof of a building 336 ft. high. If the equation of motion of the object is $s = -16t^2 + 96t$, find the following:
    (a) The instantaneous velocity of the object at 2 sec.
    (b) How high the object will go
    (c) How long it will take for the object to reach the ground

19. Find the equation of the tangent line to the curve

$$y = 2x^2 - 3x + 1 \quad \text{at} \quad x = 4$$

20. Find the point at which the slope of the tangent line to the curve $y = 2x^2 - 3x + 1$ is 5.

21. At what point on the curve $y = -2x^2 + 4x$ is the tangent line parallel to the x-axis?

22. Find the elasticity of demand for

$$x(p) = 100 + 3p - p^2 + \frac{32}{p} \quad \text{at} \quad p = 4$$

23. Find the elasticity of cost for

$$C(x) = 200 + 5x + \frac{36}{x} \quad \text{at} \quad x = 3$$

24. Find the marginal propensity to consume and the marginal propensity to save for the consumption function:

$$C = \frac{i-2}{2i+5} \quad \text{at} \quad i = 100$$

25. Find dy/dx if $x^3 + y^2 - 2xy = 0$.

26. Find the equation of the tangent line to the curve

$$2x^2 - y^2 + xy = 0 \quad \text{at} \quad x = 1$$

27. Find $f'''(x)$ given that $f(x) = 1 - 2x + \frac{1}{x}$.

28. A bomber is flying parallel to the ground at an altitude of 2 km. and at a constant speed of 6 km./min. If the bomber flies directly over a tank, at what rate is the line-of-sight distance between the bomber and the tank changing 30 seconds later?

29. A stone dropped into a still pond of water sends out concentric circular ripples. If the radius of the outermost ripple increases at the rate of 1 m./sec., find how quickly the area of the disturbed surface is increasing at the end of 10 seconds.

30. Two roads cross each other at right angles. One person starts from the junction point along one road with a constant speed of 1.5 m./sec., and another person walks toward the junction point from a distance of 12.5 m. on the other road with a uniform speed of 1.2 m./sec. Find the rate of increase of the distance between the two people at the end of 5 seconds.

31. A spherical iron ball is heated, and its radius is found to increase at the rate of 0.25 cm. per minute. Find the rate at which the surface area of the sphere and its volume increase when the radius of the sphere has become 12 cm.

32. Water is flowing into a vertical circular cylinder of diameter 4 ft. at the rate of 10 cm. ft./min. Find how quickly the surface of the water rises in the tank.

Chapter Three

Curve Sketching, Extrema

The simplest way to understand the behavior of a function is to look at its graph (assuming, of course, that the graph is available). In Chapter 1 we indicated that a continuous function can usually be visualized as a function whose graph can be drawn over an interval without removing pencil from paper. Thus, if we know that the given function is continuous, we can sketch its graph by plotting enough points and then connecting the successive points to form a curve. It is obvious that this procedure alone produces a graph which may not be very accurate. We shall see in this chapter how the signs of the first and second derivatives are used to find the behavior of a function and thus determine the shape of the graph of a function. We shall be able to find the extreme (high and low) points on the graph of a function. This last concept will be used to solve some real-life problems.

3.1 INCREASING, AND DECREASING FUNCTIONS

Let us consider the function $y = f(x)$ whose graph is given in Figure 1. We note from the graph that the function is *increasing* (the graph is rising) in the intervals

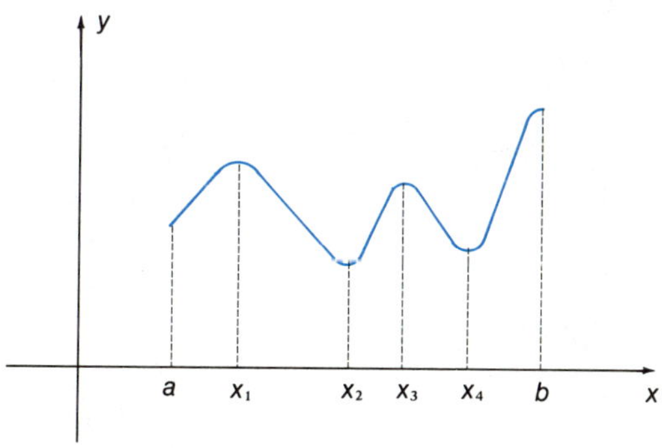

Figure 1

(a, x_1), (x_2, x_3), and (x_4, b) and that the function is *decreasing* (the graph is falling) in the intervals (x_1, x_2), and (x_3, x_4). Moreover, the slope of the tangent line (the derivative) is positive over the intervals where the function is increasing and is negative over the intervals where the function is decreasing. Before going further, we pause to define carefully the terms increasing function and decreasing function.

Definition 1. A function f is said to be *increasing* over an interval if, for any two points x_1, x_2 of the interval such that $x_1 < x_2$, we have $f(x_1) < f(x_2)$. Similarly, f is *decreasing* over the interval if $x_1 < x_2$ implies that $f(x_1) > f(x_2)$.

The following theorem shows how the derivative helps in determining whether a function is increasing or decreasing.

THEOREM. Let f be a differentiable function.

(a) If $f'(x) > 0$ for all x on an open interval I, then f is increasing on I.

(b) If $f'(x) < 0$ for all x on an open interval I, then f is decreasing on I.

We hasten to point out that the converse of the above theorem, say, "If a function f is differentiable and increasing on an interval I, then $f'(x) > 0$ on I," is not true.

For example, let us consider the function f given by

$$f(x) = x^3 \qquad (-1 < x < 1)$$

We have

$$f'(x) = 3x^2$$

Since $f'(x) > 0$ for all $x \neq 0$, the function is increasing over $(-1, 0) \cup (0, 1)$. In fact, f is increasing over $(-1, 1)$: if $x < 0$, then $x^3 < 0$, and if $x > 0$, then $x^3 > 0$ (see Figure 2). However, $f'(0) = 0$ (not positive). From the preceding discussion, we conclude that the points where the derivative is zero or where the derivative is undefined require separate consideration.

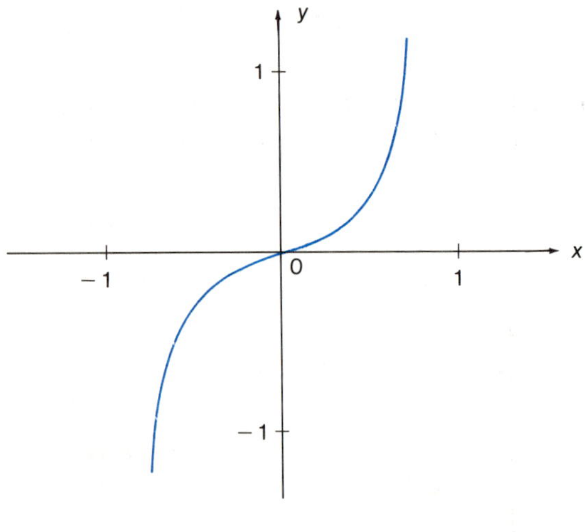

Figure 2

In conclusion, the sign of $f'(x)$ gives us the following information about the graph of the original function f.

Sign of $f'(x)$	*Graph of f*
(1) $f'(x) > 0$ on an interval (a, b)	f is increasing on (a, b); graph of f is rising
(2) $f'(x) < 0$ on an interval (a, b)	f is decreasing on (a, b); graph of f is falling

EXAMPLE 1. Find the intervals on which the graph of the function f, defined by

$$f(x) = x^2 - 2x + 3$$

is increasing and decreasing. Sketch the graph of f.

SOLUTION. We have

$$f'(x) = 2x - 2 = 2(x - 1)$$

Hence, $f'(x) > 0$ for $x > 1$ and $f'(x) < 0$ for $x < 1$. Thus, f is increasing for $x > 1$ and decreasing for $x < 1$. By using this information and plotting a few points $(x, f(x))$, we obtain the graph of f shown in Figure 3. The graph is a parabola which opens upward. The vertex of the parabola (see Chapter 1) is the point (1, 2).

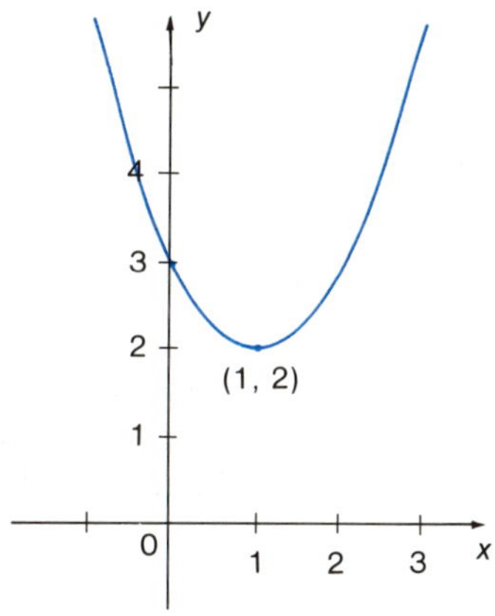

Figure 3

EXAMPLE 2. Find the intervals in which the function f defined by

$$f(x) = x^3 - 6x^2 + 9x - 7$$

is increasing and decreasing. Sketch the graph of f.

SOLUTION. We have

$$f(x) = x^3 - 6x^2 + 9x - 7$$

Differentiating, we obtain

$$f'(x) = 3x^2 - 12x + 9 = 3(x^2 - 4x + 3)$$

or

$$f'(x) = 3(x-1)(x-3) \tag{1}$$

Now, $f'(x)$ is zero only for the values $x = 1$ and $x = 3$. These two points where the derivative is zero divide the x-axis into three intervals:

$$I_1 = \{x \mid -\infty < x < 1\}$$
$$I_2 = \{x \mid 1 < x < 3\}$$
$$I_3 = \{x \mid 3 < x < \infty\}$$

By examining the sign of each factor in (1), we can determine the sign of the derivative in each of the intervals. This is done systematically in Figure 4.

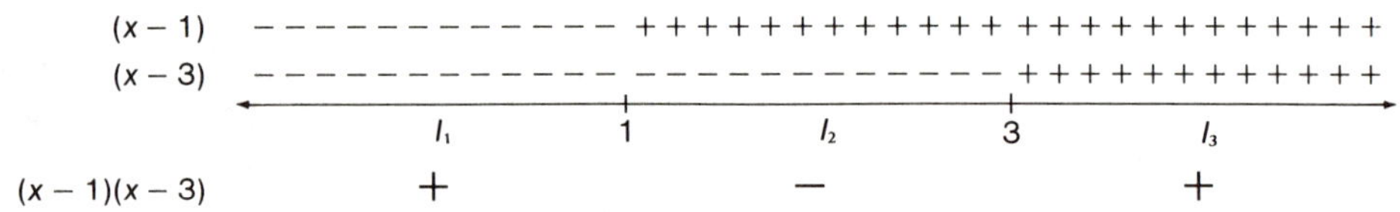

Figure 4

From Figure 4, we conclude that:

(1) $f'(x) > 0$ on I_1; f is increasing on $I_1 = (-\infty, 1)$
(2) $f'(x) < 0$ on I_2; f is decreasing on $I_2 = (1, 3)$
(3) $f'(x) > 0$ on I_3; f is increasing on $I_3 = (3, \infty)$

Since, for $x = 1, f(1) = -3$ and for $x = 3, f(3) = -7$; we conclude that, as we draw the graph from left to right, it rises (increases) to the point $(1, -3)$, then falls (decreases) to the point $(3, -7)$, and rises thereafter. The graph of f is sketched in Figure 5.

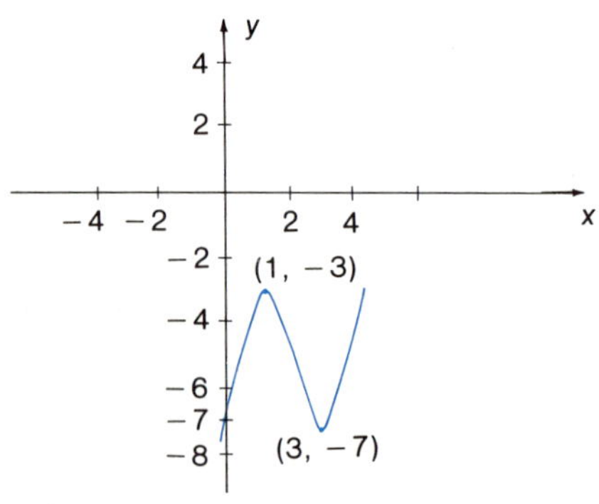

Figure 5

EXAMPLE 3. A manufacturer can sell x units of a product per day if the unit price is $p = 8 - 0.01x$. Determine whether the total revenue is increasing or decreasing when x is 200, 300, 500, and 600. Sketch a graph.

SOLUTION. The total revenue is

$$R(x) = xp(x)$$

or

$$R(x) = 8x - 0.01x^2 \tag{2}$$

Taking the derivative on both sides of Equation (2), we obtain

$$\frac{dR}{dx} = 8 - 0.02x$$

At $x = 200$, $\left.\frac{dR}{dx}\right|_{x=200} = 8 - (0.02)(200) = 4$

Since this result is a positive number, the total revenue is increasing at $x = 200$. Similarly, for $x = 300$, we have

$$\left.\frac{dR}{dx}\right|_{x=300} = 8 - (0.02)(300) = 2$$

Hence, $R(x)$ is increasing at $x = 300$. At $x = 500$,

$$\left.\frac{dR}{dx}\right|_{x=500} = 8 - (0.02)(500) = -2$$

a negative number, and so the total revenue is decreasing. Finally, at $x = 600$, the total revenue is decreasing, since

$$\left.\frac{dR}{dx}\right|_{x=600} = 8 - (0.02)(600) = -4$$

In fact, we find that

$$\frac{dR}{dx} > 0 \qquad \text{whenever} \qquad x < 400$$

and

$$\frac{dR}{dx} < 0 \qquad \text{whenever} \qquad x > 400$$

Thus, the total revenue is increasing on any interval where $x < 400$ and is decreasing when $x > 400$. A rough sketch of the graph of R is shown in Figure 6.

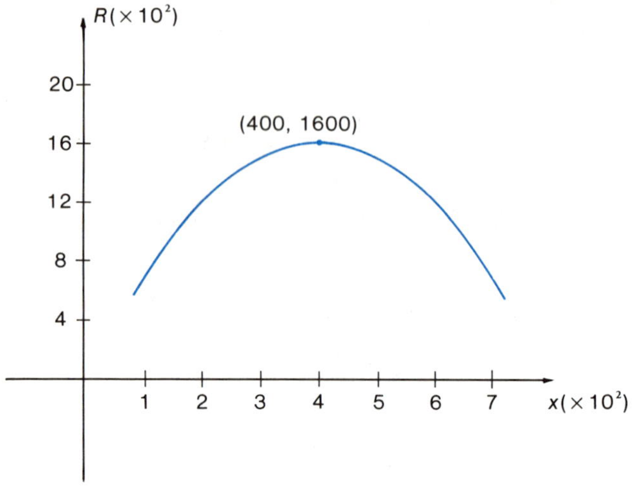

Figure 6

EXERCISES 3.1

In Problems 1 through 16, find the intervals where f is increasing and decreasing, and then sketch the graph of f.

1. $f(x) = 2x - 1$

2. $f(x) = -3x + 4$

3. $f(x) = -\frac{1}{2}x - 1$

4. $f(x) = 4x^2$

5. $f(x) = -2x^2$

6. $f(x) = x^2 - 2x + 1$

7. $f(x) = x^2 + 6x + 3$

8. $f(x) = -3x^2 + 2x$

9. $f(x) = x^2 - 3x + \frac{1}{4}$

10. $f(x) = -\frac{1}{2}x^2 + 4x - 1$

11. $f(x) = x^3 - 3x^2 - 9x + 15$

12. $f(x) = x^3 - 12x$

13. $f(x) = 3x^5 - 5x^3$

14. $f(x) = 4x^3 + 3x^2 - 6x - 1$

15. $f(x) = 2x^4 - x^2$

16. $f(x) = 4x^3 - 15x^2 - 18x + 2$

17. Suppose that the rate R of an autocatalytic reaction is given by

$$R = kx(4 - x) \qquad 0 \le x \le 4$$

where k is a positive constant and x is the amount of substance produced. Find the intervals where the rate R is increasing and where it is decreasing. Sketch the graph.

18. For the function $f(x) = x^4 - 2x^3 - 36x^2 + x + 1$, find the intervals over which f' (not f) is increasing and decreasing. Sketch the graph of $y = f'(x)$.

19. A dress manufacturer has found that the number x of dresses sold per month is related to the price p dollars per dress by the equation

$$p(x) = 2 - \left(\frac{x}{300}\right)^2$$

Determine whether the marginal revenue is increasing or decreasing when x is 200, 300, 500, and 600. Sketch a graph of the marginal revenue.

20. A manufacturer sells x units of a product at a dollar price of

$$p(x) = 200 - 2x - x^2 \text{ per unit sold.}$$

If the total cost for these x units is

$$C(x) = x^3 - 50x^2 + 416x + 300$$

find the demand intervals over which the profit $P(x)$ is (a) increasing, (b) decreasing. [*Hint:* $P(x) = R(x) - C(x) = x\,p(x) - C(x)$.]

21. Let $x(p) = 120 - 3p$ be the demand function of a company. Then $R(p) = p \cdot x(p)$ expresses revenue as a function of price.

(a) Where is the revenue (1) increasing, (2) decreasing?

(b) Compute the elasticity of demand. Where is the demand (1) inelastic, (2) elastic? Compare your answer with (a).

22. For the demand function $x(p)$, prove that

(a) The revenue is increasing whenever the demand is inelastic.

(b) The revenue is decreasing whenever the demand is elastic.

3.2 EXTREME VALUES OF A FUNCTION

In many practical problems, we are interested in determining the maximum or minimum values of a function. Geometrically, of course, these values correspond to high and low points on a function's graph. Presently we shall see how the derivative of a function is helpful in locating these values. We begin our discussion with the following definition.

Definition 1. Let f be a function. A point c for which

$$f'(c) = 0$$

is called a *critical number* in the domain of f, and the point $(c, f(c))$ is called a *critical point* of the graph of f.

By our definition, the slope of the tangent line to the graph of $y = f(x)$ at a critical point $(c, f(c))$ is zero. Consequently, the tangent line at the critical point is parallel to the x-axis. The function whose graph is given in Figure 1 has three critical points: $(x_1, f(x_1))$, $(x_2, f(x_2))$, and $(x_3, f(x_3))$.

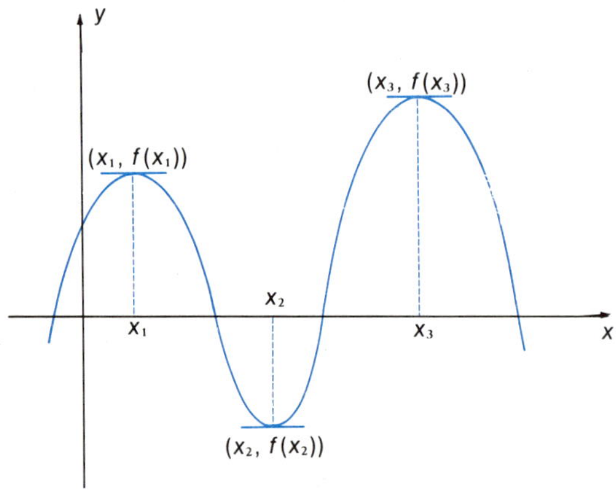

Figure 1

To compute the critical numbers of a function f, we need to solve the equation

$$f'(x) = 0$$

for x.

EXAMPLE 1. Find the critical numbers of

(a) $f(x) = 2x^3 - 3x^2 - 36x + 5$ (b) $g(x) = 5 + \dfrac{1}{x}$

SOLUTION.

(a) Differentiating, we get

$$f'(x) = 6x^2 - 6x - 36$$

so the critical numbers must satisfy the equation

$$6x^2 - 6x - 36 = 0 \quad \text{or} \quad x^2 - x - 6 = 0 \quad \text{or} \quad (x + 2)(x - 3) = 0$$

Thus, $x = -2$ and $x = 3$ are the critical numbers of the function f.

(a) Differentiating, we get

$$g'(x) = -\frac{1}{x^2}$$

Since $-1/x^2$ is never zero, the function g has no critical numbers.

RELATIVE EXTREMUM

Definition 2. (*Relative extremum*). A function f has a *relative maximum* value at c if there is an open interval I in the domain of f containing c, such that

$$f(c) \geq f(x) \qquad \text{for all} \qquad x \in I$$

Similarly, f has a *relative minimum* value at c if there is an open interval I in the domain of f containing c, such that

$$f(c) \leq f(x) \qquad \text{for all} \quad x \in I$$

The term *relative extremum* covers both situations.

In Figure 1 the function f has relative maxima at the points x_1 and x_3 and a relative minimum at x_2.

Definition 3. (*Absolute extremum*). A function f is said to have an *absolute maximum* at a if $f(a) \geq f(x)$ for all x in the domain of f. Similarly, f has an *absolute minimum* at b if $f(b) \leq f(x)$ for all x in the domain of f.

Let us discuss some of the preceding concepts in the following example.

EXAMPLE 2. Find the extreme values of the following functions:

(a) $f(x) = x^2 \qquad (-\infty < x < \infty)$

(b) $f(x) = x^2 \qquad (-2 \leq x \leq 2)$

(c) $f(x) = x^2 \qquad (-1 \leq x < 3)$

SOLUTION. The graphs of the functions are given in Figure 2.

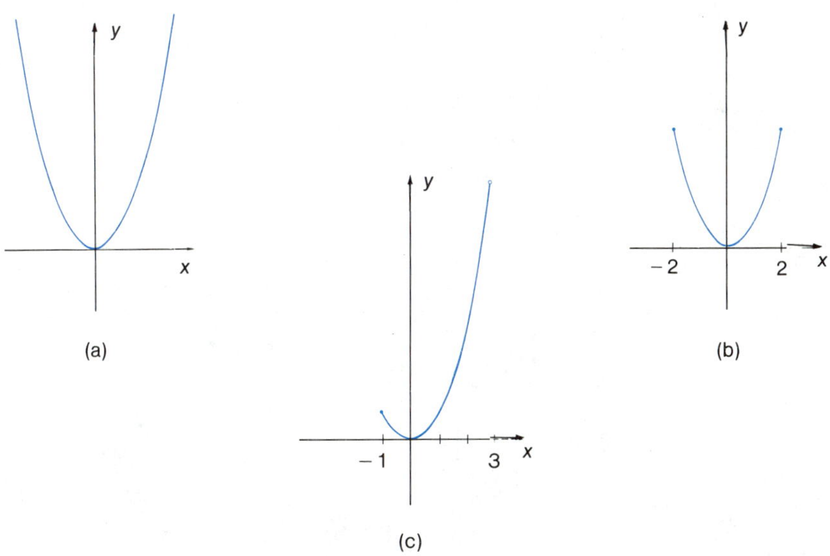

Figure 2

The relative and, in fact, the absolute minimum value for (a), (b), and (c) is at $x = 0$, since $f(0) = 0 \leq x^2 = f(x)$ for all x in the domain of each of the functions. Obviously, the function $f(x) = x^2$ $(-\infty < x < \infty)$ has no relative or absolute maxima. The function

$$f(x) = x^2 \qquad (-2 \leq x \leq 2)$$

has an absolute maximum value at $x = -2$ and at $x = 2$. The function

$$f(x) = x^2 \qquad (-1 \leq x < 3)$$

has no relative or absolute maximum value. (Why?)

We shall now state a theorem which gives a connection between the critical numbers and the extreme values of a function.

THEOREM 1. Let f be differentiable at c, and suppose that f has a relative extremum at c; then

$$f'(c) = 0$$

In other words, c is a critical number of f.

We advise the reader not to jump to the conclusion that a local extremum always exists when the derivative is zero. The theorem does not say that. The reader should carefully reread the theorem. Let us see why such care is necessary by considering some examples.

EXAMPLE 3. Find the critical numbers and the extreme values for the function

$$f(x) = x^3 \qquad (-\infty < x < \infty)$$

SOLUTION. The graph of this function is sketched in Figure 3. Solving the equation

$$f'(x) = 3x^2 = 0$$

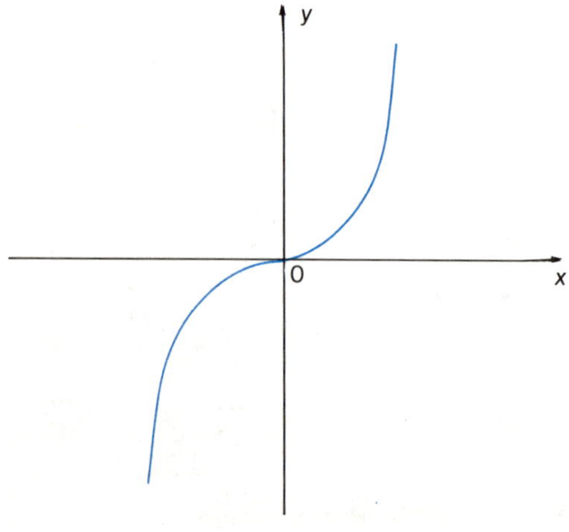

Figure 3

to find all the critical numbers of f, we find that $x = 0$ is the only critical number.

We determine that $f(0) = 0$ is *not* a relative extreme value. Since for $h > 0, f(-h) = -h^3 < 0$ and $f(h) = h^3 > 0$, $f(0)$ is neither greater nor less than both $f(-h)$ and $f(h)$.

As the graph shows, this function f is an increasing function. It has no extreme values.

The next example shows that a function may have an extremum, either relative or absolute, without the derivative being zero at that point.

EXAMPLE 4. Find the extreme values of the function

$$f(x) = |x| \qquad (-\infty < x < \infty)$$

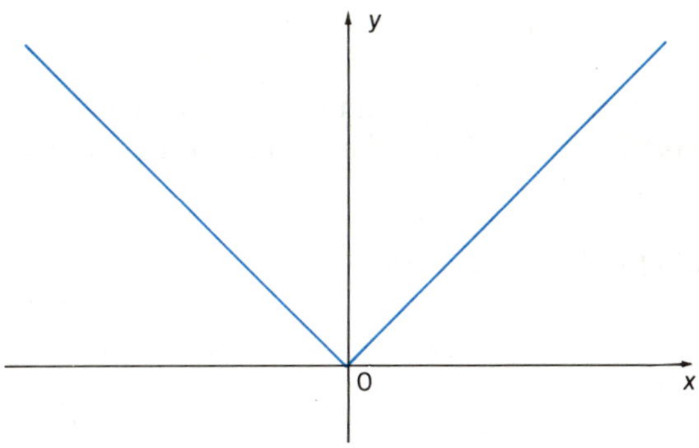

Figure 4

SOLUTION. The graph of this function is sketched in Figure 4. Clearly, $f(0) = 0$ is a relative minimum value (in fact, an absolute minimum value). However, $f'(0)$ is not equal to zero, because $f'(0)$ does not exist! (See Example 5, Section 2.1.)

The above examples illustrate that the derivative does not always tell us where the extreme values are. Rather, it tells us where the extreme values are not. If the derivative f' is positive (or negative) at a point c, then f is increasing (or decreasing) on an interval containing the point c. Thus, there cannot be an extreme value of f at c.

Consequently, if we wish to locate the extreme values of a function, only three kinds of points need be examined:

(1) Points where the derivative is zero;

(2) Points where the derivative does not exist;

(3) Endpoints of the intervals over which the function is defined.

Remark. Example 3 illustrates that a function need not have an extreme value. The following theorem shows that an important class of functions always has maximum and minimum values.

THEOREM 2. If a function f is continuous on a closed interval, then f has a maximum and a minimum value on that interval.

We arrive at the following procedure for locating the absolute extrema for a function which is continuous on a closed interval $[a, b]$ and differentiable at all (except perhaps *finitely* many) points on $[a, b]$.

PROCEDURE

Step 1. Find critical values of f'. That is, solve the equation $f'(x) = 0$.

Step 2. Find those values of x in (a, b) where $f'(x)$ does not exist.

Step 3. Evaluate f at the values obtained in steps 1 and 2.

Step 4. Evaluate f at a and b.

Step 5. The largest number of the numbers computed in steps 3 and 4 is the maximum of $f(x)$ in $[a, b]$, and the smallest is the minimum.

EXAMPLE 5. Find the absolute extrema for the function

$$f(x) = x^3 - 6x^2 + 9x - 7 \qquad (0 \le x \le 5)$$

SOLUTION. We have

(1) $$f'(x) = 3x^2 - 12x + 9 = 3(x-1)(x-3)$$

and the critical numbers are obtained from the equation $f'(x) = 0$ or $3(x-1)(x-3) = 0$, namely, $x = 1$ and $x = 3$.

(2) The function f is differentiable everywhere in $(0, 5)$.

(3) (a) $f(1) = (1)^3 - 6(1)^2 + 9(1) - 7 = -3$
(b) $f(3) = (3)^3 - 6(3)^2 + 9(3) - 7 = -7$

(4) Evaluating f at the endpoints 0 and 5, we obtain

$f(0) = -7$
$f(5) = (5)^3 - 6(5)^2 + 9(5) - 7 = 13$

(5) Therefore, the function f has absolute maximum value at $x = 5$ and absolute minimum value at $x = 0$ and $x = 3$.

In Example 5, suppose we modify the function so that it is defined for every real number x. That is,

$$f(x) = x^3 - 6x^2 + 9x - 7 \qquad (-\infty < x < \infty)$$

Then this equation has neither an absolute maximum nor an absolute minimum. However, it does have relative extrema. The candidates for relative extrema are the critical values $[f'(x) = 3(x-1)(x-3)]$, $x = 1$ and $x = 3$.

We observe (see Example 2, Section 3.1) that the sign of the derivative f' changes from positive to negative at $x = 1$. Consequently, the function f is increasing to the left of $x = 1$ and is decreasing to the right of $x = 1$. Therefore, $f(1) = -3$ is a relative maximum value of the function. Similarly, the sign of f' changes from negative to positive at $x = 3$, so that the function f is decreasing to the left of 3 and increasing to the right of 3. Hence, $f(3) = -7$ is a relative minimum value of the function.

So, in general, we know that the candidates for relative extrema of a continuous function are the critical values (values where $f'(x) = 0$) or the values of x where $f'(x)$ does not exist. We apply the following *first derivative test* to classify such points.

THEOREM 3. (*First derivative test*). Let c be a number in the domain of the function f such that $f'(c) = 0$ or $f'(c)$ does not exist. Then

(1) If the sign of f' changes from positive to negative at $x = c$, then $f(c)$ is a relative maximum.

(2) If the sign of f' changes from negative to positive at $x = c$, then $f(c)$ is a relative minimum.

(3) If the sign of f' does not change at $x = c$, then $f(c)$ is not a relative extremum.

At this stage, since we have mentioned the *first derivative test*, we take the liberty of stating the *second derivative test* also. Although we shall discuss the second derivative test in Section 3.4, we shall not hesitate to use it now, because, whenever the second derivative test is applicable, it is much simpler than the first derivative test for classifying the relative extrema.

THEOREM 4. (*Second derivative test*). Let a function f have a continuous second derivative at a critical number c.

(1) If $f''(c) > 0$, then f has a relative minimum at c.

(2) If $f''(c) < 0$, then f has a relative maximum at c.

(3) If $f''(c) = 0$, then the test is not applicable.

EXAMPLE 6. Find the relative maxima and minima for the function

$$f(x) = x^3 - 6x^2 + 9x - 7$$

SOLUTION. We have

$$f'(x) = 3x^2 - 12x + 9 = 3(x-1)(x-3)$$

The critical values are $x = 1$ and $x = 3$. Now

$$f''(x) = 6x - 12$$

We have

$$f''(1) = 6(1) - 12 = -6 < 0$$

Hence, $f(1)$ is a relative maximum. Similarly,

$$f''(3) = 6(3) - 12 = 6 > 0$$

Hence, $f(3)$ is a relative minimum.

EXAMPLE 7. Find the relative maxima and minima for the function

$$f(x) = x + \frac{9}{x}$$

SOLUTION. We have

$$f'(x) = 1 - \frac{9}{x^2}$$

Since $f(x)$ is undefined for $x = 0$, the function cannot have a relative maximum or a relative minimum value at 0.

Now, from the equation

$$f'(x) = 1 - \frac{9}{x^2} = 0$$

we have

$$x^2 - 9 = 0$$

and the critical numbers are $x = 3$ and $x = -3$. Since

$$f''(x) = \frac{18}{x^3}$$

we have

$$f''(3) = \frac{18}{(3)^3} = \frac{2}{3} > 0$$

Thus, by Theorem 3, $x = 3$ yields a relative minimum value for f. This minimum value is given by

$$f(3) = 3 + \frac{9}{3} = 6$$

Also, we find that

$$f''(-3) = \frac{18}{(-3)^3} = -\frac{2}{3} < 0$$

Thus, $x = -3$ yields a relative maximum value for f. This maximum value is given by

$$f(-3) = (-3) + \frac{9}{(-3)} = -6 \qquad \text{(See Figure 5)}$$

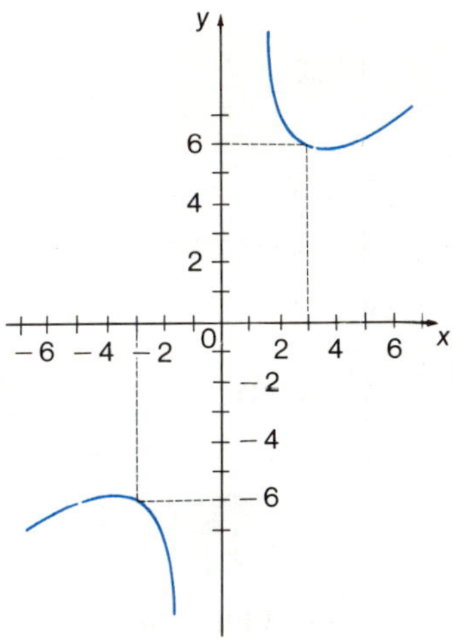

Figure 5

The obvious question is: what do we do if $f''(x)$ is zero at a critical value? This is a rare occurrence, but it can happen. In this situation we apply the first derivative test.

EXAMPLE 8. Find the relative maxima and the relative minima for the functions

$$\text{(a)}\quad f(x) = x^3, \qquad \text{(b)}\quad g(x) = x^4$$

SOLUTION. (a) Computing the first two derivatives, we find that

$$f'(x) = 3x^2 \text{ and } f''(x) = 6x$$

There is only one critical value, namely, $x = 0$, but at that critical value the second derivative $f''(0) = 0$. Thus, the second derivative test does not apply. So we apply the first derivative test. We notice that the sign of f' is positive to the left of zero and positive to the right of zero. Hence, by Theorem 2 (part (3)), the function $f(x) = x^3$ has neither a relative maximum nor a relative minimum.

(b) Computing the first two derivatives, we have

$$g'(x) = 4x^3 \text{ and } g''(x) = 12x^2$$

As in (a), the only critical value is $x = 0$, and, since $g''(0) = 0$, the second derivative test does not apply. However, the sign of g' changes from negative to positive at $x = 0$. Consequently, $g(0)$ is a relative minimum.

EXERCISES 3.2

1. Consider the function whose graph is shown in Figure 6. State whether the function has a relative maximum, relative minimum, absolute maximum, or absolute minimum at each of the points x_1, x_2, x_3, x_4, x_5, x_6, x_7, and x_8.

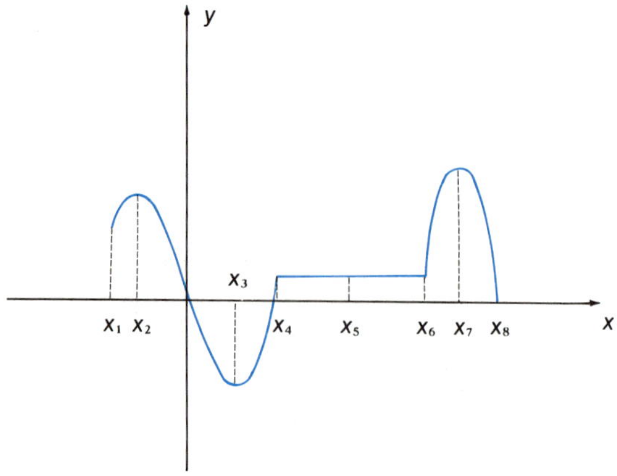

Figure 6

In Problems 2 through 20, find the absolute maximum and absolute minimum (if any exist) for the given function in the domain: $-2 \le x \le 4$.

2. $f(x) = x^2 + 4x - 5$

3. $f(x) = 5 - 4x - x^2$

4. $f(x) = 2x^3 - 6x^2 + 9$

5. $f(x) = x^3 + 3x^2 + 3x$

6. $f(x) = x^3 - 3x + 2$

7. $f(x) = x^3 + 3x + 3$

8. $f(x) = 2x^4 + x^2$

9. $f(x) = x^3 + 3x^2 - 12x + 7$

10. $f(x) = 2x^4 - x^2$

11. $f(x) = x^3 - 3$

12. $f(x) = x^4 + 3x^2 - 6$

13. $f(x) = \dfrac{x}{x+1}$

14. $f(x) = \dfrac{1}{x^2+1}$

15. $f(x) = \dfrac{1+x^2}{x^2}$

16. $f(x) = \dfrac{1-x^2}{x^3}$

17. $f(x) = (x^2 - 4)^2$

18. $f(x) = x^3(x + 1)$

19. $f(x) = (x - 2)^2(x + 4)$

20. $f(x) = x^{2/3}$

In Problems 21 through 40, find all relative extrema for the given function.

21. $f(x) = x^2 - 6x + 11$

22. $f(x) = 6 + 2x - x^2$

23. $g(x) = \frac{1}{3}(x^2 + 8x + 7)$

24. $h(x) = -\frac{1}{4}(x^2 + 6x + 17)$

25. $F(x) = 2 - 3x - x^2$

26. $G(x) = x^2 - 5x + 3$

27. $y = \frac{1}{12}(x^3 - 12x^2 + 21x + 14)$

28. $y = \frac{1}{10}(130 + 15x - 6x^2 - x^3)$

29. $y = \frac{1}{8}(24 - 36x - 12x^2 - x^3)$

30. $y = \frac{1}{2}(x^3 - 27x + 6)$

31. $y = x^4 - 2x^2 - 1$

32. $y = x^4 + x^2 - 25$

33. $f(x) = \frac{1}{x}$

34. $f(x) = \frac{x-1}{x+1}$

35. $f(x) = \frac{2}{x} - 2x$

36. $f(x) = x + \frac{9}{x}$

37. $f(x) = \frac{8}{3 + x^2}$

***38.** $f(x) = \frac{4x}{1 - x^2}$

***39.** $f(x) = x^{1/3}(x-1)^{4/3}$

***40.** $f(x) = (x+2)^{2/3}(x-1)^{1/3}$

41. Consider the cubic polynomial

$$f(x) = x^3 + ax^2 + bx + c$$

(a) For what values of a and b will f have critical numbers at $x = 4$ and $x = -2$?

(b) Determine the nature of the critical points.

42. Consider the function $f(x) = 3x^2 - x + 1$. Show that $f(x) > 0$ for all x. [*Hint*: Determine the minimum value of $f(x)$.]

43. If $f(x) = -x^2 + x - 4$, show that $f(x) < 0$ for all x.

49. Given that the selling price p is related to the profit P by the equation

$$P = 5000p - 125p^2$$

(a) For what range of values of p is the profit increasing?

(b) For what range of values of p is the profit decreasing?

(c) Determine the value of p that would yield maximum profit.

3.3 APPLICATION OF MAXIMA AND MINIMA

Although many students understand the concept involved in the theory of maxima and minima for a curve, they often find it difficult to apply the concept in solving word problems. In this section we shall see the applications of maxima-minima theory, also called *optimization theory*, to problems in various fields. First we will give a number of examples, and then we will state some general principles that are useful in attacking any problem involving extreme values.

EXAMPLE 1. Using a 120-ft. length of fencing, a farmer wishes to contain a cow in a rectangular plot of land along the bank of a river. What should the dimensions of the rectangle be to provide the cow with maximum grazing ground? (Assume that no fence is needed along the river, which is flowing along a straight edge.)

SOLUTION. We illustrate in steps a method for solving problems such as this one.

Step 1. Draw a figure showing the relevant features (see Figure 1).

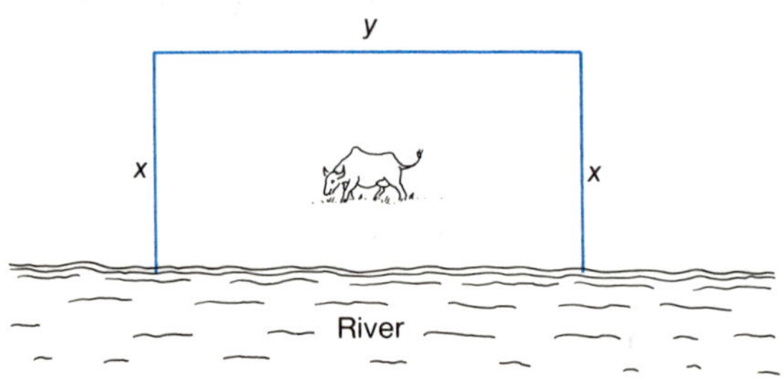

Figure 1

Step 2. Using Figure 1, find a relationship between the variables involved, and express the quantity to be maximized or minimized (in this case, the area) as a function of one of the variables. Since the length of fencing is 120 ft., we have

$$120 = 2x + y \tag{1}$$

and the area of the rectangle is

$$A = xy \tag{2}$$

Solving for y in Equation (1), we obtain

$$y = 120 - 2x \tag{3}$$

Substitution of (3) in (2) yields

$$A = x(120 - 2x)$$

or

$$A = 120x - 2x^2 \tag{4}$$

Thus, the area is expressed in terms of a single variable x.

Step 3. Compute the derivative of $A(x)$:

$$\frac{dA}{dx} = 120 - 4x \tag{5}$$

Step 4. Locate the critical numbers by solving the equation

$$\frac{dA}{dx} = 0$$

Thus $$120 - 4x = 0 \quad \text{or } x = 30$$

Step 5. Apply the second derivative test to determine which of the critical points yields the maximum or minimum value of the function. In this case there exists only one critical number, namely, $x = 30$. From Equation (5), we find that

$$\frac{d^2A}{dx^2} = -4 < 0 \tag{6}$$

and, by Theorem 3, the function A has a maximum value at $x = 30$. (Note again that if you obtain more than one critical point, select those that render the problem meaningful.) To solve for y, we use Equation (3):

$$y = 120 - 2(30) = 60$$

In conclusion, the farmer encloses maximum area if $x = 30$ and $y = 60$. The area enclosed is

$$A = (30)(60) = 1800 \text{ sq./ft.}$$

EXAMPLE 2. Find the two positive numbers having maximum product whose sum is 30.

SOLUTION. In this case we do not have the aid of a figure. So we proceed with Step 2 and find the product as a function of a single variable.

Let the two numbers be x and y. We have

$$x + y = 30 \tag{7}$$

and the product P is

$$P = xy \tag{8}$$

Using equation (7), we express one variable in terms of the other, say

$$y = 30 - x \tag{9}$$

Substitution of (9) in (8) yields the function P, where

$$P(x) = x(30 - x) = 30x - x^2 \tag{10}$$

The derivative of P with respect to x is

$$\frac{dP}{dx} = 30 - 2x \tag{11}$$

Therefore,

$$\frac{dP}{dx} = 0 \qquad \text{when} \qquad x = 15$$

Now, from Equation (11), we have

$$\frac{d^2P}{dx^2} = -2 < 0 \tag{12}$$

and by Theorem 3, the function P has a maximum value when $x = 15$. From Equation (7), we obtain $y = 15$.

EXAMPLE 3. A soup manufacturing company wishes to pack 25 cu. in. of mushroom soup in a can having the form of a right circular cylinder. Find the dimensions of the can if the surface area is to be a minimum.

SOLUTION. Again, we start by drawing a figure (see Figure 2). Clearly, the surface area of the metal used is

$$A = 2\pi r^2 + 2\pi rh \tag{13}$$

where $2\pi r^2$ is the area of the top and the bottom. Here the area is a function of r and h. To obtain A as a function of a single variable, we use the formula for the volume of the can:

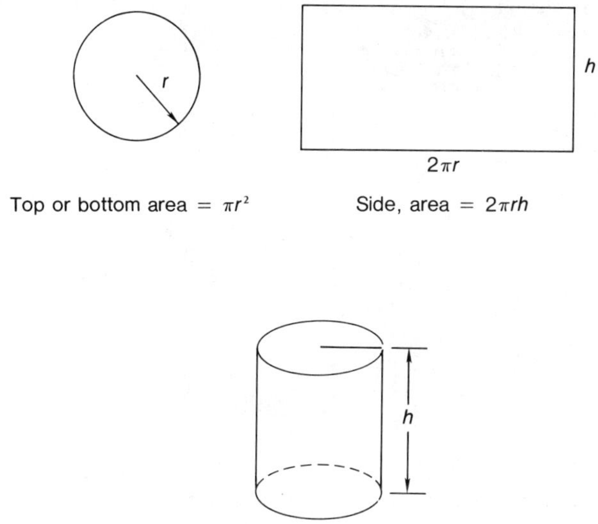

Figure 2

$$V = \pi r^2 h = 25 \tag{14}$$

Solving for h, we have

$$h = \frac{25}{\pi r^2} \qquad (r \neq 0) \tag{15}$$

Substitution of (15) in (13) yields the function A, where

$$A(r) = 2\pi r^2 + 2\pi r \frac{25}{\pi r^2} \qquad \text{or} \qquad A(r) = 2\pi r^2 + \frac{50}{r}$$

Therefore, we have expressed A as a function of a single variable r. Of course, we could have expressed A as a function of h only. The student should do this and show that the end result is the same.

Next, we compute the derivative of A

$$\frac{dA}{dr} = 4\pi r - \frac{50}{r^2}$$

Therefore, from the Equation $dA/dr = 0$, we have

$$4\pi r - \frac{50}{r^2} = 0 \quad \text{or} \quad 4\pi r^3 - 50 = 0$$

and

$$r^3 = \frac{50}{4\pi} = \frac{25}{2\pi} \quad \text{or} \quad r = \sqrt[3]{\frac{25}{2\pi}}$$

Applying the second derivative test, we find that

$$\frac{d^2A}{dr^2} = 4\pi + \frac{100}{r^3}$$

and at $r = \sqrt[3]{25/2\pi}$,

$$\frac{d^2A}{dr^2} = 4\pi + \frac{100}{(\sqrt[3]{25/2\pi}\,)^3} = 4\pi + (100)\left(\frac{2\pi}{25}\right) = 12\pi \qquad (>0)$$

By Theorem 3, therefore, A has a minimum value at $r = \sqrt[3]{25/2\pi}$. The corresponding value of h is obtained from Equation (15):

$$h = \frac{25}{\pi(25/2\pi)^{2/3}} = \frac{25}{\pi}\left(\frac{2\pi}{25}\right)^{2/3}$$

We now give an example of an apparently similar problem for which an extremum does not exist.

EXAMPLE 4. A cylindrical package having 18 cu. in. volume is to be wrapped in a rectangular sheet of paper. What should the dimensions of the cylinder be to minimize the surface area of the paper?

SOLUTION. Let the dimensions of the rectangular paper be x by y (see Figure 3). Now, the volume of the cylinder is

$$V = \pi r^2 x = 18 \tag{16}$$

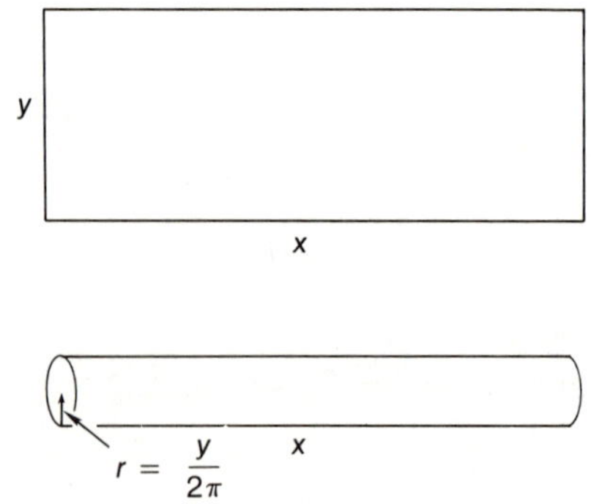

Figure 3

Since $r = y/2\pi$, we have

$$\pi\left(\frac{y}{2\pi}\right)^2 x = 18$$

or

$$x = \frac{18(4\pi)}{y^2} \tag{17}$$

The area A of the paper is

$$A = xy \tag{18}$$

Substituting (17) in (18), we obtain A in terms of a single variable.

$$A(y) = y\frac{72\pi}{y^2} = \frac{72\pi}{y} \tag{19}$$

The derivative of A with respect to y is

$$\frac{dA}{dy} = -\frac{72\pi}{y^2}$$

Since dA/dy is never zero, A has no critical number.

Thus, we check for those points where the derivative does not exist. In this case dA/dy does not exist at $y = 0$. But when $y = 0$, we have $r = y/2\pi = 0$. Thus, from (16), we obtain $V = 0 \neq 18$! Consequently, $y = 0$ does not minimize A. We naturally then turn to the endpoints (in any) of the interval for the variable y. In all physical problems there is a natural interval for the variables involved. In this case, we know that y cannot be negative. Hence, the interval for y is $I = \{y \mid y \geq 0\}$. This interval has only one endpoint, namely, $y = 0$. But we have already seen that $y = 0$ does not minimize A. Consequently, A has no minimum value.

EXAMPLE 5. An experiment shows that the rate of an autocatalytic reaction is proportional to the amount of substance produced times the amount of the original substance. For what value of the substance produced will the rate of the reaction be maximal?

SOLUTION. Let c denote the amount of the original substance at the beginning of the reaction. At a later time, if the amount of the new substance formed is x, then the amount of original substance left is $c - x$. Therefore, if R is the rate of the reaction, then

$$R = kx(c - x) = ckx - kx^2 \qquad (0 \leq x \leq c) \tag{20}$$

where $k > 0$ is a constant of proportionality. For the rate of the reaction to be a maximum, we must have for some value of x

$$\frac{dR}{dx} = ck - 2kx = 0 \tag{21}$$

Equation (21) is satisfied if $x = c/2$. Thus, when the concentration of the new substance formed is one-half the amount of the original substance at the beginning of the reaction, the rate of the reaction is maximal.

We now show by application of the second derivative test that the rate of the reaction is a maximum when $x = c/2$. From Equation (21) we have

$$\frac{d^2R}{dx^2} = -2k$$

Since $k > 0$, $d^2R/dx^2 < 0$, and, by Theorem 3, R attains a maximum at $x = c/2$. Notice that R attains a minimum at the endpoints $x = 0$ and $x = c$.

EXAMPLE 6. Suppose the response R of the body is found to be related to the dosage x of the drug administered according to the equation

$$R = kx^2 - \frac{x^3}{3} \qquad (k > 0) \tag{22}$$

The rate of change of R is

$$\frac{dR}{dx} = 2kx - x^2 \tag{23}$$

Equation (23) tells us that the body is responding as long as $0 < x < 2k$. For what dosage of the drug is the rate of change dR/dx a maximum?

SOLUTION. The function we wish to maximize is dR/dx. Set

$$y = \frac{dR}{dx}$$

Then y has critical points for the values of x for which $dy/dx = 0$:

$$\frac{dy}{dx} = \frac{d}{dx}\left(\frac{dR}{dx}\right) = \frac{d^2R}{dx^2} = 2k - 2x = 0 \tag{24}$$

Equation (24) is satisfied if $x = k$. Application of the second derivative test yields

$$\frac{d^2y}{dx^2} = \frac{d^3R}{dx^3} = -2 < 0$$

The second derivative test indicates that for $x = k$, the function dR/dx attains a maximum. In other words, when the concentration of the drug in the body is 50% of the amount administered, the greatest changes in the reaction will occur. This type of analysis is used to determine the optimum amount of the drug to be administered to give the greatest changes in the reaction for a small change in dosage.

EXAMPLE 7. A Florida orange grower finds that an orange tree produces (on the average) 400 oranges per year if no more that 16 trees are planted in a unit area. For each additional tree planted per unit area, the grower finds that the yield decreases by 20 oranges per tree. How many trees should the grower plant per unit area to maximize the yield?

SOLUTION. If x is the number of trees planted per unit area, then the number N of oranges per tree depends on x; that is, $N = N(x)$. Since each tree produces 20 fewer oranges for every additional tree over 16, we have for $x \geq 16$

$$N = 400 - 20(x - 16)$$

and the total number T of oranges produced per unit area is

$$T = xN = x[400 - 20(x - 16)]$$

or

$$T = 720x - 20x^2$$

The first derivative condition for maximum yields

$$\frac{dT}{dx} = 720 - 40x = 0 \tag{25}$$

Solving Equation (25), we obtain $x = 18$; that is, we find that T has a critical point at $x = 18$. Applying the second derivative test, we determine that

$$\frac{d^2T}{dx^2} = -40$$

which is a negative number, and so, by Theorem 3, the yield is maximum if 18 trees are planted per unit area.

(What assumptions are we making about the function T in order to take its derivative?)

EXAMPLE 8. A firm manufactures swim suits and sells its product at \$24 per unit. The total cost C (in dollars) of producing x swim suits is given by

$$C = 150 + \frac{39}{10}x + \frac{3}{1000}x^2 \tag{26}$$

Write the profit P as a function of x and determine the number of swim suits that the firm should produce and sell to achieve a maximum profit.

SOLUTION. The firm's profit is its revenue minus its total cost; that is,

$$P = R(x) - C(x) \tag{27}$$

The total revenue from the sale of x swim suits at \$24 per swim suit is

$$R(x) = 24x \tag{28}$$

Substituting (26) and (28) into (27), we find that

$$P = 24x - \left(150 + \frac{39}{10}x + \frac{3}{1000}x^2\right) \tag{29}$$

To achieve maximum profit, we must have

$$\frac{dP}{dx} = 24 - \frac{39}{10} - \frac{3}{500}x = 0 \tag{30}$$

Solving Equation (30) for x, we obtain

$$\frac{201}{10} - \frac{3}{500}x = 0$$

Thus, the critical number for P is $x = 3350$. Applying the second derivative test, we find that

$$\frac{d^2P}{dx^2} = -\frac{3}{500}$$

which is a negative number, and so, by Theorem 3, a production level of 3350 swim suits yields a maximum profit. The profit in dollars at this level of production is

$$P = 24(3350) - \left[150 + \frac{39}{10}(3350) + \frac{3}{1000}(3350)^2\right] = \$33{,}517.50$$

Note that if the firm should increase (or decrease) its output by one or more swim suits, the firm's profit would decrease.

It is of interest to observe that the marginal revenue is $dR/dx = 24$ and that the marginal cost is also

$$\left.\frac{dC}{dx}\right|_{x=3350} = \left.\frac{39}{10} + \frac{3}{500}x\right|_{x=3350} = 24$$

at a production level of 3350 swim suits.

EXERCISES 3.3

1. Find two numbers whose sum is 28 and whose product is the largest possible among those pairs of numbers whose sum is equal to 28.
2. Do problem 1 when the sum of the two numbers is 37.
3. Find positive numbers x and y whose sum is 48 and such that xy^2 is a maximum.
4. Do problem 3 when $x + y = 45$.
5. Can you find two positive numbers such that $x + y = 8$ and (a) $x^3 + y^3$ is a minimum and (b) $x^3 + y^3$ is a maximum?
6. A rectangle is to be constructed having a perimeter of 60 meters. Find the dimensions of the rectangle of maximum area.
7. A rectangle is to be constructed having a given perimeter P. Find the dimensions of the rectangle of maximum area.

8. A closed box with square base is to be constructed to hold 64 cu. in. Find the dimensions of the box of minimum surface area.

9. Do problem 8 if the box has no top.

10. A square piece of aluminum is 18 inches on each side. An open rectangular container is to be made by cutting a square from each corner and bending up the sides (See Figure 4). Find the dimensions of the container with maximum volume.

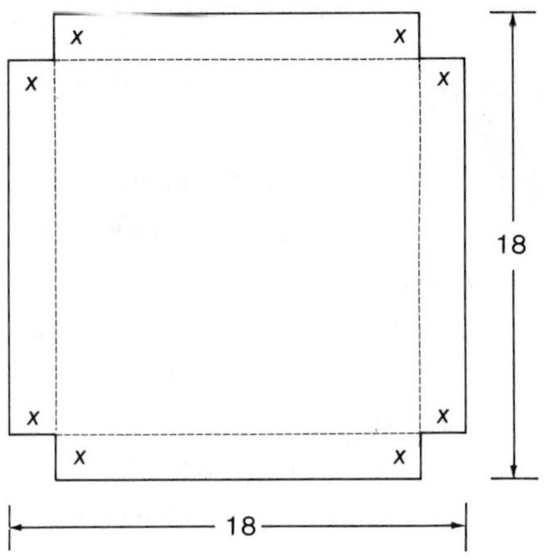

Figure 4

11. Do problem 10 if the given piece of aluminum is rectangular and is 48 in. $\times$ 90 in.

12. Find the dimensions of the rectangle of the largest area that can be inscribed in a semicircle having diameter $2r$.

13. A piece of wire 20 in. long is to be cut into two pieces, one to form a circle and the other a square. How should the wire be cut in order that the sum of the two areas enclosed by the wire be minimal?

14. An open cylindrical tank with circular base is to be constructed of sheet metal so as to contain a volume πa^3 of water. Find the height and the radius of the base so that the quantity of sheet metal required may be minimal.

15. A 288-cu. ft. pool is to have a square top. The sides are to be built of glass and the bottom of mosaic. The cost per unit area of glass is three times the cost of mosaic. Find the dimensions of the pool having minimum cost.

16. Two cars are traveling along two roads which cross each other at right angles at A. Both cars are traveling toward A at 30 ft./sec. Initially, their distances from A are 1500 ft. and 2100 ft., respectively. At what time is the distance between the two cars a minimum? Find this distance.

17. A rectangular box with square bottom and top is to contain 1000 cu. ft. The cost of material per square foot for the bottom is 25¢; for the top, 15¢; and for the sides, 20¢. The labor charge for making the box is \$3.00. Find the dimensions of the box when the cost is minimal.

18. Find the points on the parabola $y = x^2 - \frac{9}{2}$ that are nearest to the origin.

19. Is there a point on the parabola of Problem 18 that is farthest from the origin?

20. Find the points of the parabola $y = x^2$ that are closest to the point (8, 2).

21. If the rate of a certain autocatalytic reaction is given by

$$r(x) = 12x(200 - x)$$

where x is the amount of the new substance formed and 200 is the amount of the original substance, for what value of x is the rate of the reaction a maximum? What is the maximum rate of the reaction?

22. If the number of bacteria (in thousands) present in a culture is given by

$$N(t) = 250t - 50t^2 + 20$$

where t is the time, at what time will the population of the bacteria be a maximum? What is the maximum population?

23. The concentration of hydrogen ion in a solution is given by

$$X = H + \frac{10^{-5}}{H}$$

For what value of H is the concentration a minimum?

24. Experiments show that the velocity of the flow of air through the respiratory system during coughing is given by

$$v = \frac{k}{\pi} r^2(r_0 - r)$$

where k is a constant, r_0 is the radius of the windpipe when there is no pressure, and r is the radius of the windpipe after pressure builds up. Find the value of r for which the velocity of the flow of air is maximum.

25. For each of the following cost functions, find the production level x for which the cost is a minimum (x is in thousands).

(a) $C = 2x^3 - 8x^2 + 10x$

(b) $C = 800 - \frac{x}{2} + \frac{1}{1000}x^2$

(c) $C = \frac{x}{2} + \frac{5 \cdot 10^5}{x} + 10^3$

26. A firm produces x units of a product. If the profit P in dollars earned in the manufacture and sale of x units is

$$P = 20x - \frac{1}{500}x^2$$

find the number x that yields maximum profit. What is the maximum profit?

27. A firm produces mink coats and sells each coat for \$1500. The cost of producing x coats per year is

$$C = 9 + 3x + 0.015x^2$$

(a) Write an expression for the profit in terms of x.

(b) Find a number x that yields maximum profit.

(c) What is the maximum annual profit?

28. A manufacturer finds that it is possible to sell x units per month of a certain item if the price is $p = 18 - 0.001x$ dollars. Find the production level that will maximize profits if the cost of producing x items is $C = 4x + 1500$ dollars. What is the price of each item at this production level?

29. A firm's production level is such that the profit is maximum. Find a relationship between the firm's marginal revenue and marginal cost.

30. Let $C(x) = 0.2x^2 + 4x + 20$ be the total cost of producing x units.

(a) Find the average cost $\bar{C}(x) = \dfrac{C(x)}{x}$.

(b) When is $\bar{C}$ minimum?

(c) What is the minimum average cost?

(d) Find the marginal cost at the production level x, when $\bar{C}$ is minimum. Compare your answer with (c).

31. Prove that $\bar{C}(x)$ is minimum, when $\bar{C}(x) = C'(x)$.

32. If 25 apple trees are planted per acre in a certain area, the average yield per tree is 450 apples. For each tree in excess of 25 per acre, the average yield is reduced by 15 apples per tree.

(a) Show that the total number of apples produced per acre is

$$T = 825x - 15x^2 \qquad (x \geq 25)$$

(b) Show that if 28 trees are planted per acre, the yield is maximum. (Remember that you cannot plant a fraction of a tree.)

33. The owner of a day nursery finds that if a monthly fee of \$75 per child is charged, 20 children will be enrolled. To increase the enrollment, the owner finds that the fee will have to be decreased by \$3 for each additional child.

(a) If x ($x \geq 20$) is the number of children enrolled, show that the total revenue is $r = 135x - 3x^2$.

(b) Show that the maximum revenue is obtained by enrolling either 22 of 23 children.

34. A group of students arrange a chartered flight from New York to London, Paris, Heidelberg, Barcelona, and back to New York with a three-day stay at each European city. The charge per person is \$499 if 100 students go on the flight. If more than 100 participate, the charge per student is reduced by an amount equal to \$4 times the number of students above 100.

(a) Find the total revenue.

(b) Find the number of students that will furnish maximum revenue.

(c) What is the cost per student if the maximum revenue is obtained?

3.4 CONCAVITY, CURVE SKETCHING

In this section we shall study the sign of the second derivative to find where the graph of a function is concave upward and where it is concave downward. We shall also look for the behavior of the graph of $y = f(x)$ when $|x|$ is very large. This will lead us to the concept of an asymptote to the graph. We shall conclude this section with a few hints for graphing a function f.

CONCAVITY

By determining the sign of the first derivative, we can tell whether the graph of a function is rising (f is increasing) or falling (f is decreasing). Suppose we know that the graph of a particular function is rising from $(a, f(a))$ to $(b, f(b))$. Then the graph may be rising and may be (1) concave upward as in Figure 1(a) or (2) concave downward as in Figure 1(b). Similar comments apply for a decreasing function. But the first derivative cannot distinguish between these possibilities. The second derivative, however can!

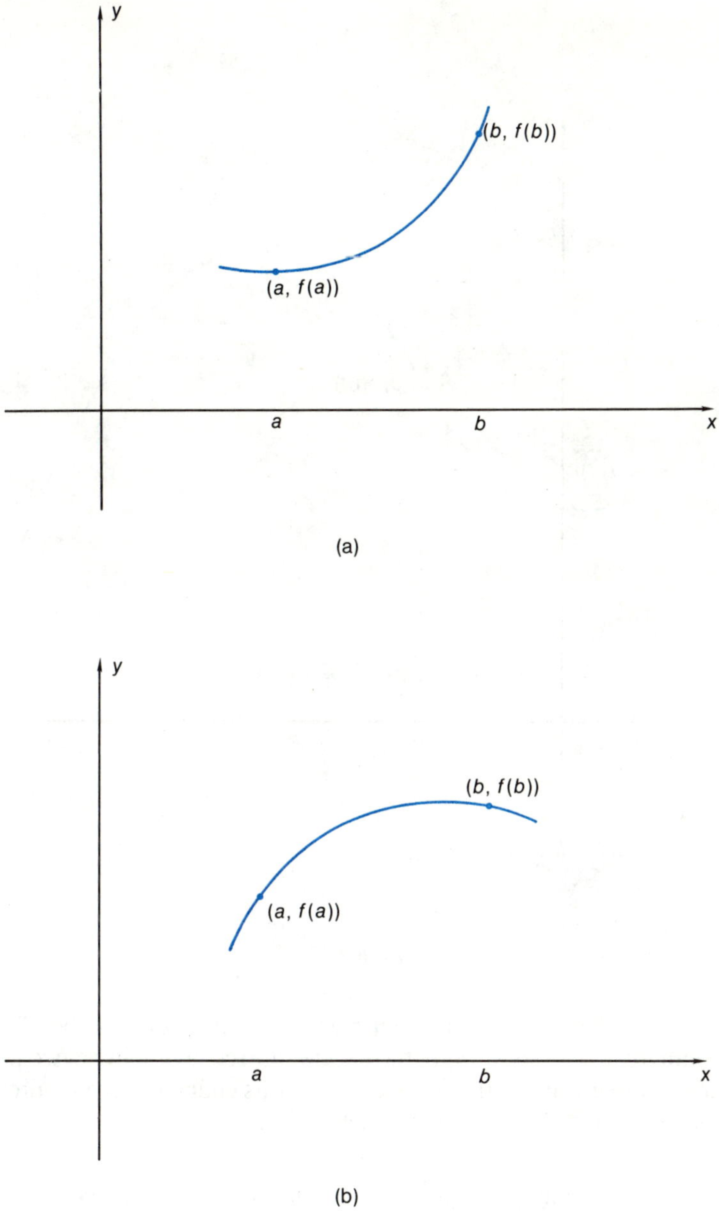

Figure 1

Consider the graph of the function f in Figure 2. As x increases from a to b, the graph of the function bends in a counterclockwise direction, and from b to c it bends in a clockwise direction. We say that the graph of the function is *concave upward* between a and b and that it is *concave downward* between b and c. At the point $x = b$, the direction of bending changes. The point $(b, f(b))$ is called the *point of inflection* of the curve.

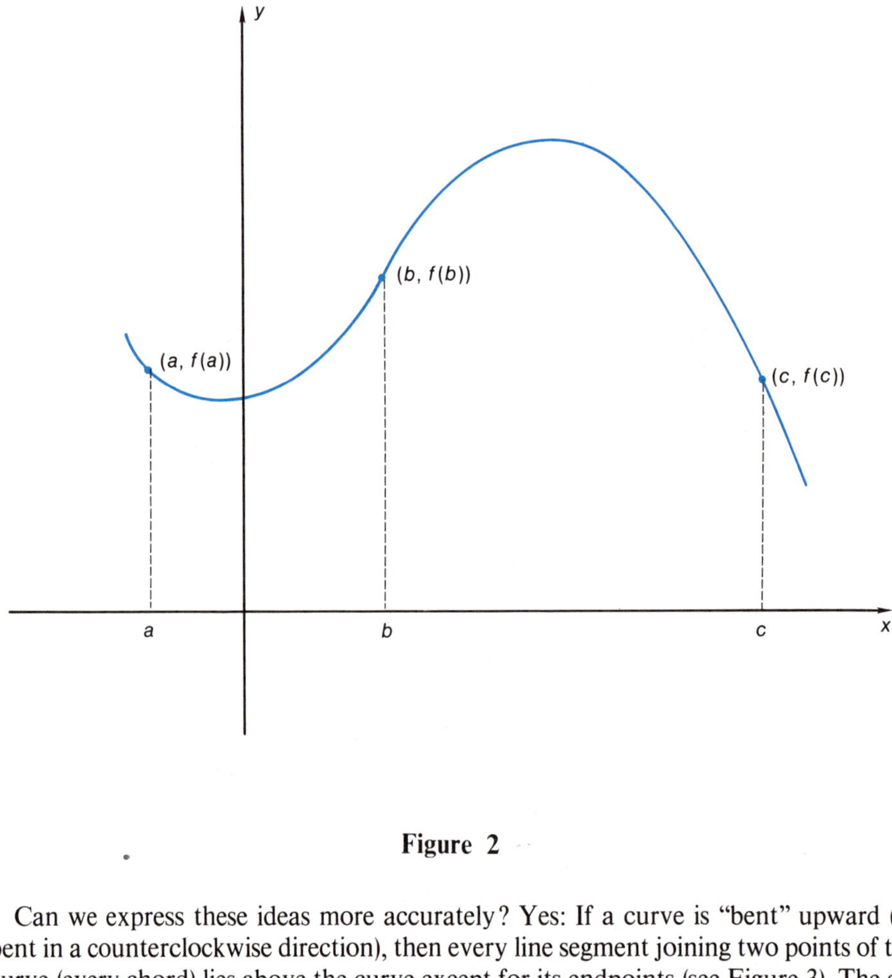

Figure 2

Can we express these ideas more accurately? Yes: If a curve is "bent" upward (or bent in a counterclockwise direction), then every line segment joining two points of the curve (every chord) lies above the curve except for its endpoints (see Figure 3). The exact meaning is given in the definition which follows.

Definition 1. A curve (or graph of a function) is said to be *concave upward* in an interval if every chord in that interval lies above the curve except for its endpoints (see Figure 3). The curve is said to be *concave downward* if every such chord, except for its endpoints, lies below the curve (see Figure 4).

There is an alternative (but equivalent) definition for concavity, as follows:

Definition 1′ We say that a curve is concave upward on an interval I if and only if the curve lies above its tangent lines for all $x \in I$. Similarly, a curve is said to be concave downward on I if and only if the curve lies below its tangent lines for all $x \in I$.

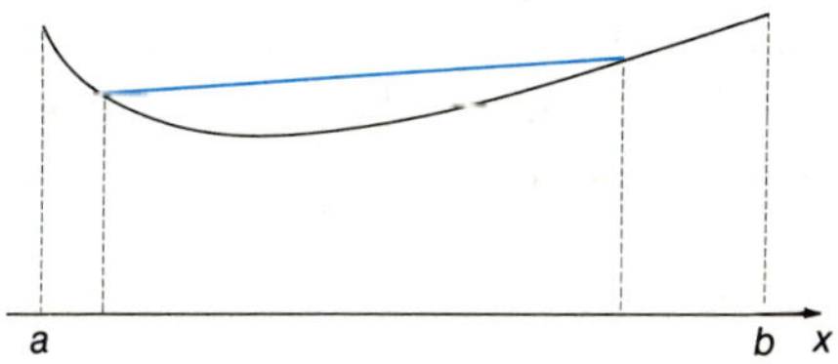

Figure 3

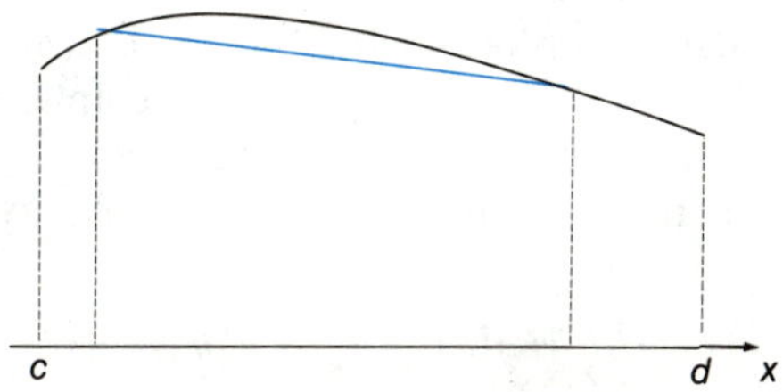

Figure 4

Let us consider a concave upward curve, shown in Figure 5.

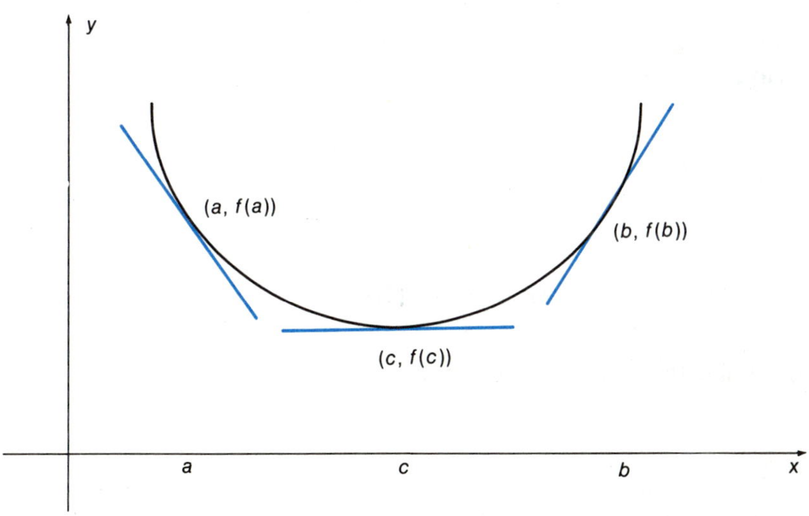

Figure 5

We observe that the slope of the tangent line at $(a, f(a))$ is negative, the slope of the tangent line at $(c, f(c))$ is zero, and the slope of the tangent line at $(b, f(b))$ is positive. This leads us to suspect that if a curve is concave upward, then the slopes of the tangent lines increase when we move from left to right on the curve. In fact, this is true in general. It can be shown that if the graph of a function f is concave upward in an interval I, then the slopes of the tangent lines increase. Similarly, if the graph of a function f is concave downward in an interval I, then the slopes of the tangent lines decrease. We shall use this result to arrive at a test for the concavity of the graph of a function.

We recall that if $f'(x) > 0$ on an interval I, then $f(x)$ is increasing on I. Let us apply the same fact to the function $f'(x)$. We conclude that if $f''(x) > 0$ on an interval I, then $f'(x)$ [slope of the tangent line at $(x, f(x))$] is increasing on the interval I. Consequently, the graph of f is concave upward on I. Similarly, if $f''(x) < 0$ on an interval J, then the graph of f is concave downward on J. We summarize these observations as the second derivative test for the concavity of the graph of f as:

THEOREM 1. (*Test for Concavity*).

(1) If $f''(x) > 0$ for $a < x < b$, then the graph of f is concave up (or upward) for $a \le x \le b$.

(2) If $f''(x) < 0$ for $c < x < d$, then the graph of f is concave down (or downward) for $c \le x \le d$.

EXAMPLE 1. Discuss the concavity of the graph of

$$y = f(x) = x^4 - 6x^3 + 12x^2 + 5x + 7$$

SOLUTION. Differentiating, we have

$$f'(x) = 4x^3 - 18x^2 + 24x + 5$$

and

$$f''(x) = 12x^2 - 36x + 24 = 12(x^2 - 3x + 2) = 12(x - 1)(x - 2)$$

We determine the sign of $f''(x)$ from Figure 6.

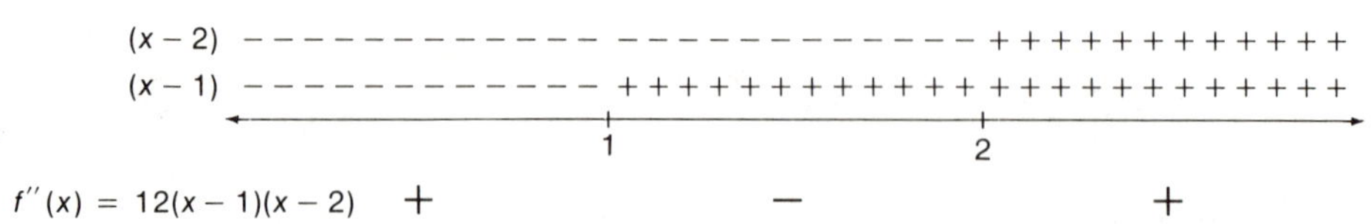

Figure 6

We observe that

$$f''(x) > 0 \qquad \text{for} \qquad x < 1$$

$$f''(x) < 0 \qquad \text{for} \qquad 1 < x < 2$$

and

$$f''(x) > 0 \qquad \text{for} \qquad x > 2$$

Hence, the graph of f is concave up in the intervals $(-\infty, 1]$ and $[2, \infty)$; and the graph of f is concave down in the interval $[1, 2]$.

There is an interesting application of concavity to the problem of locating the relative extrema. Suppose Figure 7 depicts the graph of a "well-behaved" and "nice" function $y = f(x)$ and $f'(a) = 0$. If the graph of f is concave upward at a, then f must have a relative minimum at a. Similarly, from Figure 8 if the graph is concave downward at a and $f'(a) = 0$, then f must have a relative maximum at a.

Concave upward

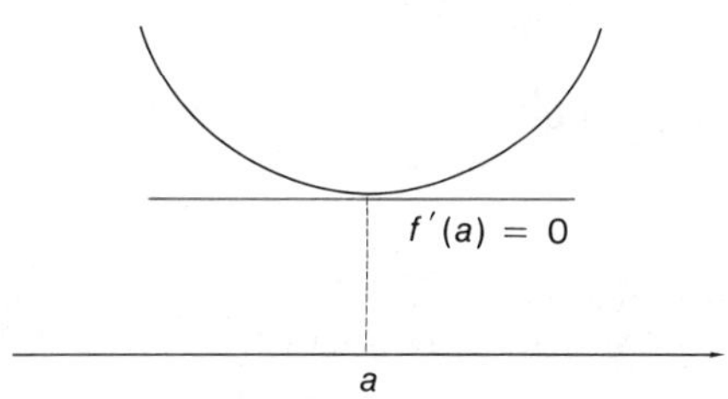

Figure 7

Concave downward

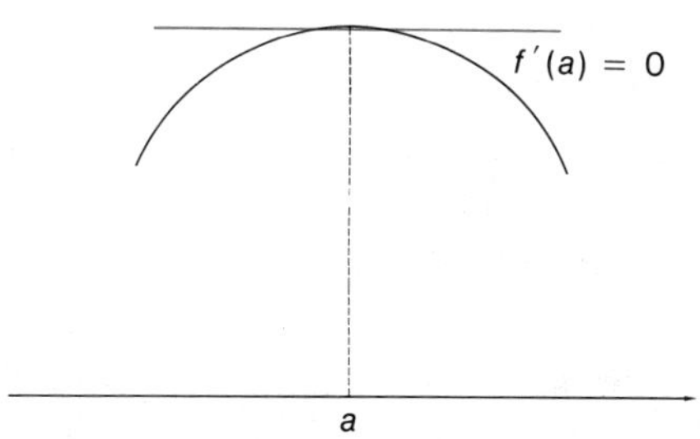

Figure 8

These observations lead us to the second derivative test for relative extrema, mentioned in Section 3.2.

THEOREM 2. (*Test for relative extrema*). Let $f''(x)$ be continuous in an interval that contains a, and $f'(a) = 0$. Then

(1) if $f''(a) > 0$, f has a relative minimum at a;

(2) if $f''(a) < 0$, f has a relative maximum at a.

For the function $f(x) = x^4 - 6x^3 + 12x^2 + 5x + 7$, of Example 1, we see that $f''(1) = 0$. Furthermore, the curve is concave upward on one side of $(1, f(1))$ and concave downward on the other side of $(1, f(1))$. In other words the concavity changes at $(1, f(1))$. This point $(1, f(1))$ is called a *point of inflection*. Similarly, the point $(2, f(2))$ is also a point of inflection. Thus, a point of inflection on a curve is that point at which the concavity of the curve changes. In terms of the derivative, we have the following definition:

Definition 2. (*Point of Inflection*) A point $(a, f(a))$ on the graph of a function f is called a point of inflection if the second derivative $f''(x)$ changes sign at a.

The following theorem enables us to find the "candidates" for points of inflection.

THEOREM 3. If $(a, f(a))$ is a point of inflection of the graph of the function f and if f'' is continuous at a, then $f''(a) = 0$.

Theorem 3 asserts that under the assumption that f'' is continuous, the points of inflection are among those points $(x, f(x))$ for which $f''(x) = 0$. However, if $f''(c) = 0$, $(c, f(c))$ is not necessarily a point of inflection, as the following example illustrates.

EXAMPLE 2. Show that (0, 0) is not a point of inflection for the curve

$$y = f(x) = x^4$$

SOLUTION. We have

$$f'(x) = 4x^3 \qquad \text{and} \qquad f''(x) = 12x^2$$

Thus, $f''(x) = 0$ if $x = 0$. However, (0, 0) is not a point of inflection since $f''(x) > 0$ for every $x \neq 0$. Therefore, the concavity of the graph of f does not change at $x = 0$. Hence (0, 0) is not an inflection point.

From Theorem 3, we conclude that the point $(a, f(a))$ is a candidate for being the point of inflection if either $f''(a) = 0$ or f'' is discontinuous at a. Thus, we follow the following *procedure to find the points of inflection*. For the function f defined by $y = f(x)$:

(1) Find $f''(x)$.

(2) Determine those values of x for which $f''(x) = 0$ or f'' is discontinuous. Let us call these values of x the *hypercritical numbers*.

(3) Check to see if $f''(x)$ changes sign at a hypercritical number. If $f''(x)$ changes sign at the hypercritical number $x = a$, then $(a, f(a))$ is a point of inflection of f provided that a is in the domain of f.

EXAMPLE 3. For the function f defined by

$$y = f(x) = x^3 - 3x^2 + 1$$

(a) Find the points of inflection.

(b) Find the equations of the tangent lines to the graph of f at the points of inflection.

SOLUTION.

(a) (1) Differentiating, we have

$$f'(x) = 3x^2 - 6x \qquad \text{and} \qquad f''(x) = 6x - 6 = 6(x - 1)$$

(2) f'' is continuous everywhere and $f''(x) = 0$ when $x = 1$. Thus, $x = 1$ is the only hypercritical number.

(3) We now check to see if f'' changes sign at the hypercritical number $x = 1$. In an interval around $x = 1$, if $x < 1$, then $f''(x) < 0$, and if $x > 1$, then $f''(x) > 0$. Therefore, when $x < 1$, the curve is concave downward and when $x > 1$, the curve is concave upward. Since $x = 1$ is in the domain of f, the only point of inflection for f is $(1, f(1)) = (1, -1)$.

(b) To find the equation of the tangent line to the graph of f at $(1, -1)$, we first find the slope of this line. The slope m is given by

$$f'(1) = 3(1)^2 - 6(1) = -3$$

Using the point-slope form of the equation of a line, we have

$$y - (-1) = -3(x - 1) \quad \text{or} \quad y + 1 = -3x + 3 \quad \text{or} \quad y = -3x + 2$$

as an equation of the tangent line to the graph of f at the point of inflection $(1, -1)$.

In the next example we shall illustrate the use of our knowledge of increasing functions, decreasing functions, critical points, relative extrema, concavity, and points of inflection to sketch the graph of a given function f.

EXAMPLE 4. Sketch the graph of

$$f(x) = x^4 - 6x^2 + 1$$

SOLUTION.

1. We first find $f'(x)$ to determine:

(1) The critical numbers that may yield the relative extrema.

(2) The intervals where f is increasing and decreasing.

$$\begin{aligned} f'(x) &= 4x^3 - 12x = 4(x^3 - 3x) \\ &= 4x(x^2 - 3) = 4x(x - \sqrt{3})(x + \sqrt{3}) \end{aligned}$$

Then

$$f'(x) = 0 \quad \text{when} \quad x = 0, \quad x = \sqrt{3}, \quad \text{and} \quad x = -\sqrt{3}$$

Therefore, 0, $\sqrt{3}$, and $-\sqrt{3}$ are critical numbers. When

$$\begin{array}{lll} x < -\sqrt{3}, & f'(x) < 0 & \text{and } f \text{ is decreasing} \\ -\sqrt{3} < x < 0, & f'(x) > 0 & \text{and } f \text{ is increasing} \\ 0 < x < \sqrt{3}, & f'(x) < 0 & \text{and } f \text{ is decreasing} \\ \sqrt{3} < x, & f'(x) > 0 & \text{and } f \text{ is increasing} \end{array}$$

2. We next find $f''(x)$ to:

(1) Test the critical numbers for relative extrema.

(2) Find the intervals where f is concave upward and concave downward.

(3) Find the points of inflection.

Since

$$\begin{aligned} f''(x) &= 12x^2 - 12 = 12(x^2 - 1) = 12(x - 1)(x + 1) \\ f''(0) &= -12 < 0 \end{aligned}$$

f has a relative maximum at $x = 0$. Since

$$f''(-\sqrt{3}) = 24 > 0$$

f has a relative minimum at $x = -\sqrt{3}$. Since

$$f''(\sqrt{3}) = 24 > 0$$

f has a relative minimum at $x = \sqrt{3}$. When

$$\begin{array}{lll} x < -1, & f''(x) > 0 & \text{and } f \text{ is concave upward} \\ -1 < x < 1, & f''(x) < 0 & \text{and } f \text{ is concave downward} \\ 1 < x, & f''(x) > 0 & \text{and } f \text{ is concave upward} \end{array}$$

The concavity of the curve changes at 1 and -1. Consequently, $(1, f(1)) = (1, -4)$ and $(-1, f(-1)) = (-1, -4)$ are the points of inflection.

3. A rough sketch of the graph of f is shown in Figure 9. We do not need to plot additional points.

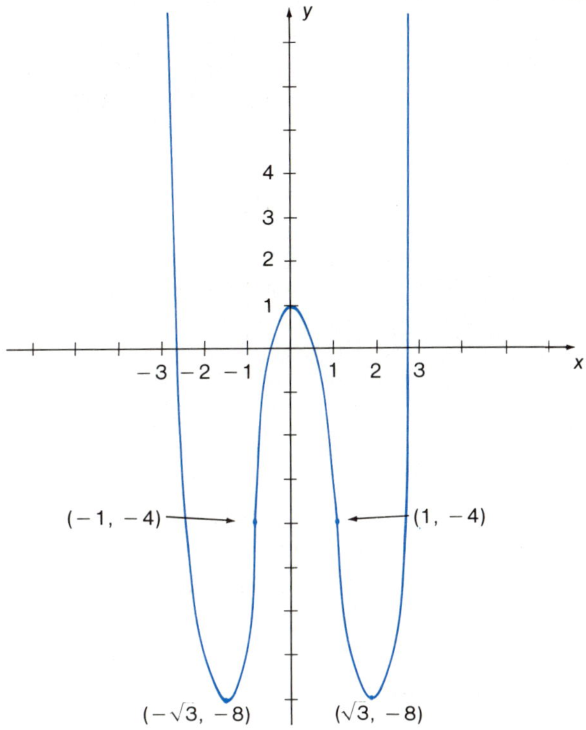

Figure 9

EXAMPLE 5. Sketch the graph of

$$f(x) = \frac{x+1}{x-1} \tag{1}$$

SOLUTION. The function f is undefined at $x = 1$. We have

$$f'(x) = \frac{(x-1)-(x+1)}{(x-1)^2} = -\frac{2}{(x-1)^2} \tag{2}$$

Since $f'(x)$ is never zero, there are no critical values. We see from (2) that for $x \neq 1$, $f'(x) < 0$. Consequently, the function f is decreasing over the intervals $(-\infty, 1) \cup (1, \infty)$. Also,

$$f''(x) = \frac{4}{(x-1)^3} \tag{3}$$

so that, for $x \neq 1$, $f''(x) > 0$ for $x > 1$ and $f''(x) < 0$ for $x < 1$. Hence, the function f is concave upward in the interval $(1, \infty)$ and concave downward in the interval $(-\infty, 1)$. The function f has no points of inflection.

We now investigate the behavior of the function (1) when $|x|$ is very large. We note that, as x gets larger and larger, $1/x$ gets closer and closer to zero. The expression "x tends to infinity" is used to indicate that x increases without bound. The following symbols are used:

$$x \to \infty \qquad \text{if } x \text{ increases without bound}$$

and

$$x \to -\infty \qquad \text{if } x \text{ decreases without bound}$$

We then have

$$\lim_{x \to \infty} \frac{1}{x} = 0$$

and

$$\lim_{x \to -\infty} \frac{1}{x} = 0$$

The limit theorems of Chapter 1 still hold when the number c is replaced by ∞ or $-\infty$. Thus,

$$\lim_{x \to \infty} f(x) = \lim_{x \to \infty} \frac{x+1}{x-1} = \lim_{x \to \infty} \frac{1 + \frac{1}{x}}{1 - \frac{1}{x}} = 1 \tag{4}$$

Similarly,

$$\lim_{x \to -\infty} f(x) = 1 \tag{5}$$

From (4) and (5), we conclude that the graph of f approaches the horizontal line $y = 1$ as $x \to \infty$ and as $x \to -\infty$. The line $y = 1$ will be called a *horizontal asymptote* to the graph of f as x approaches ∞ and $-\infty$.

We also note that the function f defined by (1) is discontinuous at $x = 1$. However,

$$\lim_{x \to 1^+} f(x) = \lim_{x \to 1^+} \frac{x+1}{x-1} = \infty$$

and

$$\lim_{x \to 1^-} f(x) = \lim_{x \to 1^-} \frac{x+1}{x-1} = -\infty$$

In this situation the line $x = 1$ will be called a *vertical asymptote* to the graph of f.

Combining everything we have discussed so far, we obtain the graph of f shown in Figure 10.

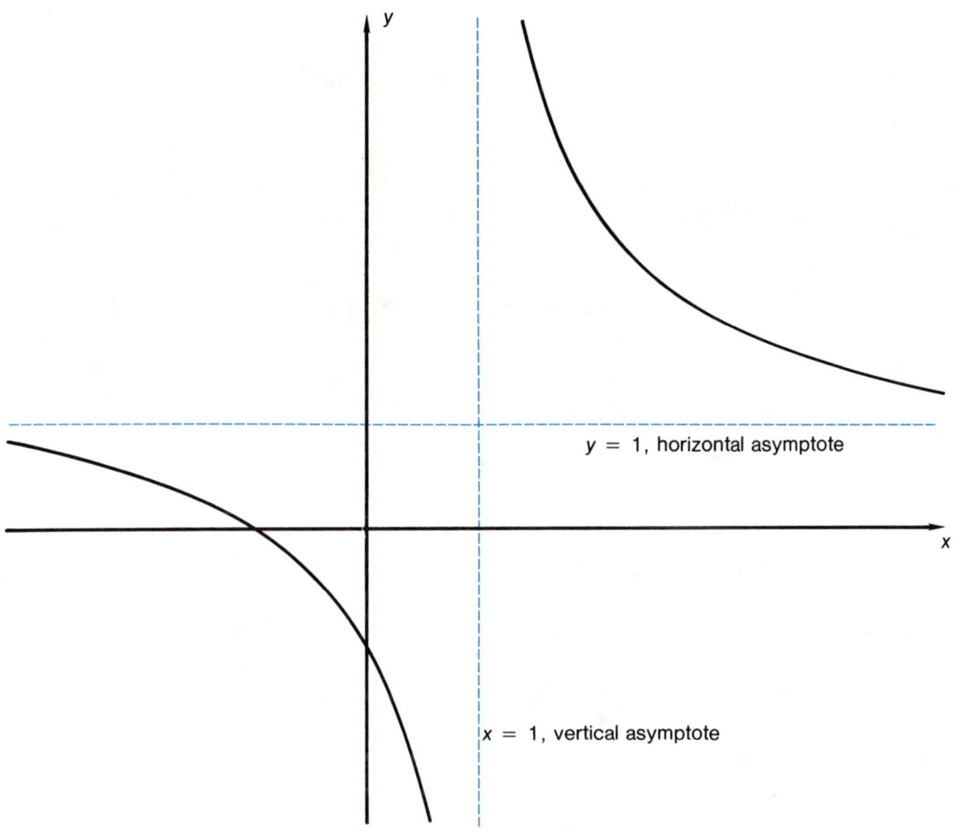

Figure 10

Following from the above example, we determine the two definitions which follow.

Definition 3. (*Horizontal Asymptote*) If $\lim_{x \to \infty} f(x) = L$, then the horizontal line $y = L$ is called a horizontal asymptote to the graph of f as $x \to \infty$. Similarly, if $\lim_{x \to -\infty} f(x) = M$, then the horizontal line $y = M$ is called a horizontal asymptote to the graph of f as $x \to -\infty$.

If $f(x) = P(x)/Q(x)$, then the following example illustrates the procedure for finding the horizontal asymptotes as $x \to \infty$.

EXAMPLE 6. Evaluate each of the following limits (if it exists):

(a) $\lim_{x\to\infty} \dfrac{2x^2+x+2}{x^2+2x-3}$ (b) $\lim_{x\to\infty} \dfrac{x^2-3x+5}{2x^3+3}$

(c) $\lim_{x\to\infty} \dfrac{2x^2+1}{3x+5}$

SOLUTION. (a) We divide the numerator and denominator by x^2.
Thus,

$$\lim_{x\to\infty} \frac{2x^2+x+2}{x^2+2x-3} = \lim_{x\to\infty} \frac{2+\dfrac{1}{x}+\dfrac{2}{x^2}}{1+\dfrac{2}{x}-\dfrac{3}{x^2}} = \frac{2}{1} = 2$$

(b) We divide the numerator and denominator by x^3 (the highest power of x that occurs).
Thus,

$$\lim_{x\to\infty} \frac{x^2-3x+5}{2x^3+3} = \lim_{x\to\infty} \frac{\dfrac{1}{x}-\dfrac{3}{x^2}+\dfrac{5}{x^3}}{2+\dfrac{3}{x^3}} = \frac{0}{2} = 0$$

(c) Dividing the numerator and denominator by x^2, we get

$$\lim_{x\to\infty} \frac{2x^2+1}{3x+5} = \lim_{x\to\infty} \frac{2+\dfrac{1}{x^2}}{\dfrac{3}{x}+\dfrac{5}{x^2}}$$

but the limit of the denominator is

$$\lim_{x\to\infty} \frac{3}{x}+\frac{5}{x^2} = 0$$

and the limit of the numerator is 2.
In this case we say the limit does not exist or the limit is equal to ∞. So, geometrically, there is no horizontal asymptote.

In general, we have

$$\lim_{x\to\infty} \frac{a_0x^n+a_1x^{n-1}+\cdots+a_n}{b_0x^m+b_1x^{m-1}+\cdots+b_m} = \begin{cases} 0 & \text{if } n<m \\ \dfrac{a_0}{b_0} & \text{if } n=m \\ \infty & \text{if } n>m \end{cases}$$

Definition 4. (*Vertical Asymptote*) The vertical line $x = a$ is called a vertical asymptote to the graph of f is $\lim_{x\to a^+} f(x) = \pm\infty$ or $\lim_{x\to a^-} f(x) = \pm\infty$ or both.

If $f(x) = P(x)/Q(x)$, where $P(x)$ and $Q(x)$ have no common factors, then the vertical line $x = a$ is a vertical asymptote if and only if $Q(a) = 0$. For example, for the function.

$$f(x) = \frac{x^2 + x + 1}{x^2 - 3x - 4} = \frac{x^2 + x + 1}{(x-4)(x+1)}$$

the lines $x = 4$ and $x = -1$ are the vertical asymptotes to the graph of f.

HINTS FOR SKETCHING CURVES.

(1) Get as much "free" information as you can by inspection.

(a) *Locate intercepts.* Set $x = 0$, so that $f(0)$ is the y-intercept. If possible, find the x-intercepts by setting $f(x) = 0$.

(b) If easily done, find where y is positive, negative, or zero.

(c) *Look for symmetry.* The graph of $y = f(x)$ is symmetric about the y-axis if $f(-x) = f(x)$. Such a function is called an *even function* (see Figure 11). The graph of $y = f(x)$ is symmetric through the origin if $f(-x) = -f(x)$. Such a function is called an *odd function* (see Figure 12).

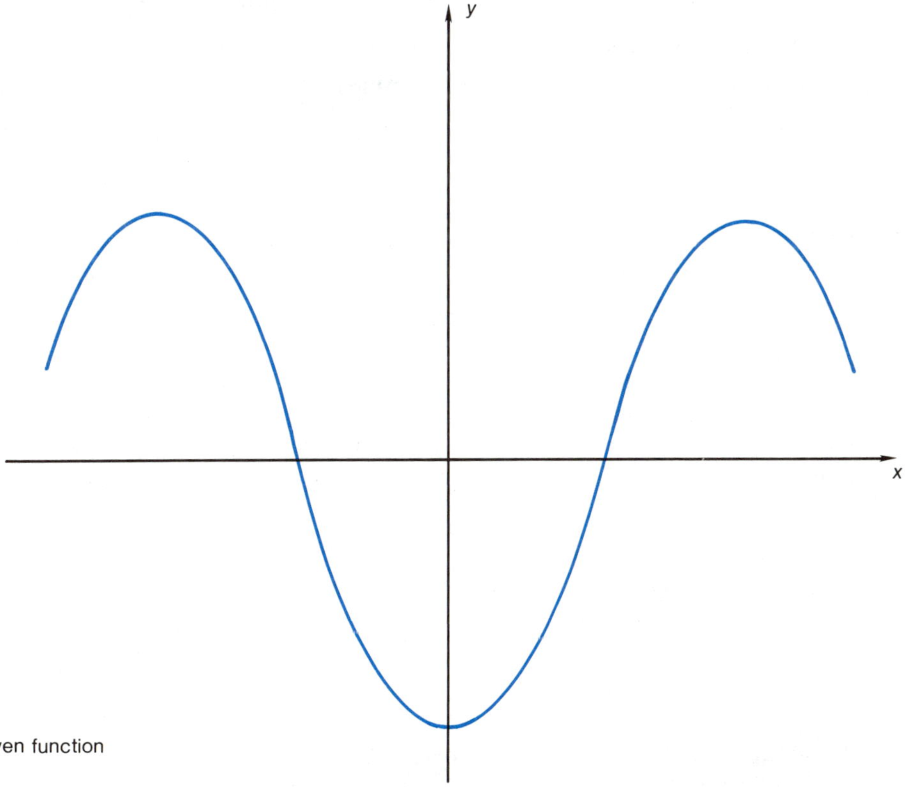

Even function

Figure 11

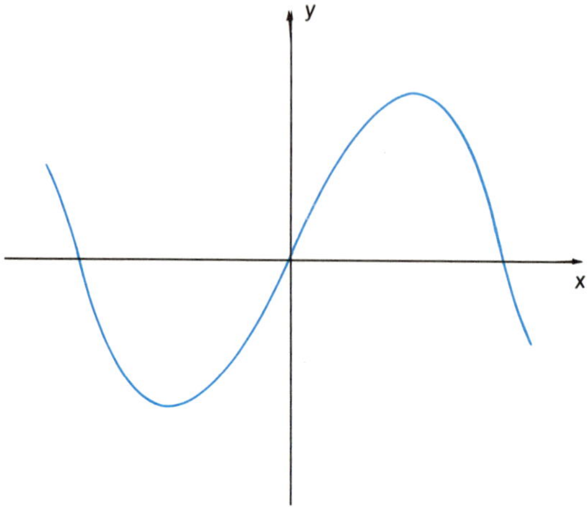

Odd function

Figure 12

(d) *Horizontal asymptotes.* Find the behavior of the function as $x \to \infty$ and $x \to -\infty$; that is, find $\lim_{x \to \infty} f(x)$ and $\lim_{x \to -\infty} f(x)$. If either of these is finite, we obtain horizontal asymptotes.

(e) *Vertical asymptotes.* If f is not defined at a, determine $\lim_{x \to a^+} f(x)$ and $\lim_{x \to a^-} f(x)$; then the line $x = a$ is a vertical asymptote if either of these limits is ∞ or $-\infty$.

(2) *Take the first derivative.* Examine the sign of f'. This will tell you where the curve is rising, falling, or level. Solve $f'(x) = 0$ to locate critical values. Find where f' does not exist.

(3) *Take the second derivative.* Examine the sign of f''. This will tell you where the curve is concave upward and downward. Find the points of inflection. Use the second derivative test to see if the critical values obtained above give relative a maximum or minimum. If the second derivative at a critical value is zero, use the first derivative test.

(4) For greater accuracy, plot a few additional points.

EXERCISES 3.4

In Problems 1 through 22, evaluate the limit (if it exists).

1. $\lim_{x \to +\infty} \dfrac{5x-1}{-x+4}$

2. $\lim_{x \to +\infty} \dfrac{3x+1}{4x-2}$

3. $\lim_{x\to+\infty} \dfrac{x^2+3x+7}{3x^2-2x-1}$

4. $\lim_{x\to+\infty} \dfrac{2x+5}{x^2+1}$

5. $\lim_{x\to+\infty} \dfrac{2x^2+4x-1}{x^3+4}$

6. $\lim_{x\to-\infty} \dfrac{\sqrt{x^2+6}}{x+6}$

7. $\lim_{x\to-\infty} \dfrac{x^2+6}{x+6}$

8. $\lim_{x\to+\infty} \dfrac{2x^3+3x^2-4}{5x^3+x+2}$

9. $\lim_{x\to+\infty} \dfrac{2-x^2}{3x+5}$

10. $\lim_{x\to+\infty} \dfrac{2x^2-3x+4}{x+4}$

11. $\lim_{x\to 2} \dfrac{x}{x-2}$

12. $\lim_{x\to 3} \dfrac{x^2}{9-x^2}$

13. $\lim_{x\to 1^+} \dfrac{x+1}{x^2-1}$

14. $\lim_{x\to 1^-} \dfrac{x+1}{x^2-1}$

15. $\lim_{x\to 2} \dfrac{-2}{(x-2)^2}$

16. $\lim_{x\to 2^+} \dfrac{x+2}{x^2-4}$

17. $\lim_{x\to 0^+} \dfrac{\sqrt{1+x^2}}{x}$

18. $\lim_{x\to 0^-} \dfrac{\sqrt{1+x^2}}{x}$

19. $\lim_{x\to 3^-} \dfrac{\sqrt{9-x^2}}{x-3}$

20. $\lim_{x\to 3^+} \dfrac{\sqrt{9-x^2}}{x-3}$

21. $\lim_{x\to 0} \left(\dfrac{1}{x}-\dfrac{1}{x^2}\right)$

22. $\lim_{x\to 3} \left(\dfrac{1}{x-3}-\dfrac{2}{x^2-9}\right)$

In Problems 23 through 30, find the horizontal and vertical asymptotes to the graphs of the given functions. Determine if the function is odd or even or neither.

23. $f(x) = \dfrac{x}{x+1}$

24. $f(x) = \dfrac{x^2}{x^2-1}$

25. $f(x) = \dfrac{6x^2+x+11}{x^2+4x+3}$

26. $f(x) = \dfrac{x^2}{x-1}$

27. $f(x) = \dfrac{1}{x-1}+3$

28. $f(x) = \dfrac{x^2+x-2}{x^2+4x-5}$

29. $f(x) = \sqrt{x}$

30. $f(x) = \dfrac{3}{x^2}$

In Problems 31 through 50, sketch the graph of the given function.

31. $f(x) = x^2-3x+4$

32. $f(x) = -2x^2+3x-2$

33. $f(x) = x^3 - 1$

34. $f(x) = -x^3 + 4$

35. $f(x) = 4x^{2/3}$

36. $f(x) = -3x^{5/3}$

37. $f(x) = \frac{1}{3}x^3 + 3x^2$

38. $f(x) = x^3 - 3x^2$

39. $f(x) = 2x^3 + 4x^2 + 2x + 1$

40. $f(x) = x^3 - 5x^2 + 8x - 4$

41. $f(x) = x^2 - 3x^3 + 3x^4$

42. $f(x) = -x^3 + 3x^2 - 3x + 7$

43. $f(x) = \dfrac{x^3}{6} - \dfrac{x^5}{20}$

44. $f(x) = \dfrac{1}{x}$

45. $f(x) = \dfrac{1}{1 - x}$

46. $f(x) = x + \dfrac{1}{x}$

47. $f(x) = -3x + \dfrac{3}{x}$

48. $f(x) = \dfrac{x}{x^2 + 3}$

49. $f(x) = \dfrac{x^2}{1 + x^2}$

50. $f(x) = \dfrac{x}{(x + 3)^2}$

51. Find the equation of the tangent line to the graph of f at each point of inflection in Problems 38 through 41 above.

52. Determine a value of k so that the function

$$f(x) = x^2 + \frac{k}{x}$$

has a point of inflection at $x = 2$.

53. If k is a positive integer, then show that

(a) $f(x) = x^{2k+1}$ has $(0, 0)$ as a point of inflection,

(b) $f(x) = x^{2k}$ does not have $(0, 0)$ as a point of inflection.

54. Consider the fourth-degree polynomial function

$$f(x) = x^4 + ax^3 + bx^2 + cx + d$$

and suppose that $3a^2 = 4b$. Show that f has no points of inflection.

55. If the cost function $C(x)$ is given by

$$C(x) = 6x - 0.03x^2 + 0.00005x^3$$

determine whether the curve is concave upward, concave downward, or has a point of inflection at $x = 100$, $x = 200$, $x = 300$, and $x = 400$.

56. Show that the vertex of the parabola

$$f(x) = ax^2 + bx + c$$

is at $x = -\dfrac{b}{2a}$. Discuss the cases (1) $a > 0$ and (2) $a < 0$.

3.5 THE DIFFERENTIAL

Let $y = f(x)$ define a differentiable function of x. In Chapter 2 we introduced the notation dy/dx to represent the derivative of f with respect to x without clarifying the meaning of the symbols dy and dx. During the nineteenth century much controversy shrouded these symbols. Historically, the symbols date back to Leibniz, who attempted to define dy and dx as infinitesimals (very small quantities) "greater than zero but less than any positive real number." Leibniz was aware that there were no such real numbers. Nevertheless, he was very successful using these symbols in solving many problems. Some years later, Cauchy[1] offered an alternative definition of the symbols. In this section we shall define these symbols and show how they are used in mathematics and the sciences.

Let us return to the definition of the derivative of a function. Suppose that f, given by $y = f(x)$, is a differentiable function of x. Let Δx (read "delta x") be the *increment*, or change, in the variable x. Then the *increment* Δy in the functional value $f(x)$ is given by

$$\Delta y = f(x + \Delta x) - f(x)$$

For a particular function f, Δy depends on both the numbers x and Δx.

We may write the derivative f' as

$$f'(x) = \lim_{\Delta x \to 0} \frac{f(x + \Delta x) - f(x)}{\Delta x} = \lim_{\Delta x \to 0} \frac{\Delta y}{\Delta x}$$

---

[1] AUGUSTIN LOUIS CAUCHY (1789-1857) was born in Paris at the outset of the French Revolution. His father was a lawyer who later became Secretary of State under Napoleon. The first eleven years of Cauchy's life were spent in the village of Arcueil, where the family had gone to escape the Reign of Terror.

At the age of 18, Cauchy entered civil engineering school, and two years later Napoleon commissioned him to participate in the construction of the harbor at Cherbourg. In 1815 he was appointed professor at the École Polytechnique and at the Sorbonne. With the revolution of 1830, he left France in voluntary exile. When Cauchy returned to Paris eight years later, he resumed his post without giving allegiance to the government.

Cauchy introduced rigorous methods in mathematical analysis. He created the theory of functions of a complex variable and produced over 500 papers on all branches of mathematics.

In many practical problems (see Example 3), we are interested in computing Δy. However, the computation of Δy may be quite difficult or tedious. Therefore, we use a close approximation of Δy, called the differential.

Definition 1. Let $y = f(x)$ define a differentiable function of x. Then we have the following:

(1) The *differential* dx of the independent variable x is defined by $dx = \Delta x$, where Δx is any real number.

(2) The *differential* dy of the dependent variable y is defined by $dy = f'(x)\,dx$.

A few remarks are in order.

Remark 1. The definition $dx = \Delta x$ is completely arbitrary and is, therefore, an independent variable whose range is the set of real numbers.

Remark 2. As we have indicated earlier, dy is used as an approximation for Δy (see the following examples), and in many cases it will be a relatively simpler matter to calculate dy rather than Δy.

Remark 3. An important result of this definition is the fact that the ratio of dy to dx is $f'(x)$.

EXAMPLE 1. Let $y = f(x) = x^2 - 3x + 2$. Compare Δy and dy for

(a) $x = 3, \quad \Delta x = -\frac{1}{2}$ (b) $x = 3, \quad \Delta x = \frac{1}{8}$

SOLUTION. Since

$$f(x) = x^2 - 3x + 2 \quad \text{and} \quad f'(x) = 2x - 3$$

then, by definition,

$$dy = f'(x)dx = (2x - 3)dx \tag{1}$$

Also,

$$\begin{aligned} \Delta y &= f(x + \Delta x) - f(x) \\ &= [(x + \Delta x)^2 - 3(x + \Delta x) + 2] - (x^2 - 3x + 2) \end{aligned}$$

or

$$\Delta y = 2x\Delta x + (\Delta x)^2 - 3\Delta x \tag{2}$$

(a) Substituting $x = 3$ and $dx = \Delta x = -\frac{1}{2}$ in Equations (1) and (2), respectively, we obtain

$$dy = (2 \cdot 3 - 3)(-\tfrac{1}{2}) = -\tfrac{3}{2}$$

$$\Delta y = 2 \cdot 3(-\tfrac{1}{2}) + (-\tfrac{1}{2})^2 - 3(-\tfrac{1}{2})$$

$$= -3 + \tfrac{1}{4} + \tfrac{3}{2} = -\tfrac{3}{2} + \tfrac{1}{4}$$

Thus,

$$|\Delta y - dy| = \tfrac{1}{4}$$

(b) Setting $x = 3$ and $\Delta x = \frac{1}{8}$, we get

$$dy = (2 \cdot 3 - 3)\tfrac{1}{8} = \tfrac{3}{8}$$

and

$$\Delta y = 2 \cdot 3 \cdot \tfrac{1}{8} + (\tfrac{1}{8})^2 - 3 \cdot \tfrac{1}{8} = \tfrac{3}{8} + \tfrac{1}{64}$$

or

$$|\Delta y - dy| = \tfrac{1}{64}$$

We observe that even in such simple cases it is easier to calculate dy than Δy. We also note in this example that dy approximates Δy better when Δx is changed from $-\frac{1}{2}$ to $\frac{1}{8}$ while $x = 3$ is kept fixed.

Let us now consider the geometrical meaning of the differential. Let $P(x, y)$ and $Q(x + \Delta x, y + \Delta y)$ be two points on the graph of $y = f(x)$ (see Figure 1). Then the slope of the tangent line at P is given by $f'(x) = dy/dx$. But the slope of the tangent

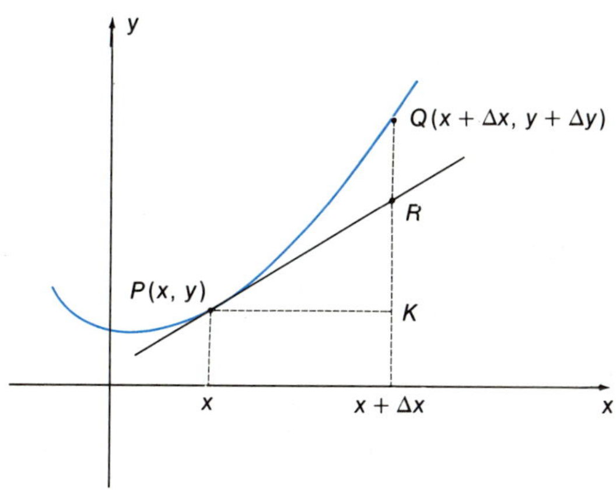

Figure 1

line is also given by the ratio RK/PK, and since $PK = \Delta x = dx$, it follows that $RK = f'(x)dx = dy$.

Now $\Delta y = KQ$. Thus, the difference between Δy and dy is represented geometrically by the segment RQ.

Applications. In many problems we come across a quantity x which is measured by a faulty instrument. As a result of this error in the increment of x, there will be a corresponding error in the computed value of y, where $y = f(x)$. The error in y (that is, Δy) is approximated by dy. If the error in y is dy, the dy/y is called the *relative error* in y and $100dy/y$ is called the *percentage error* in y.

EXAMPLE 2. Calculate approximately the fourth root of 626.

SOLUTION. Since we know that the fourth root of 625 is 5, we shall approximate the amount by which $x^{1/4}$ increases when x is increased by 1. Let

$$y = x^{1/4} \quad \text{and} \quad y + \Delta y = \sqrt[4]{x + \Delta x}$$

Then

$$dy = \tfrac{1}{4}x^{-3/4}dx \quad \text{or} \quad dy = \frac{dx}{4x^{3/4}}$$

If $x = 625$ and $dx = \Delta x = 1$, then

$$y = (625)^{1/4} = 5 \quad \text{and} \quad dy = \frac{1}{4(625)^{3/4}} = \frac{1}{4(125)} = \frac{1}{500} = 0.002$$

Now

$$(626)^{1/4} = y + \Delta y \approx y + dy = 5 + 0.002 = 5.002$$

EXAMPLE 3. A student in a physics laboratory measures the sides of a cube as 2 ft. (see Figure 2) with a possible error of 0.002 ft. What is the greatest possible error in the volume computed by the student?

SOLUTION. The possible error of 0.002 ft. in the side x of the cube signifies that the value 2 ft. is subject to a correction which may amount to as much as 0.002 ft. in either direction. In other words, the error in x may be as much as $dx = \Delta x = \pm 0.002$. Then, since Δx is small, the computed value of the volume

$$V = f(x) = x^3$$

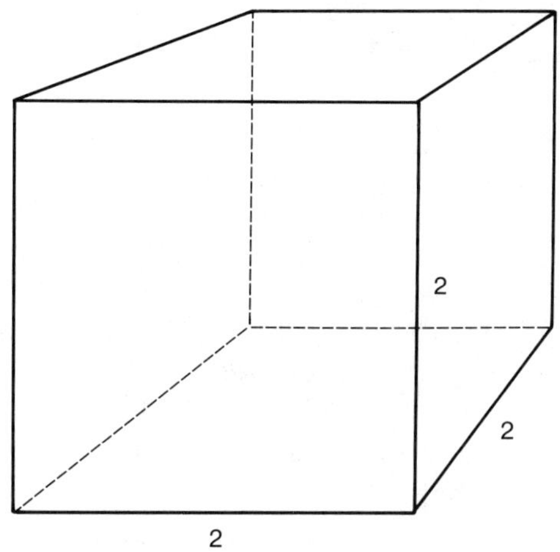

Figure 2

is subject to an approximate error given by

$$dV = f'(x)dx = 3x^2dx$$

Here $x = 2$ and $\Delta x = \pm 0.002$. Therefore,

$$dV = \pm 3(2)^2(0.002) = \pm 0.024$$

Thus, the computed volume of 8 cu. ft. may be too large or too small by 0.024 cu. ft., approximately.

The relative error in this case is

$$\frac{dV}{V} = \frac{0.024}{8} = 0.003$$

The relative error in x is

$$\frac{dx}{x} = \frac{0.002}{2} = 0.001$$

The percentage error in the side of the cube and its volume will be 0.1 and 0.3 respectively.

EXERCISES 3.5

In Problems 1 through 6, find dy, Δy, and $|\Delta y - dy|$, when $x = 4$ and $\Delta x = 0.5$.

1. $y = x^3$

2. $y = 2x^3 - 7x$

3. $y = x^2 + 5x + 3$

4. $y = \sqrt{x}$

5. $y = \dfrac{2x}{3 + x}$

6. $y = \dfrac{1}{\sqrt{x}}$

In Problems 7 through 15, use differentials to find an approximate value for the given quantity.

7. $\sqrt{10}$

8. $\sqrt{122}$

9. $\sqrt[5]{33}$

10. $\sqrt{0.0143}$

11. $\sqrt{255}$

12. $\sqrt[3]{7.5}$

12. $\sqrt[3]{25}$

14. $(1.98)^5$

15. $\dfrac{1}{\sqrt[5]{31}}$

16. The diameter of a circle is 4 ft. By approximately how much does the area decrease if the radius decreases by 1/30 in.?

17. The diameter of a sphere is 6 in. with a possible error of 0.05 in. What is the greatest possible error in the computed area and the relative possible error?

18. A cubic box is to hold 1000 cu. ft. What is the allowable error in the edge of the cube if the error in the volume is not to exceed 2 cu. ft.?

19. If the area of a circle increases at a constant rate, show that the rate of increase of the perimeter varies inversely as the radius.

20. A spherical shell has an inner radius of 10 cm. and a thickness of $\frac{1}{4}$cm. Find the approximate volume of the shell.

21. The total cost of producing x units is C dollars, where

$$C = \tfrac{1}{100}x^2 + 5x + 200$$

(a) Find the differential dC.

(b) When the level of production changes from 100 to 101, what is the value of dC?

(c) Determine the error involved when dC is used as an approximation for ΔC.

(d) Determine the relative and percentage errors.

22. The total cost C (in dollars) of producing x shirts is

$$C = \frac{x^3}{1500} - \frac{2}{5}x^2 + 45x$$

and each shirt is sold at \$10.
(a) Find the total profit P.
(b) Find the differential dP.
(c) When the production level changes from $x = 350$ to $x = 355$, what will be an approximate change in P?

23. Use differentials to estimate the number of sales lost when the price is increased from \$5 to \$6 if the demand function is

(a) $x(p) = 5000\left(\frac{100-p}{p^2+5}\right)$, (b) $x(p) = 5000\left(\frac{100-p}{p+45}\right)$

24. Profits for the Kookie Krumbal Co. are given by

$$P(x) = -80 + 10x - \frac{x^2}{210}$$

Estimate the change in profits if the sales are increased:
(a) from 30 to 31 units, (b) from 135 to 140 units.

25. For the revenue function

$$R(x) = 200x + 50x^2 - x^3,$$

use differentials to find the estimate of the change in revenue if the number of units is increased from $x = 30$ to $x = 31$. Find the actual change.

26. Do problem 25 when $R(x) = 150x + 10x^2 - \frac{3}{x}$.

CHAPTER 3 REVIEW

Part 1 (ORAL)

Define or discuss the following:

1. Increasing function
2. Decreasing function

3. Critical number
4. Relative maximum, absolute maximum
5. Relative minimum, absolute minimum
6. Concave upward, concave downward
7. Point of inflection
8. Hypercritical number
9. Differential

Part 2 (WRITTEN)

In problems 1 through 20, find $f'(x)$ and $f''(x)$. Find the intervals over which f is increasing, decreasing, concave upward, and concave downward. Locate relative extrema. Find the points of inflection. Graph each function.

1. $f(x) = 3x^2 - 7x + 2$

2. $f(x) = 2x^2 + x + 1$

3. $f(x) = x(3 - x)$

4. $f(x) = x^3 - 27x$

5. $f(x) = x^2(x + 9)$

6. $f(x) = x^3 - 3x^2 + 3x - 9$

7. $f(x) = \sqrt{x + 3}$

8. $f(x) = x^4 - x^2$

9. $f(x) = 3x^{1/3}$

10. $f(x) = \dfrac{1}{x + 1}$

11. $f(x) = 2x + \dfrac{1}{x}$

12. $f(x) = \sqrt{4 - x^2}$

13. $f(x) = \dfrac{-2}{x + 3}$

14. $f(x) = -3x + \dfrac{3}{x}$

15. $f(x) = \dfrac{4}{x^2}$

16. $f(x) = \dfrac{1}{5}x^5 - 1$

17. $f(x) = 4x^3 + 3x^2 - 6x - 1$

18. $f(x) = x^4 - 14x^2 + 24x + 5$

19. $f(x) = 2x^4 + x^2 + 2x$

20. $f(x) = \dfrac{2x - 1}{x + 3}$

21. Find the relative and absolute extrema (if any) for the function $f(x) = 2x + 3$, when

(a) $x \in (0, 1)$ (b) $x \in [1, 3)$ (c) $x \in (-\infty, \infty)$

22. Repeat problem 21 for $f(x) = 1/x$.

23. Find two positive numbers whose sum is 50 and whose product is maximum.

24. Find two numbers x and y that have a sum of 80 and such that $x^3 + y^3$ is minimum.

25. Of all the rectangles with an area of 400 sq. ft., find the following:
(a) The one with the smallest perimeter
(b) The one with the shortest diagonal

26. Find the dimensions of the most economical can (with closed top) if it is to hold 40 cu. in. of a liquid.

27. A rectangular field along the bank of a river is to have an area of 1800 sq. m. What must the dimensions of the field be to use the least amount of fencing material in enclosing the field? (Assume that no fence is needed along the river, which is flowing along a straight line.)

28. A figure consists of a rectangle surmounted by a semicircle. If the perimeter of the figure is $(8 + 2\pi)$ in., and the area is maximum, find the dimensions of the rectangle and the radius of the semicircle.

29. Two sources of light are 10 ft. apart. Assuming that the intensity of illumination varies inversely as the square of the distance from the source, find the point of maximum illumination on the straight line joining the two sources if one source has six times the candle power of the other.

30. P and Q are two points on a straight coast at a distance of 10 mi. L is a point in the sea, where PL is perpendicular to PQ, and $PL = 3$ mi. A person wishes to go from L to Q as quickly as possible partly by rowing on the sea and partly by running along the coast. The person is able to row at 4 mph and run at 5 mph. How far from P should the person land?

31. Let the reaction R of the body to a dose d of a drug be given by

$$R(d) = d^2\left(\frac{c}{2} - \frac{d}{3}\right)$$

where $c > 0$ is the maximum amount of the drug that can be administered. Find the dose for which the rate of increase of R is maximum. In other words, find d when $R'(d)$ is maximum.

In problems 32 through 37, find the horizontal and vertical asymptotes to the graphs of the given function.

32. $f(x) = \dfrac{2x + 1}{x + 5}$

33. $f(x) = x^3 + 2x$

34. $f(x) = \dfrac{x^2 + 1}{x^2 - 1}$

35. $f(x) = \dfrac{2x^2 + 3}{x^2 - 16}$

36. $f(x) = 2 + \dfrac{1}{x + 1}$

37. $f(x) = \dfrac{2x^2 + 3x + 1}{x^2 - 3x - 10}$

38. Let $f(x) = \sqrt{x^2 + x + 1} - \sqrt{x^2 + 1}$. Show that $\lim_{x \to \infty} f(x) = \frac{1}{2}$.

[*Hint:* Rationalize first; i.e., multiply and divide $f(x)$ by $\sqrt{x^2 + x + 1} + \sqrt{x^2 + 1}$.]

39. Suppose $f(x)$ is the attainment for x hours of practice. Psychologist Thurston suggested the "learning curve," which expresses the relationship between attainment and practice as:

$$f(x) = \frac{a(x + b)}{x + c}$$

where a, b, and c are positive constants with $b < c$. Sketch the graph of $f(x)$. What does a represent?

40. Suppose the cost function is given by

$$C(x) = x^2 + 2x + 9 \qquad (0 < x \leq 20)$$

For what value of x is the average cost a minimum?

41. If the cost function is given by

$$C(x) = 6x - 0.03x^2 + 0.00005x^3$$

determine where the curve is concave upward and where it is concave downward. Find the inflection points.

42. In problem 41, find when the average cost is a minimum.

43. If $y = 8 - \frac{1}{3}x^3$, find dy and Δy when $x = 3$ and $\Delta x = 0.5$.

44. Compute $\sqrt[5]{31}$ approximately.

45. A cubical box is to have a capacity of 1000 cu. in. What is the allowable error in measuring the edge of the cube if the error in the volume is not to exceed 3 cu. in.?

Chapter Four

Exponential and Logarithmic Functions

In this Chapter we shall discuss two of the most important functions in mathematics: the exponential function and the logarithmic function. Until now we concentrated our attention on functions which can be constructed by taking combinations of sums, differences, products, ratios, powers, and roots of polynomial functions. While these functions are certainly important, they fail to describe certain real-life situations, such as the growth of population, decay of radioactive samples, continuously compounded interest, and several others. To fill this gap, we introduce the exponential function and the logarithmic function. We also find the derivatives of these functions and discuss some applications.

4.1 THE EXPONENTIAL FUNCTION.

Suppose b is a positive number and m and n are positive integers; then we recall that

$$b^{m/n} = (\sqrt[n]{b})^m \tag{1}$$

$$b^0 = 1 \tag{2}$$

$$b^{-m/n} = \frac{1}{b^{m/n}} \tag{3}$$

This indicates that for $b > 0$, the function $f(x) = b^x$ is defined for rational values of x. We do require b to be positive, since if b is not positive then there are values of x for which b^x is meaningless or is not a real number. For example, $0^{-2} = 1/0^2$ is meaningless and $(-1)^{1/2}$ is not a real number.

The function $f(x) = b^x$ can be extended to produce a continuous function whose domain consists of all real numbers. But how do we interpret expressions like $3^{\sqrt{2}}$ and

3^π? To do so we make use of the fact that any irrational number can be approximated as closely as desired by a rational number. For example, let us consider 3^π. The number π is an irrational number and, to eight decimal places, $\pi = 3.14159265$. Since the numbers

$$3.1 = \frac{31}{10}, \; 3.14 = \frac{314}{10^2}, \; 3.141 = \frac{3141}{10^3}, \; 3.1415 = \frac{31415}{10^4}, \text{ etc.}$$

are rational, we can use a calculator to find:

$$\begin{aligned} 3^{3.1} &= 30.1353257 \\ 3^{3.14} &= 31.48913565 \\ 3^{3.141} &= 31.52374901 \\ 3^{3.1415} &= 31.54106996 \\ &\text{etc.} \end{aligned}$$

The more decimal places we take, the nearer the exponent x gets to π. It can be proved that as x approaches π, 3^x approaches a unique number, which is defined as 3^π. In fact, this property holds for any number b^x, where b is a positive number and x is any irrational number. We use this idea to define an *exponential function.*

Definition 1. (*Exponential function*) If $b > 0$ and $b \neq 1$, then the function f defined by $f(x) = b^x$ is called an *exponential function*. The number b is called the *base* of the function. The domain of f is the set of all real numbers, and the range of f is the set of all positive numbers.

Note that we require that $b \neq 1$ because in this case $f(x) = 1^x = 1$, which represents a constant function.

We illustrate the behavior of the function $f(x) = b^x$ in the following example.

EXAMPLE 1. Sketch the graph of each of the following functions:

(a) $f(x) = 2^x$ (b) $f(x) = (\frac{1}{2})^x$

SOLUTION.

(a) Assigning certain values for x and finding the corresponding values $f(x)$, we obtain the following table:

x	-2	-1	0	1	2
$f(x)$	$\frac{1}{4}$	$\frac{1}{2}$	1	2	4

It can be shown that the exponential function is continuous. Thus, locating the points in the preceding table, we construct a graph of f as shown in Figure 1.

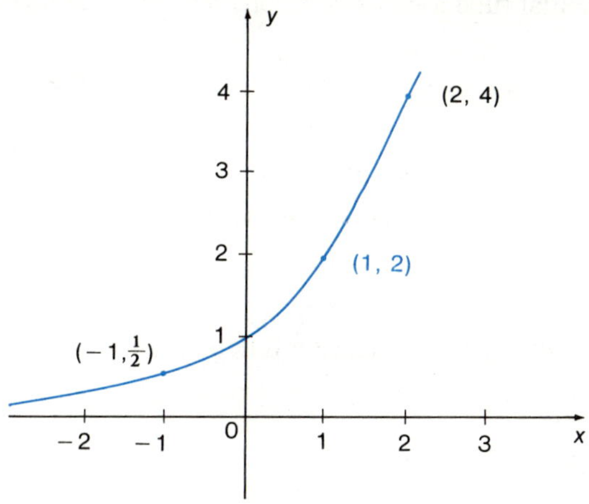

Figure 1

(b) First we construct the following table:

x	-2	-1	0	1	2
$f(x)$	4	2	1	$\frac{1}{2}$	$\frac{1}{4}$

A graph of the function $f(x) = (\frac{1}{2})^x$ is shown in Figure 2.

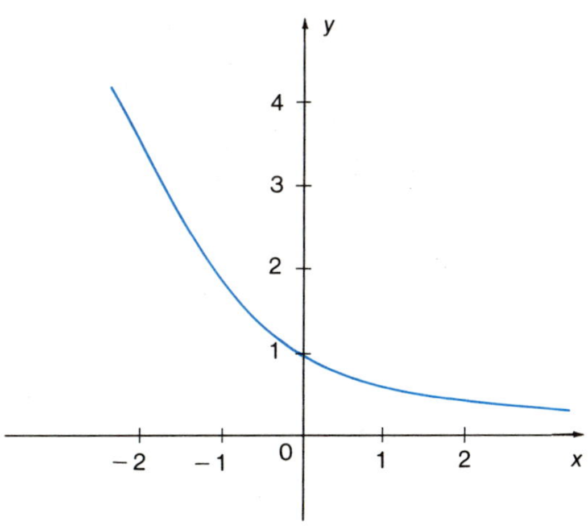

Figure 2

In general, the exponential function $f(x) = b^x$ has the following properties:

1. The exponential function is continuous over the real line.

2. The graph of f intersects the y-axis at $y = 1$; i.e., $b^0 = 1$.

3. If $b > 1$, then f is an increasing function, and

$$f(x) \to +\infty \quad \text{as} \quad x \to +\infty$$
$$f(x) \to 0 \quad \text{as} \quad x \to -\infty$$

4. If $0 < b < 1$, then f is a decreasing function, and

$$f(x) \to +\infty \quad \text{as} \quad x \to -\infty$$
$$f(x) \to 0 \quad \text{as} \quad x \to +\infty$$

5. $b^x \cdot b^y = b^{x+y}$

6. $(b^x)^y = b^{xy}$

7. $\dfrac{b^x}{b^y} = b^{x-y}$

8. $(ab)^x = a^x b^x$

9. $\left(\dfrac{a}{b}\right)^x = \dfrac{a^x}{b^x}$

10. $b^{-x} = \dfrac{1}{b^x}$

The general shape of the graphs of $f(x) = b^x$ for $b > 1$ and $0 < b < 1$ are shown in Figures 3 and 4 respectively.

Graph of $f(x) = b^x$ $(b > 1)$

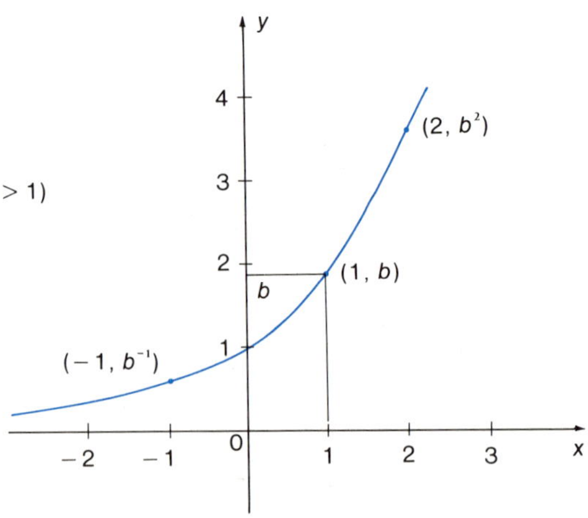

Figure 3

Graph of $f(x) = b^x$ $(0 < b < 1)$

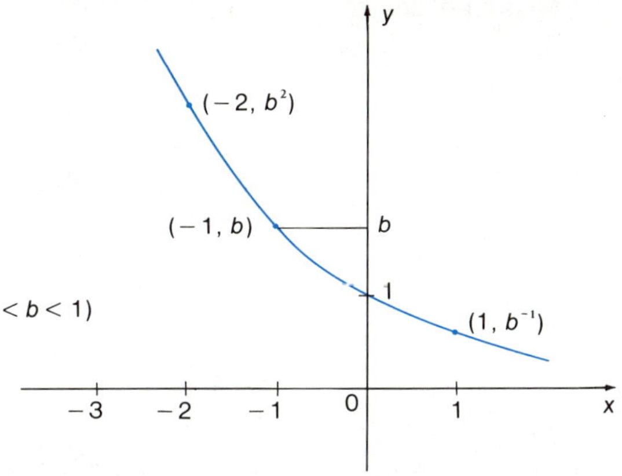

Figure 4

Before continuing, let us investigate an important limit.

EXAMPLE 2. Using a calculator, approximate $\lim_{x \to 0} (1 + x)^{1/x}$.

SOLUTION. We use a calculator and make the following table:

x	$1 + x$	$(1 + x)^{1/x}$
0.1	1.1	2.59374246
0.01	1.01	2.704813815
0.001	1.001	2.716923842
0.0001	1.0001	2.718145918
0.00001	1.00001	2.718254646
0.000001	1.000001	2.718281828
0.0000001	1.0000001	2.718281828
−0.1	0.9	2.867971989
−0.01	0.99	2.731999012
−0.001	0.999	2.719642125
−0.0001	0.9999	2.718417746
−0.00001	0.99999	2.718281828
−0.000001	0.999999	2.718281828
−0.0000001	0.9999999	2.718281828

It appears from the table that the limit exists and has a value near 2.718281828. In fact, it can be proved that the limit does indeed exist. The limit is so important that it was given a special symbol, e, by Euler. He showed that e is an irrational number. A twenty-decimal place approximation to e is 2.71828182845904523536.

COMPOUND INTEREST

Suppose P dollars are deposited in a bank paying interest at a rate i ($100i\%$) compounded annually.

After 1 year the principal (P) has earned Pi dollars interest, and, hence, there are

$$S_1 = P + Pi = P(1+i)$$

dollars on deposit. If interest is compounded, this means that during the second year interest is paid on S_1. Hence, at the end of the second year the amount is

$$S_2 = S_1 + S_1 i = S_1(1+i) = P(1+i)(1+i) = P(1+i)^2$$

Thus, the amount S (or S_n) on deposit after n years is given by:

$$S = P(1+i)^n \tag{1}$$

For example, if \$5000 is deposited for 10 years in an account paying 7% interest compounded annually, then the total amount on deposit in the account after 10 years will be ($P = 5000, i = 0.07, n = 10$):

$$S = 5000(1+0.07)^{10} = \$9835.76 \tag{2}$$

Here we have used a calculator and rounded off the result to the nearest cent.

Next, suppose we shop around and deposit \$5000 for 10 years in another bank that also pays 7% yearly but compounds the interest quarterly (every 3 months). The phrase "7% compounded quarterly" means that each 3 months, the previous balance earns interest at the rate 7/4%. Thus, in equation (1) we use $i = 0.07/4 = 0.0175$, but n is replaced by $4(10) = 40$ because ther are 40 periods of 3 months each in a period of 10 years. We will have:

$$S = 5000(1+0.0175)^{40} = \$10{,}007.99 \tag{3}$$

in the bank after 10 years.

We observe that by looking for a "better deal" (comparing (3) and (2)), we have gained

$$10{,}007.99 - 9835.76 = \$172.23$$

Now suppose that interest $100i\%$ is quoted on a yearly basis but is compounded q times a year ($q = 4$ for interest compounded every 3 months; $q = 12$ if interest is compounded monthly). Then equation (1) is replaced by

$$S = P(1 + \tfrac{i}{q})^{qn} \tag{4}$$

where n is the number of years, q represents how many times a year interest is compounded, and S is the amount after n years.

The present custom is to compound interest daily, or even continuously. If the interest is compounded daily, then, of course, we substitute $q = 365$ in equation (4), but what does the continuous compounding of interest mean?

If we let $y = \frac{1}{q}$ in equation (4), we have

$$S = P(1 + iy)^{n/y} \tag{5}$$

Now, as q gets larger and larger, y gets closer and closer to zero. The total amount:

$$S = \lim_{x \to 0} P(1 + iy)^{n/y} = P \lim_{x \to 0} (1 + iy)^{n/y} \tag{6}$$

is referred to as the final amount if P dollars are deposited at the rate i *compounded continuously* over a period of n years.

We recall from Example 2, that

$$e = \lim_{x \to 0} (1 + x)^{1/x}$$

Consequently, if we set $x = iy$ in (6), then when $y \to 0$, it is clear that $x \to 0$. Hence,

$$\begin{aligned} S &= P \lim_{y \to 0} (1 + iy)^{n/y} \\ &= P \lim_{x \to 0} (1 + x)^{ni/x} \\ &= P \lim_{x \to 0} [(1 + x)^{1/x}]^{ni} \end{aligned}$$

or

$$S = Pe^{ni} \tag{7}$$

Thus, if P dollars are invested at $100i\%$ compounded continuously, then at the end of n years, the account will have a total of Pe^{ni} dollars.

EXAMPLE 3. A principal of \$5000 is deposited in a bank at an annual rate of 7% for 10 years. How much money will the depositor have in the bank if the interest is compounded (a) yearly, (b) quarterly, (c) continuously.

SOLUTION.

(a) From (2), $S = \$9835.76$

(b) From (3), $S = \$10{,}007.99$

(c) $S = 5000e^{(0.07)(10)}$

$= 5000e^{0.7}$

$= \$10{,}068.76$

Here again, we use a calculator with a "raise to the power" key and round off to the nearest cent.

The exponential function with base $b = e$ is ofter denoted by *exp*, and we write

$$f(x) = \exp(x) = e^x$$

EXERCISES 4.1

In Problems 1 through 16, sketch a graph of the given exponential function.

1. $f(x) = 3^x$

2. $f(x) = (\frac{1}{3})^x$

3. $f(x) = 4^x$

4. $f(x) = (\frac{1}{4})^x$

5. $f(x) = 3^{x+1}$

6. $f(x) = (\frac{1}{4})^{-x}$

7. $f(x) = 2 \cdot 3^x$

8. $f(x) = 5^x$

9. $f(x) = 2^{x-1}$

10. $f(x) = 3 \cdot 2^x$

11. $f(x) = 3^{x-1}$

12. $f(x) = 2^{-x^2}$

13. $f(x)\ 3^{-x^2}$

14. $f(x) = e^x$ [*Hint*: Use tables to compute e^x.]

15. $f(x) = e^{-x}$

16. $f(x) = e^{2x}$

17. If $f(x) = b^x$, show that $f(x_1 - x_2) = \dfrac{f(x_1)}{f(x_2)}$.

In Problems 18 through 25, the graph of the exponential function $y = b^x$ contains the given point. Determine the base b.

18. $(2, \frac{1}{9})$

19. $(-2, \frac{1}{9})$

20. $(3, \frac{1}{8})$ **21.** $(-4, 16)$

22. $(-2, 16)$ **23.** $(-1, 5)$

24. $(\frac{1}{2}, 2)$ **25.** $(-\frac{1}{2}, 2)$

In Problems 26 through 40, use a calculator.

26. The popualtion $P(t)$ of Osmaria is given by

$$P(t) = 5{,}000{,}000e^{0.01t}$$

where t represents time measured in years. Find: (a) $P(0)$, (b) $P(2)$, (c) $P(5)$, (d) $P(100)$.

27. Suppose the amount $A(t)$ of a radioactive substance present at time t is given by

$$A(t) = 100e^{-0.02t}$$

where t is measured in years and $A(t)$ is measured in grams. Find: (a) $A(0)$, (b) $A(5)$, (c) $A(50)$.

28. Suppose the sales of a new model of television set are approximated by

$$S(x) = 4000 - 2000e^{-x}$$

where x is the number of years the new model television set has been on the market and $S(x)$ represents the sales in thousands per year. Find: (a) $S(0)$, (b) $S(2)$, (c) $S(10)$. Sketch the graph of $S(x)$.

29. A manufacturer's annual profit $P(x)$ from the sales of a certain appliance is given by

$$P(x) = 100{,}000 + 50{,}000e^{-0.2x}$$

where x is the number of years the appliance has been on the market. Find the manufacturer's profit for $x = 1, 5$, and 10 (to the nearest dollar).

30. Suppose that the percentage of television viewers who remember what products are regularly advertised during the daily "Today Show" is approximated by

$$y = 70 - 30e^{-x}$$

where x is the number of times they have seen the show. Calculate the percentages that correspond to $x = 1, 2, 3$, and 4. Sketch the graph of the function.

In Problems 31 through 38, find the amount due the investor after 8 years on an investment of \$6000 at the stated rate of interest.

31. 6% (simple interest) **32.** 6% compounded yearly

33. 6% compounded quarterly

34. 6% compounded semiannually

35. 6% compounded monthly

36. 6% compounded daily

37. 6% compounded continuously

38. 5.5% compounded continuously

39. How much principal should be deposited in a bank that pays 6.5% interest compounded (a) quarterly, (b) continuously, so as to have $10,000 accumulated at the end of 10 years?

40. Tom needs money to keep his television store operating through a recession. If money is worth 8% compounded quarterly and he signs a note promising to pay the bank $6000 two years from today, how much will the bank lend him today?

4.2 THE LOGARITHMIC FUNCTION

The invention of logarithms by John Napier[1] was a milestone in the advancement of mathematics. Whereas most mathematical concepts evolve from previous ideas, the logarithm is an exception. Napier invented logarithms to simplify the computation of cumbersome algebraic operations, though his original work was unwieldy. However, together with Henry Briggs (1561-1631), he constructed tables for $\log_{10} x$. These tables were completed and published by Briggs after Napier's death. The invention of logarithms paved the way for the famous astronomer Johann Kepler (1571-1630) to complete the calculations that led to the third law of planetary motion. Having received Napier's tables, Kepler refused to use them until he had derived the formula responsible for their construction. Kepler invented other formulas and constructed his own tables, which he published in 1624. Napier's invention led to the logarithmic and exponential functions which today play important roles in the sciences.

[1] JOHN NAPIER (1550-1617) was born at Merchiston, near Edinburgh, Scotland. His early education was received at St. Andrews, and he completed his studies in Europe. In 1572 he married, and he had two children. His wife, Elizabeth, died seven years later. He remarried and fathered ten more children. Theology and mathematics occupied Napier. He took an active part in church politics and was the author of a theological treatise.

His work with equations involved imaginary roots, which he referred to as "a great algebraic secret." Napier devised several mechanical aids for calculations including the Ragologia (familiarly known as "Napier's bones"). He spent many years preparing the Tables of Logarithms, which were published in 1614. This invention was welcomed by the mathematical world. Napier also wrote the *Constructio* and *Descripto*, which explained logarithms.

Although logarithms were originally invented as an aid to arithmetic computations, the availability of calculators and computers has rendered logarithms of little use for this purpose. However, we now study logarithmic functions because they provide us with another function which can be used as a mathematical model for real-life situations. We have the following definition:

Definition 1. Let $b > 0$ and $b \neq 1$. We say that y is the logarithm of x to the base b if $b^y = x$. We symbolize this statement by writing $y = \log_b x$.

It is clear from the definition that $y = \log_b x$ is equivalent to $x = b^y$.

For example, since

$$25 = 5^2$$

we may write this statement as

$$2 = \log_5 25$$

Similarly,

$$10^4 = 10000$$

and we may write

$$4 = \log_{10} 10000$$

Thus, a statement in exponential form can be written in the logarithmic form and vice versa:

Exponential Form	Logarithmic Form
$2^3 = 8$	$\log_2 8 = 3$
$(\frac{1}{4})^2 = \frac{1}{16}$	$\log_{1/4} \frac{1}{16} = 2$
$(4)^{-2} = \frac{1}{16}$	$\log_4 \frac{1}{16} = -2$
$(27)^{1/3} = 3$	$\log_{27} 3 = \frac{1}{3}$

EXAMPLE 1. Find the unknown x, y, or b in each of the following:

(a) $y = \log_2 8$ (b) $4 = \log_3 x$

(c) $\frac{1}{2} = \log_b 3$ (d) $y = \log_2 \sqrt{2}$

SOLUTION.

(a) The equation $y = \log_2 8$ is equivalent to

$$2^y = 8 \qquad \text{or} \qquad 2^y = 2^3$$

Thus

$$y = 3$$

(b) The equation $4 = \log_3 x$ is equivalent to $3^4 = x$.
Thus

$$x = 81$$

(c) The equation $\frac{1}{2} = \log_b 3$ is equivalent to $b^{1/2} = 3$.
Thus

$$b = 9$$

(d) The equation $y = \log_2 \sqrt{2}$ is equivalent to $2^y = \sqrt{2}$.
Thus

$$y = \frac{1}{2}$$

The usefulness of logarithmic functions comes from the properties of logarithms given in the following theorem:

THEOREM 1. If a, b, and c are positive numbers and $b \neq 1$, then:

(1) $\log_b (ac) = \log_b a + \log_b c$

(2) $\log_b \frac{a}{c} = \log_b a - \log_b c$

(3) $\log_b 1 = 0$

(4) $\log_b b = 1$

(5) $\log_b a^r = r\log_b a \qquad$ if r is a real number

(6) $a = b^{\log_b a}$

These properties may be easily proved by writing each property in the exponential form and using the appropriate law of exponents.

For example, property (1) may be proved as follows:

Let $x = \log_b a$ and $y = \log_b c$. In exponential form, we have

$$a = b^x \qquad \text{and} \qquad c = b^y$$

Thus,

$$ac = b^x \cdot b^y = b^{x+y}$$

Finally, we rewrite $ac = b^{x+y}$ in the logarithmic form as:

$$\log_b ac = x + y = \log_b a + \log_b c$$

The verification of the other properties is left to the reader.

EXAMPLE 2. Express the given expression in terms of logarithms of x, y, and z:

$$\log_b \left(\frac{x^2 z^{2/3}}{y^4}\right)$$

SOLUTION. Using Theorem 1, we have

$$\begin{aligned}\log_b \left(\frac{x^2 z^{2/3}}{y^4}\right) &= \log_b (x^2 z^{2/3}) - \log_b y^4 \\ &= \log_b x^2 + \log_b z^{2/3} - \log_b y^4 \\ &= 2\log_b x + \frac{2}{3}\log_b z - 4\log_b y\end{aligned}$$

EXAMPLE 3. Simplify each of the following expressions.

(a) $\log_2 4^2$ (b) $\log_3 (9^{1/2} \cdot 3^{1/4})$

SOLUTION.

(a) $\log_2 4^2 = \log_2 (2^2)^2 = \log_2 2^4 = 4\log_2 2 = 4$

(b) $\log_3 (9^{1/2} \cdot 3^{1/4}) = \log_3 9^{1/2} + \log_3 3^{1/4} = \log_3 (3^2)^{1/2} + \log_3 3^{1/4}$
$= \log_3 3 + \frac{1}{4}\log_3 3 = 1 + \frac{1}{4} = \frac{5}{4}$

EXAMPLE 4. Given that $\log_{10} 2 = 0.301$ and $\log_{10} 3 = 0.477$, compute the value of each of the following expressions.

(a) $\log_{10} 6$ (b) $\log_{10} 200$ (c) $\log_{10} \sqrt{1.5}$

SOLUTION.

(a) $\log_{10} 6 = \log_{10} (2 \cdot 3) = \log_{10} 2 + \log_{10} 3 = 0.301 + 0.477 = 0.788$

(b) $\log_{10} 200 = \log_{10} (2 \cdot 10^2) = \log_{10} 2 + \log_{10} 10^2$

$$= 0.301 + 2\log_{10} 10 = 0.301 + 2 = 2.301$$

(c) $\log_{10} \sqrt{1.5} = \log_{10} (\frac{3}{2})^{1/2} = \frac{1}{2}\log_{10} (\frac{3}{2})$

$$= \tfrac{1}{2}[\log_{10} 3 - \log_{10} 2] = \tfrac{1}{2}[0.477 - 0.301] = 0.088$$

EXAMPLE 5. Solve each of the following equations for x:

(a) $\log_2 (2x + 1) = 4$ (b) $\log_2 x + \log_2 (3x - 2) = 3$

SOLUTION.

(a) The equation $\log_2 (2x + 1) = 4$ is equivalent to

$$2x + 1 = 2^4 \qquad \text{or} \qquad x = \tfrac{15}{2}$$

(b) We have from Theorem 1,

$$\log_2 x + \log_2 (3x - 2) = \log_2 x(3x - 2)$$

Thus,

$$\log_2 (3x^2 - 2x) = 3 \qquad \text{or} \qquad 3x^2 - 2x = 2^3$$

or

$$3x^2 - 2x - 8 = 0 \qquad \text{or} \qquad (3x + 4)(x - 2) = 0$$

Hence,

$$x = 2 \qquad \text{or} \qquad x = -\tfrac{4}{3}$$

Since $x = -\frac{4}{3}$ does not satisfy the original equation, the only solution is $x = 2$.

Two numbers, 10 and e (e is an irrational number approximately equal to 2.71828), are each used extensively as bases of logarithms. The base e is primarily used in calculus and the sciences. The logarithms to base e are called *natural logarithms* and are denoted by $\ln x$ (read "ell en x"). That is, $\ln x$ is an *abbreviation* for *$\log_e x$*. The base 10 is used for computations. The logarithms to base 10 are called *common logarithms* and are denoted by $\log x$ instead of $\log_{10} x$. Tables 3 and 4 at the back of the book are given for common logarithms and natural logarithms, respectively.

The following theorem states the relationship between the common and natural logarithms.

THEOREM 2. For any positive number x,

$$\ln x \approx 2.3026 \log x$$

EXAMPLE 6. Solve for x (leave the answer in common logarithms):

$$5^{2x-1} = 7$$

SOLUTION. Since equal numbers have equal logarithms, we have

$$\log 5^{2x-1} = \log 7$$

or

$$(2x - 1)\log 5 = \log 7$$

or

$$(2x - 1) = \frac{\log 7}{\log 5}$$

or

$$x = \frac{1}{2}\left[\frac{\log 7}{\log 5} + 1\right]$$

EXAMPLE 7. Suppose that $P(t)$, the population of bacteria at time t, is given by

$$P(t) = 1000e^{kt} \tag{1}$$

where k is a constant. After 10 hours the population of bacteria is observed to be 3000. What was the population after 5 hours?

SOLUTION. We are given that when $t = 10$, $P = 3000$. Substituting in equation (1), we have

$$3000 = 1000e^{10k}$$

or

$$3 = e^{10k}$$

or

$$k = \frac{1}{10}\ln 3$$

Therefore, from (1), we have

$$P(t) = 1000e^{((1/10)\ln 3)t} \tag{2}$$

To find the population after 5 hours, we set $t = 5$ in (2) and obtain

$$\begin{aligned} P(5) &= 1000e^{((1/10)\ln 3)5} \\ &= 1000e^{(1/2)\ln 3} = 1000e^{\ln\sqrt{3}} = 1000\sqrt{3} \\ &\approx 1732. \end{aligned}$$

Suppose we wish to sketch the graph of $y = \log_2 x$. Since this statement is equivalent to $x = 2^y$, we sketch the graph of this exponential function to obtain the graph of $y = \log_2 x$. We notice that the variable x is isolated in the equation $x = 2^y$; so, to sketch its graph, we choose various values of y and compute the corresponding values of x. The graph of $y = \log_b x$ if $b > 1$ has the general shape given in Figure 1, whereas the graph of $y = \log_b x$ if $0 < b < 1$ has the general shape given in Figure 2.

Graph of $y = \log_b x$ $(b > 1)$

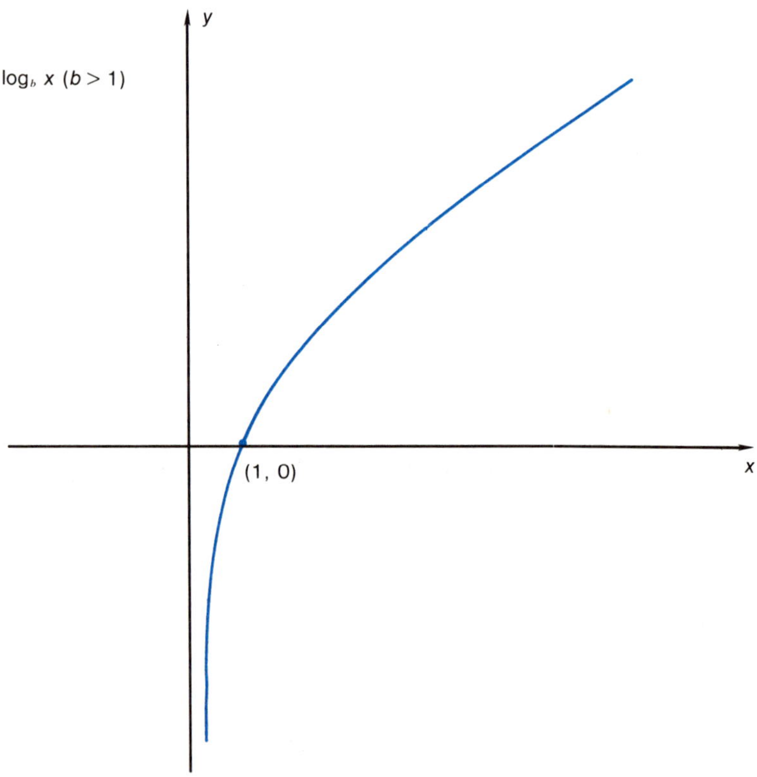

Figure 1

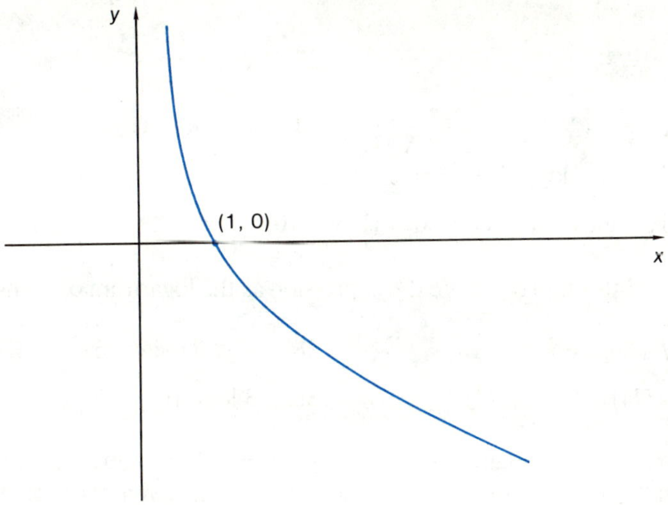

Graph of $y = \log_b x$ $(0 < b < 1)$

Figure 2

EXERCISES 4.2

In problems 1 through 10, change each equation to logarithmic form.

1. $2^5 = 32$

2. $6^{-2} = \dfrac{1}{36}$

3. $\left(\dfrac{1}{2}\right)^4 = \dfrac{1}{16}$

4. $9^0 = 1$

5. $(100)^{3/2} = 1000$

6. $(27)^{-2/3} = \dfrac{1}{9}$

7. $10^{-5} = 0.00001$

8. $(16)^{-3/4} = 0.125$

9. $a^x = y$

10. $N^u = v$

In problems 11 through 18, change each equation to exponential form.

11. $\log_{10} 100 = 2$

12. $\log_7 343 = 3$

13. $\log_{2/3} \dfrac{4}{9} = 2$

14. $\log_{1/3} 243 = -5$

15. $\log_6 1 = 0$

16. $\log_2 x = 4$

17. $\ln 5 = x$

18. $\ln (a + b) = y$

In problems 19 through 26, solve for x.

19. $\log_3 x = 4$

20. $\log_{16} x = -\frac{3}{2}$

21. $\log_{64} x = \frac{5}{6}$

22. $\log_x 8 = -3$

23. $\log_4 x^2 = -2$

24. $\log_6 (x-1) = 2$

25. $\log_2 (7-x) - \log_2 (2+x) = 2$

26. $\log_7 3x + \log_7 (2x-1) = \log_7 (16x-10)$

In problems 27 through 30, write the expression as the logarithm of a single number.

27. $\log_2 7 + \log_2 3$

28. $\log_2 8 - \log_2 6$

29. $\log_{10} (5/4) - \log_{10} (2/3)$

30. $3 \log_5 10 - 2 \log_5 15$

In problems 31 through 38, use $\log_{10} 3 = 0.4771$, $\log_{10} 5 = 0.699$, and $\log_{10} 7 = 0.8451$ to compute the value of each of the given expressions.

31. $\log_{10} 35$

32. $\log_{10} 15$

33. $\log_{10} 1.5$

34. $\log_{10} 500$

35. $\log_{10} \sqrt{7}$

36. $\log_{10} 0.003$

36. $\dfrac{\log_{10} 5}{\log_{10} 7}$

38. $\log_{10} \frac{5}{7}$

39. Sketch the graph of $y = \log_3 x$.

40. Sketch the graph of $y = \log_{1/3} x$.

In problems 41 through 48, express the logarithm in terms of logarithms in x, y, and z.

41. $\log_b xyz$

42. $\log_b \sqrt{xyz}$

43. $\log_b x^2y^3z^4$

44. $\log_b \dfrac{x^2y^3}{z^5}$

45. $\log_b x^{1/2}y^{1/3}z^{1/4}$

46. $\log_b (\sqrt{x}\sqrt[3]{yz})$

47. $\log_b \sqrt{x\sqrt{yz}}$

48. $\log_b (xy \log_b b^{\sqrt{yz}})$

In problems 49 through 54, solve for x (leave the answer in common logarithms).

49. $2^x = 9$

50. $3^x = 5$

51. $4^{x-1} = 27$

52. $5^{1-x} = 16$

53. $3^{1-x} = 5^{2-3x}$

54. $\dfrac{4^x + 4^{-x}}{3} = 2$

55. Population P of Bloomville in t years is given by

$$P = P_0e^{kt}$$

where P_0 is the population at $t = 0$ and k is a constant. Suppose that the present population of Bloomville is 150,000. Twenty years ago the population was 50,000. Find the population 10 years from now.

56. The population P of a colony of ants in t weeks is given by

$$P = 2000e^{kt}$$

where k is a constant. If in 24 weeks the population is 25,000, find the population in: (a) 10 weeks measured from $t = 0$; (b) 35 weeks form $t = 0$.

57. A certain radioactive substance decomposes according to the law

$$m = ae^{-0.36t}$$

where m is the number of grams of the substance present at t years. Find the half-life of the substance. (*Hint:* Find t, when $m = a/2$).

58. A certain radioactive substance decays so that an amount of m grams of the substance is present after t years, where

$$m = 1000e^{0.04t}$$

How long will it be until an amount of only 600 grams of the substance is present?

4.3 DERIVATIVES OF EXPONENTIAL AND LOGARITHMIC FUNCTIONS

In this section we shall give the formulas for the derivatives of exponential and logarithmic functions.

$$\text{If } y = e^x, \quad \text{then} \quad \frac{dy}{dx} = e^x \tag{1}$$

That is, $\frac{d}{dx}(e^x) = e^x$. Thus, the exponential function with base e is its own derivative. By using the chain rule, we conclude that if u is a differentiable function of x, then

$$\frac{d}{dx}e^u = e^u\frac{du}{dx} \tag{2}$$

For example, to compute the derivative of $y = e^{3x^2}$, we set $u = 3x^2$, so that $\frac{du}{dx} = 6x$. Thus,

$$\frac{dy}{dx} = e^u\frac{du}{dx} = e^{3x^2}(6x)$$

If a base other than e is used for the exponential function, then its derivative is a little more complicated. In fact, if

$$y = 2^x$$

then it can be shown that

$$\frac{dy}{dx} = 2^x \ln x$$

In general, we have ($a > 0$):

$$\text{If } y = a^x, \quad \text{then} \quad \frac{dy}{dx} = a^x \ln a \tag{3}$$

and, by the chain rule,

$$\frac{d}{dx}(a^u) = a^u(\ln a)\left(\frac{du}{dx}\right) \tag{4}$$

EXAMPLE 1. Find $\dfrac{dy}{dx}$.

(a) $y = e^{-3x}$ (b) $y = e^{x^2+x-4}$ (c) $y = 4e^{1/x}$

SOLUTION.

(a) Let $u = -3x$; then $du/dx = -3$. Using the chain rule, we have $y = e^u$, so that $dy/dx = (dy/du) \cdot (du/dx) = e^u(-3) = e^{-3x}(-3) = -3e^{-3x}$.

(b) Let $u = x^2 + x - 4$; then $du/dx = 2x + 1$, and $y = e^u$. Thus, by the chain rule we have $dy/dx = (dy/du) \cdot (du/dx) = e^u(2x + 1) = (2x + 1)e^{x^2+x-4}$.

(c) Let $u = 1/x$; then $du/dx = -1/x^2$. Therefore, $dy/dx = 4e^{1/x}(-1/x^2) = -(4/x^2)e^{1/x}$.

EXAMPLE 2. Find $\dfrac{dy}{dx}$, if

(a) $y = 3^x$ (b) $y = 5^{3x^2-4x}$ (c) $3^{x-1} \cdot 2^x$

SOLUTION.

(a) $\dfrac{dy}{dx} = 3^x \ln 3$ (by (3))

(b) By (3) and the chain rule:

$$\frac{dy}{dx} = 5^{3x^2-4x}(\ln 5)(6x - 4)$$

(c) $$\frac{dy}{dx} = 3^{x-1}\frac{d}{dx}(2^x) + 2^x\frac{d}{dx}(3^{x-1})$$

$$= 3^{x-1}2^x\ln 2 + 2^x3^{x-1}\ln 3$$

$$= 3^{x-1}2^x(\ln 2 + \ln 3)$$

$$= 3^{x-1}2^x\ln 6$$

The exponential functions are very useful in depicting growth patterns such as in the sale of a product or in population growth. Generally, the growth in sales of a product (or in population) is slow in the beginning, then it becomes rapid, and then it gradually levels off. Such growth can be approximated by the functions of the type

$$f(x) = \frac{a}{b + ce^{kx}}$$

where x is the time and a, b, c, and k are appropriate constants.

EXAMPLE 3. The marketing department of a toothpaste company estimates that the sales of its new "Shiny Toothpaste" for the next few years is given by the function

$$f(x) = \frac{5000}{1 + 100e^{-0.2x}}$$

where x is the time in years. Find the rate of growth of sales of Shiny Toothpaste when $x = 5$.

SOLUTION. The rate of growth at time x is given by $f'(x)$. Thus,

$$f'(x) = \frac{(1 + 100e^{-0.2x})(0) - 5000(-20e^{-0.2x})}{(1 + 100e^{-0.2x})^2}$$

$$= \frac{100{,}000e^{-0.2x}}{(1 + 100e^{-0.2x})^2}$$

Then (using a calculator), we find that

$$f'(5) = \frac{100{,}000e^{-1}}{(1 + 100e^{-1})^2} \approx 25.76$$

The next rule tells us how to find the derivative of a natural logarithmic fuction.

$$\text{If } y = \ln x, \text{ then } dy/dx = 1/x \qquad (5)$$

If u is a differentiable function of x, we can use the chain rule to extend this result and obtain

$$\frac{d}{dx}(\ln u) = \frac{1}{u} \cdot \frac{du}{dx} \tag{6}$$

EXAMPLE 4. Find $f'(x)$, if

(a) $f(x) = \ln\sqrt{x^2+1}$ (b) $f(x) = \ln\dfrac{(x^2+1)^{20}(3x-1)^3}{(2x^2+3)^4}$

SOLUTION.

(a) Letting $u = \sqrt{x^2+1}$, then we have $\dfrac{du}{dx} = \dfrac{x}{\sqrt{x^2+1}}$. Therefore,

$$f'(x) = \frac{1}{\sqrt{x^2+1}} \cdot \frac{x}{\sqrt{x^2+1}} = \frac{x}{x^2+1}$$

(b) First we simplify the function by using the properties of logarithms. Thus,

$$\ln\frac{(x^2+1)^{20}(3x-1)^3}{(2x^2+3)^4} = 20\ln(x^2+1) + 3\ln(3x-1) - 4\ln(2x^2+3)$$

Therefore,

$$\begin{aligned} f'(x) &= \frac{d}{dx}[20\ln(x^2+1) + 3\ln(3x-1) - 4\ln(2x^2+3)] \\ &= 20\frac{d}{dx}[\ln(x^2+1)] + 3\frac{d}{dx}[\ln(3x-1)] - 4\frac{d}{dx}[\ln(2x^2+3)] \\ &= 20\frac{2x}{x^2+1} + 3 \cdot \frac{3}{3x-1} - 4 \cdot \frac{4x}{2x^2+3} \\ &= \frac{40x}{x^2+1} + \frac{9}{3x-1} - \frac{16x}{2x^2+3} \end{aligned}$$

We can make use of logarithms to differentiate complicated functions, as in the following example.

EXAMPLE 5. If $y = \dfrac{(x^2+3)^{10}(x-1)^{1/2}}{(x^2+3x+5)^{80}}$, find $\dfrac{dy}{dx}$

SOLUTION. We have

$$\ln y = \ln\frac{(x^2+3)^{10}(x-1)^{1/2}}{(x^2+3x+5)^{80}}$$

$$= 10\ln(x^2+3) + \frac{1}{2}\ln(x-1) - 80\ln(x^2+3x+5)$$

Differentiation (implicit) yields

$$\frac{1}{y}\frac{dy}{dx} = 10\frac{1}{x^2+3}\frac{d}{dx}(x^2+3) + \frac{1}{2}\frac{1}{x-1}\frac{d}{dx}(x-1)$$

$$-80\frac{1}{x^2+3x+5}\frac{d}{dx}(x^2+3x+5)$$

$$= \frac{20x}{x^2+3} + \frac{1}{2(x-1)} - \frac{80(2x+3)}{x^2+3x+5}$$

Substituting for y and solving for dy/dx, we obtain

$$\frac{dy}{dx} = \frac{(x^2+3)^{10}(x-1)^{1/2}}{(x^2+3x+5)^{80}}\left[\frac{20x}{x^2+3} + \frac{1}{2(x-1)} - \frac{80(2x+3)}{x^2+3x+5}\right]$$

The process used in Example 5 is called "logarithmic differentiation."

EXERCISES 4.3

In Problems 1 through 20, find $f'(x)$.

1. $f(x) = e^{4x}$

2. $f(x) = 4e^{4x^2+5}$

3. $f(x) = e^{2/x^2}$

4. $f(x) = e^{3x^2-4x+1}$

5. $f(x) = x^3e^{2x}$

6. $f(x) = (2x^2+3)e^x$

7. $f(x) = e^{-2x^2}$

8. $f(x) = (e^{-2x})^2$

9. $f(x) = \ln(x^2+4)$

10. $f(x) = \ln(e^x+1)$

11. $f(x) = e^{x\ln x}$

14. $f(x) = \dfrac{e^x-1}{e^x+1}$

13. $f(x) = \ln(e^{2x^2+4}+1)$

14. $f(x) = (e^{x^2}+4)e^{5x^2-1}$

15. $f(x) = 5^x$

16. $f(x) = 5^{-x}$

17. $f(x) = 2^{x-1}5^x$

18. $f(x) = \dfrac{2^{x-1}}{5^x}$

19. $f(x) = e^{2x} + (2x)^e$

20. $f(x) = \ln(x+x^2+1)^2$

In Problems 21 through 23, use the rules of logarithms to simplify the expression and find $f'(x)$.

21. $f(x) = \ln \dfrac{(x+1)^{16}(2x^2+x)^8}{\sqrt{x^2+4}}$

22. $f(x) = \ln \dfrac{(x^2+5)^7(x^3+1)^{1/3}}{(x^2+1)^{5/4}}$

*23. $f(x) = \ln \dfrac{(e^{2x}+6)^7\sqrt{x+4}}{(e^{-x}+e^x)^5}$

24. Find the equation of the tangent line to the curve $y = e^x$ at the point $(1, e)$.

25. Find the equation of the tangent line to the curve $y = \ln x$ at the point where $x = 1$.

26. If the revenue for a product is given by $R(x) = 500e^{-0.1x}$, find the marginal revenue when $x = 8$.

27. Do Example 3 if $f(x) = \dfrac{200}{1+e^{-x}}$

28. Do Example 3 if $f(x) = \dfrac{300}{6+20e^{-2x}}$

In Problems 29 through 32, use the procedure of Example 5 to find dy/dx.

29. $y = \dfrac{(3x^2+1)^5(x^2+1)^{3/2}}{(x^2+4)^{10}}$

30. $y = \dfrac{(x^2+x+1)^{60}(x^2-5x)^4}{\sqrt{x^2-4x+10}}$

31. $y = x^x$

*32. $y = x^{x^x}$

33. Find relative and absolute extrema for $f(x) = e^{-x}$.

34. Do Problem 33 for $f(x) = x^2 \ln x^2$.

CHAPTER 4 REVIEW

Part 1. (ORAL)

Define or discuss each of the following:

1. Exponential function
2. Properties of exponential functions
3. Logarithmic function
4. Properties of logarithmic functions
5. The number e

6. Natural logarithm

Part 2. (WRITTEN)

1. Sketch the graphs of each of the following functions:

(a) $f(x) = 2^{-x}$ (b) $f(x) = (\frac{1}{2})^{2-x}$ (c) $f(x) = 3^x$

(d) $f(x) = \log_2 x$ (e) $f(x) = \log_{1/2} x$

2. Let $f(x) = 2^x - 2^{-x}$ and $g(x) = 2^x + 2^{-x}$. Find

(a) $f(x) + g(x)$ (b) $f(x) - g(x)$ (c) $f(x) \cdot g(x)$

(d) $[f(x)]^2 - [g(x)]^2$ (e) $f(x^2) \cdot g(x^2)$

3. Find the number:

(a) $\log_5 125$ (b) $\log_{\sqrt{8}} 64$ (c) $\log_3 \frac{1}{27}$

(d) $3^{\log_3 7}$

4. Express $\log\sqrt{x^{-3}y^4/z^5}$ in terms of logarithms of x, y, and z.

5. Express $3\log x - \frac{3}{2}\log xy^4 + \frac{4}{3}\log z$ as one logarithm.

In Problems 6 through 22, find $f'(x)$.

6. $f(x) = e^{3x}$

7. $f(x) = e^{-5x}$

8. $f(x) = 2e^{x^2-x}$

9. $f(x) = xe^{-x}$

10. $f(x) = xe^{1/x}$

11. $f(x) = x\ln x$

12. $f(x) = \left(x^2 - \frac{1}{x}\right)e^{x+1}$

13. $f(x) = \left(x + \frac{1}{x}\right)\ln(x+3)$

14. $f(x) = 3^x$

15. $f(x) = \dfrac{e^x - e^{-x}}{e^x + e^{-x}}$

16. $f(x) = 4^{-x}$

17. $f(x) = e^x + x^e + e^e$

18. $f(x) = \log_5 x$

19. $f(x) = \log_2(x^3 - x^2)$

20. $f(x) = \log_{10} \dfrac{(x-1)^5(x+2)^7}{(x-2)^4}$

21. $f(x) = \log_{10} \dfrac{(x^2-5x+3)^{1/2}}{x(x-5)}$

22. $f(x) = \log_{10} \dfrac{(e^x - 5x)\sqrt{x^2-3}}{(x^2-3)^4}$

In Problems 23 through 28, use logarithms to find dy/dx.

23. $y = (x-1)^4(x^2+5x)(x+3)^5$

24. $y = \dfrac{(x^2-2x+3)^{40}(x-1)^{1/2}}{(x+5)^{3/4}}$

25. $y = \dfrac{(x^2-x-1)^{1/2}(x^2-4x+1)^{-3/5}}{(5x^2-2x+1)^6}$

26. $y = x^{x+1}$

27. $y = 3^x + x^3 + x^x$

28. $y = x^{xe^x}$

In Problems 29 through 32, suppose \$12,000 is deposited in a bank. Find the amount on deposit in the bank after 10 years, at the given interest rate:

29. 8% compounded annually.

30. 8% compounded semiannually.

31. 8% compounded quarterly.

32. 8% compounded continuously.

33. Find an equation of the tangent line to the graph of $y = xe^x$ at $(1, e)$.

34. Find an equation of the tangent line to the graph of $y = x\ln x$ at: (a) (e, e), (b) $(1, 0)$.

35. Suppose a population $P(t)$ at time t (in hours) of a bacteria is given by $P(t) = 5000(2^{-t/3})$. Find the population when: (a) $t = 0$, (b) $t = 3$, and (c) $t = 9$ hours. Also find the rate of change of population with respect to time at: (a) $t = 0$, (b) $t = 3$, and (c) $t = 9$.

36. Sketch the graph of $f(x) = \dfrac{e^x + e^{-x}}{2}$.
[Follow the procedure outlined in Chapter 3.]

Chapter Five

The Integral

Elementary calculus is often divided into two major branches: differential calculus and integral calculus. Differential calculus deals with the concept of the derivative. Integral calculus deals with another basic concept—that of the integral. *We have seen that the derivative, from a geometrical point of view, is related to the slope of a tangent line to a curve; and we have used this notion to study problems in the sciences and in economic theory. From a geometrical point of view, the integral is related to the area of a region. In this chapter we shall study some elementary aspects of the integral and develop techniques for calculating areas of regions bounded by curves. We shall also show how the integral is related to certain problems arising in the sciences and economic theory. First we introduce the notion of antiderivatives, and then we provide a definition of the definite integral. We shall relate the problem of tangents to the problem of areas by turning to perhaps the most important theorem in calculus—the* Fundamental Theorem of Calculus.

5.1 THE ANTIDERIVATIVE

So far, our problem has been to find the derivative of a given function. We now consider the reverse problem: given the derivative, how do we find the function it came from? For example, suppose we know the marginal cost; can we find the cost function?

In general, suppose we are given $f(x)$. We then look for a second function $F(x)$ such that

$$\frac{d}{dx} F(x) = f(x)$$

If we find such a function $F(x)$, then $F(x)$ is called an *antiderivative* of $f(x)$.

The process involved in finding $F(x)$ for a given function $f(x)$ is called *antidifferentiation.*

EXAMPLE 1. If $f(x) = 12x^2$, find an antiderivative.

SOLUTION. We attempt the trial and error method. We recall that differentiation lowers the power by 1. Hence, in looking for an antiderivative, we raise the power by 1 and try $F(x) = 12x^3$. However, the derivative of $12x^3$ is $36x^2$, which is three times the result (namely, $12x^2$) we want. To remedy this problem we divide $12x^3$ by 3 and obtain

$$F(x) = 4x^3$$

It is easy to see that $F'(x) = \frac{d}{dx}(4x^3) = 12x^2 = f(x)$. Thus, $4x^3$ is an antiderivative of $12x^2$. Is $F(x)$ the only antiderivative of $f(x)$? No! If we differentiate any one of the functions

$$F_1(x) = 4x^3 + 17, \quad F_2(x) = 4x^3 - 1776, \quad \text{or} \quad F_3(x) = 4x^3 + \sqrt[3]{98}$$

we also get $12x^2$. Thus, any one of these functions is an antiderivative of $f(x) = 12x^2$. In fact, for any constant C, $F(x) = 4x^3 + C$ is an antiderivative of $12x^2$, since

$$\frac{d}{dx}(4x^3 + C) = \frac{d}{dx}(4x^3) + \frac{d}{dx}(C) = 12x^2 + 0 = 12x^2 = f(x)$$

There still remains another question: Is there another antiderivative $G(x)$ of $12x^2$ which is not of the form $4x^3$ + (some constant)? The answer is no. It can be proved that if $F(x)$ and $G(x)$ are antiderivatives of the same function $f(x)$, then there is a constant C such that

$$F(x) = G(x) + C$$

Hence, if we can find one antiderivative, then we can find them all by adding to it different constants.

EXAMPLE 2. Show that $F(x) = \ln|x|$ is an antiderivative of $f(x) = \frac{1}{x}$.

SOLUTION.

CASE 1. If $x > 0$, then $|x| = x$, and

$$\frac{d}{dx}\ln|x| = \frac{d}{dx}\ln x = \frac{1}{x}$$

CASE 2. If $x < 0$, then $|x| = -x$, and

$$\frac{d}{dx} \ln|x| = \frac{d}{dx} \ln(-x) = \frac{1}{-x}(-1) = \frac{1}{x}$$

Therefore,

$$\frac{d}{dx} \ln|x| = \frac{1}{x}$$

in both cases, and consequently $\ln|x|$ is an antiderivative of $\frac{1}{x}$. In fact, for any constant C, $\ln|x| + C$ is an antiderivative of $\frac{1}{x}$.

The set of all antiderivatives of a function f is called the *indefinite integral* of f. We denote the indefinite integral of f by the symbols

$$\int f(x)dx$$

read "the (indefinite) integral of $f(x)$ with respect to x." We call

$\int$ the integral[1] sign

and

f the integrand

Thus, if $F(x)$ is an antiderivative of $f(x)$, then

$$\int f(x)dx = F(x) + C$$

where C is an arbitrary constant.

The process of finding the (indefinite) integral is called *integration*.

We now compile a list of important integrals. To verifty that each of the formulas is true, one merely has to show that the derivative of the right side is the intergrand. Let a, b, n, and C be constants; then:

$$\int dx = \int 1 \cdot dx = x + C \tag{1}$$

$$\int adx = ax + C \tag{2}$$

[1] Historically, the symbol $\int$ has evolved from the letter S, which in turn evolved from the summation symbol Σ. The notation Σ for the integral sign was first used by Euler, and in the sections which follow we shall indicate the relationship between $\int$ and Σ.

$$\int x^n dx = \frac{x^{n+1}}{n+1} + C, \quad n \neq -1 \tag{3}$$

$$\int ax^n dx = \frac{a}{n+1} x^{n+1} + C, \quad n \neq -1 \tag{4}$$

$$\int e^x dx = e^x + C \tag{5}$$

$$\int e^{bx} dx = \frac{e^{bx}}{b} + C, \quad b \neq 0 \tag{6}$$

$$\int ae^{bx} dx = \frac{a}{b} e^{bx} + C, \quad b \neq 0 \tag{7}$$

$$\int \frac{1}{x} dx = \ln x + C \tag{8}$$

$$\int \frac{a}{x} dx = a\ln x + C \tag{9}$$

From Example 2, we notice that we should write $|x|$ on the right side of equations (8) and (9); but we can drop the absolute value signs because, whenever we use the logarithm function, it is understood that the variable x is positive.

We must mention that any letter (variable) can be used in place of x in any of these formulas. The variable present merely instructs us to write the antiderivative in terms of that variable. For example, equation (3) can be written as:

$$\int y^n dy = \frac{y^{n+1}}{n+1} + C, \quad n \neq -1 \tag{10}$$

or

$$\int u^n du = \frac{u^{n+1}}{n+1} + C, \quad n \neq -1 \tag{11}$$

The last form is extremely important, as we will learn in Section 4.

EXAMPLE 3. Find the indefinite integral of

$$f(x) = 32x^3 + 15x^2 + \frac{12}{x^3}$$

SOLUTION. We apply formula (4) to each piece in turn. Thus,

$$\int 32x^3 dx = \frac{32}{3+1} x^{3+1} + C_1 = 8x^4 + C_1$$

$$\int 15x^2 dx = \frac{15}{2+1} x^{2+1} + C_2 = 5x^3 + C_2$$

$$\int \frac{12}{x^3}\,dx = \int 12x^{-3}dx = \frac{12}{-3+1}\,x^{-3+1} + C_3 = -6x^{-2} + C_3 = -\frac{6}{x^2} + C_3$$

If we add these three indefinite integrals, we have

$$\int (32x^3 + 15x^2 + \frac{12}{x^3})dx = 8x^4 + 5x^3 - \frac{6}{x^2} + C$$

where $C = C_1 + C_2 + C_3$.

Why can we add the three pieces as we did? Answer: Since the derivative of the sum is equal to the sum of the derivatives, the same is true for integration.

Indeed, for any two functions $f(x)$ and $g(x)$, we have

$$\int [f(x) + g(x)]dx = \int f(x)dx + \int g(x)dx \tag{12}$$

Similarly, if k is any constant, then

$$\int k[f(x)]dx = k\int f(x)dx \tag{13}$$

EXAMPLE 4. Calculate each of the following indefinite integrals:

(a) $\int (2+x)\,dx$ (b) $\int \sqrt{x}\,dx$

(c) $\int \frac{x+3}{x^3}\,dx, \quad x \neq 0$ (d) $\int \frac{x^2-4}{x-2}\,dx, \quad x \neq 2$

SOLUTION.

(a)
$$\begin{aligned}\int (2+x)\,dx &= \int 2\,dx + \int x\,dx \\ &= 2\int dx + \int x\,dx = 2x + C_1 + \frac{x^2}{2} + C_2 \\ &= 2x + \frac{x^2}{2} + C\end{aligned}$$

(b)
$$\int \sqrt{x}\,dx = \int x^{1/2}\,dx = \frac{x^{(1/2)+1}}{\frac{1}{2}+1} + C = \frac{2}{3}x^{3/2} + C$$

(c)
$$\begin{aligned}\int \frac{x+3}{x^3}\,dx &= \int\left(\frac{1}{x^2} + \frac{3}{x^3}\right)dx, \quad x \neq 0 \\ &= \int (x^{-2} + 3x^{-3})\,dx = \int x^{-2}\,dx + 3\int x^{-3}\,dx \\ &= \frac{x^{-1}}{-1} + 3\left(\frac{x^{-2}}{-2}\right) + C = -\frac{2x+3}{2x^2} + C\end{aligned}$$

(d) $$\int \frac{x^2-4}{x-2}\,dx = \int \frac{(x-2)(x+2)}{x-2}\,dx, \quad x \neq 2$$

$$= \int (x+2)\,dx = \int x\,dx + 2\int dx = \frac{x^2}{2} + 2x + C$$

EXAMPLE 5. Calculate each of the following indefinite integrals:

(a) $\int e^{3x}\,dx$

(b) $\int 5e^{-2x}\,dx$

(c) $\int \left(e^{(1/2)x} - \frac{1}{x^3} + \frac{1}{x}\right)dx$

(d) $\int \sqrt{x}(1+x^2)\,dx$

SOLUTION. (a) $$\int e^{3x}\,dx = \frac{e^{3x}}{3} + C$$

(b) $$\int 5e^{-2x}\,dx = \frac{5}{-2}\,e^{-2x} + C = -\frac{5}{2}e^{-2x} + C$$

(c) $$\int \left(e^{(1/2)x} - \frac{1}{x^3} + \frac{1}{x}\right)dx = \int e^{(1/2)x}\,dx - \int \frac{1}{x^3}\,dx + \int \frac{1}{x}\,dx$$

$$= \frac{e^{(1/2)x}}{\frac{1}{2}} - \int x^{-3}\,dx + \ln x + C$$

$$= 2e^{(1/2)x} - \frac{x^{-2}}{-2} + \ln x + C$$

$$= 2e^{(1/2)x} + \frac{1}{2x^2} + \ln x + C$$

(d) $$\int \sqrt{x}(1+x^2)\,dx = \int x^{1/2}(1+x^2)\,dx$$

$$= \int (x^{1/2} + x^{5/2})\,dx$$

$$= \int x^{1/2}\,dx + \int x^{5/2}\,dx$$

$$= \frac{x^{(1/2)+1}}{\frac{1}{2}+1} + \frac{x^{(5/2)+1}}{\frac{5}{2}+1} + C$$

$$= \frac{2}{3}x^{3/2} + \frac{2}{7}x^{7/2} + C$$

EXAMPLE 6. Suppose that the American Gadget Company has estimated that the marginal cost at a level of production of x gadgets is given by

$$C'(x) = 3 + \frac{8}{\sqrt{x}}$$

and the fixed cost is \$20,000. Find the cost function $C(x)$.

SOLUTION. Since

$$C'(x) = 3 + \frac{8}{\sqrt{x}} = 3 + 8x^{-1/2}$$

the antiderivative of $C'(x)$ is $C(x)$ given by

$$\begin{aligned} C(x) &= \int (3 + 8x^{-1/2})dx \\ &= 3x + 8(2x^{1/2}) + K \qquad (K \text{ is a constant}) \\ &= 3x + 16x^{1/2} + K \end{aligned}$$

We find K by using the fact that the fixed cost is \$20,000; that is, if $x = 0$, then $C(x) = 20{,}000$. Consequently,

$$20{,}000 = 3(0) + 16(0) + K \quad \text{or} \quad K = 20{,}000$$

Hence, the cost function is

$$C(x) = 3x + 16x^{1/2} + 20{,}000$$

EXAMPLE 7. Suppose that a curve passes through the point $P(-1, 1)$ and that at each x, the tangent line has slope $3x^2 + 5$. Find the equation of the curve.

SOLUTION. We have

$$f'(x) = 3x^2 + 5$$

so that

$$\begin{aligned} f(x) &= \int f'(x)dx = \int (3x^2 + 5)dx \\ &= \int 3x^2 dx + \int 5dx \\ &= \frac{3x^3}{3} + 5x + C \end{aligned}$$

or

$$f(x) = x^3 + 5x + C$$

or

$$y = x^3 + 5x + C \tag{14}$$

Since the curve passes through $(-1, 1)$, we must have $y = 1$ and $x = -1$. Putting these values in (14), we get

$$1 = (-1)^3 + 5(-1) + C$$

or

$$1 = -1 - 5 + C$$

Hence, $C = 7$, and the equation of the curve is

$$y = x^3 + 5x + 7$$

EXERCISES 5.1

In Problems 1 through 30, calculate the given indefinite integral.

1. $\int \frac{1}{2}\,dx$
2. $\int \frac{1}{2}x\,dx$
3. $\int 3x^3\,dx$
4. $\int (x + x^3)\,dx$
5. $\int (x^2 + 2x + 1)\,dx$
6. $\int x(x - 2)\,dx$
7. $\int \frac{1}{x^2}\,dx, \quad x \neq 0$
8. $\int \frac{3}{x^3}\,dx, \quad x \neq 0$
9. $\int (x - 1)^2\,dx$
10. $\int \frac{x + 4}{x^4}\,dx, \quad x \neq 0$
11. $\int 2\sqrt{x}\,dx$
12. $\int \frac{1}{\sqrt{x}}\,dx, \quad x > 0$
13. $\int x^{-1/3}\,dx, \quad x \neq 0$
14. $\int (3x^{1/3} + 2x^{-1/3})\,dx, \quad x \neq 0$
15. $\int \frac{x^2 - 1}{x + 1}\,dx, \quad x \neq -1$
16. $\int x^{1/3}(2 + x^{1/3})\,dx$
17. $\int \frac{x^3 - 2}{\sqrt{x}}\,dx, \quad x > 0$
18. $\int \frac{x^3 - 27}{x - 3}\,dx, \quad x \neq 3$
19. $\int \frac{x - 4}{\sqrt{x} + 2}\,dx, \quad x \geq 0$
20. $\int \frac{x + 1}{x^{1/3} + 1}\,dx, \quad x \neq -1$
21. $\int e^{4x}\,dx$
22. $\int e^{-5x}\,dx$
23. $\int e^{2x/3}\,dx$
24. $\int e^{-3x/4}\,dx$
25. $\int (e^{2x} + \frac{1}{x})\,dx$
26. $\int \frac{1 + xe^{-x}}{x}\,dx$
27. $\int (2\pi r + 8r^3)\,dr$
28. $\int \left(\frac{1}{y^2} - \frac{8}{y^3}\right)\,dy$

29. $\int 6\left(u+\frac{1}{u}\right)^2 du$

30. $\int \left(z-\frac{5}{z^2}\right)^2 dz$

In Problems 31 through 36, find the equation of the curve that passes through the given point, and at each point on the curve the tangent line that has the given slope.

31. slope $= 3x^2 + 4$, $P(1, 3)$

32. slope $= \frac{4}{x^2} - x$, $P(2, 5)$

33. slope $= 5x + \frac{5}{x}$, $P(1, \frac{9}{2})$

34. slope $= -6x^2 + \frac{10}{x^3}$, $P(-1, -7)$

35. slope $= 10e^{5x} - 4x^3$, $P(0, -5)$

36. slope $= 5e^{-3x} + 12x$, $P(0, 6)$

37. Find $f(x)$, if $f'(x) = 2x - 1$ and $f(2) = 4$.

38. Find $f(x)$, if $f''(x) = 6x - 10$, $f'(0) = 6$, and $f(1) = 5$.

In Problems 39 and 40, find the velocity $v(t)$ and the distance traveled $s(t)$ at the end of t seconds if a particle moves with the given acceleration $a(t)$. [Recall, $a(t) = v'(t)$ and $v(t) = s'(t)$.]

39. $a(t) = 6t - \frac{15}{2}\sqrt{t}$, $v(1) = \frac{1}{2}$, $s(0) = 3$.

40. $a(t) = 12t^2 - 30t + 14$, $v(1) = 3$, $s(1) = 4$.

41. Give examples of functions f and g having antiderivatives such that

$$\int f(x)g(x)dx \neq \left(\int f(x)dx\right)\left(\int g(x)dx\right)$$

42. Suppose $F(x)$ and $G(x)$ are two antiderivatives of the function $f(x)$. What can you say about the graphs of $F(x)$ and $G(x)$?

43. The marginal cost of the ABC Shirt Company for producing x shirts is given by

$$C'(x) = \frac{12}{\sqrt[3]{x}}$$

and the fixed cost is \$10,000. Find the cost function $C(x)$. Find also the total cost of producing 2000 shirts.

44. Find the cost function $C(x)$ if the marginal cost of producing x items is given by

$$C'(x) = 2x^2 + 5x - 3$$

It is given that the fixed cost is \$5000.

45. The marginal profit for the Pizza Parlor is given by

$$P'(x) = 3x + 5000$$

where x (in thousands) is the number of pizzas sold. If there is a loss of \$200 if no pizzas are sold, find the profit function $P(x)$.

46. Suppose the marginal revenue is $R'(x) = 500 - \frac{1}{4}x^2$. Find the revenue function $R(x)$. [*Hint:* What is $R(0)$?]

47. Find the cost function if the marginal cost is

$$C'(x) = \frac{1}{2}x^2 - \frac{1}{10}x + 50$$

and the total cost of producing 60 units is \$200,000.

48. Suppose the marginal cost for the BT Corporation to produce T-shirts is

$$C'(x) = 3 - 0.004x$$

and the fixed cost is \$500. If each T-shirt sells for \$7, find the profit function.

5.2 THE DEFINITE INTEGRAL

In this section we shall introduce the concept of the *definite integral*, and in the next section we shall relate this concept to antidifferentiation.

What is meant by the area of a region in the plane? If the region is bounded by straight line segments, it is easy to calculate its area by dividing it into rectangles and triangles. Thus, the area $A(R)$ of the region R in Figure 1 is given by

$$A(R) = A(R_1) + A(R_2) + A(R_3) + A(R_4)$$

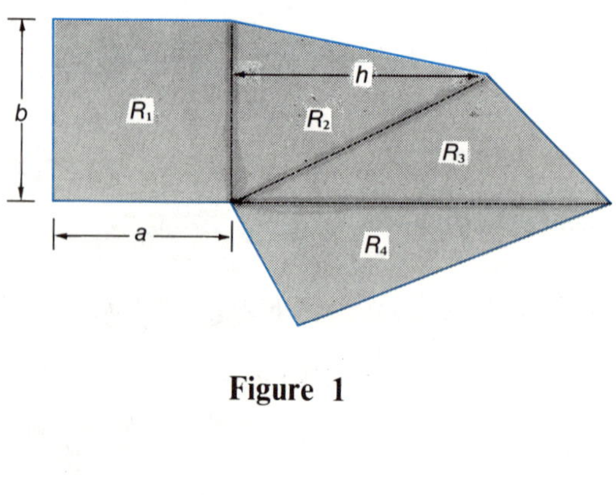

Figure 1

where the area of the rectangular region R_1 with length a and height b is

$$A(R_1) = ab$$

the area of the triangular region R_2 with base b and height h is

$$A(R_2) = \frac{1}{2}bh$$

and so on.

The problem of determining areas is more complicated if the region has curvilinear boundaries (see Figures 2 and 3). This problem dates back to the time of Archimedes (287-212 B.C.). He was successful in solving the problem in some simple cases; however, his method could not be applied generally. It was not until the time of Newton and Leibniz that the area problem was solved through the use of calculus. In this section we shall find that the area of a region in a plane can be represented as the limit of a sequence.

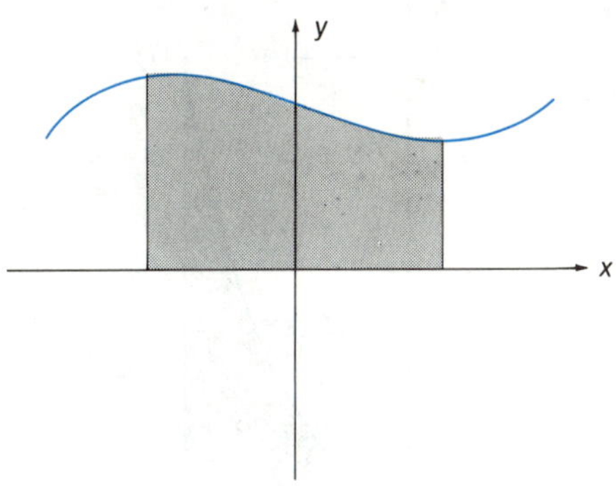

Figure 2

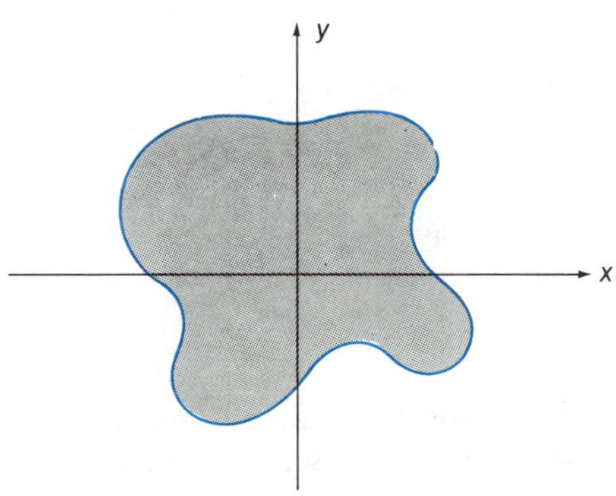

Figure 3

The main idea is to approximate the area by using rectangles: first, a set of rectangles enclosed in R for which the area is less than $A(R)$; and, second, a set of rectangles that contain R for which the area is greater than $A(R)$.

EXAMPLE 1. Approximate the area $A(R)$ of the region R bounded by the graph of $f(x) = x^2$ and the lines $x = 0$, $x = 2$, and $y = 0$.

SOLUTION. We shade the required region as shown in Figure 4. We see that R is contained in a rectangular region whose area is 8 square units. Then assuming that R has an area, $0 < A(R) < 8$, and we have an approximation of the area of the region R.

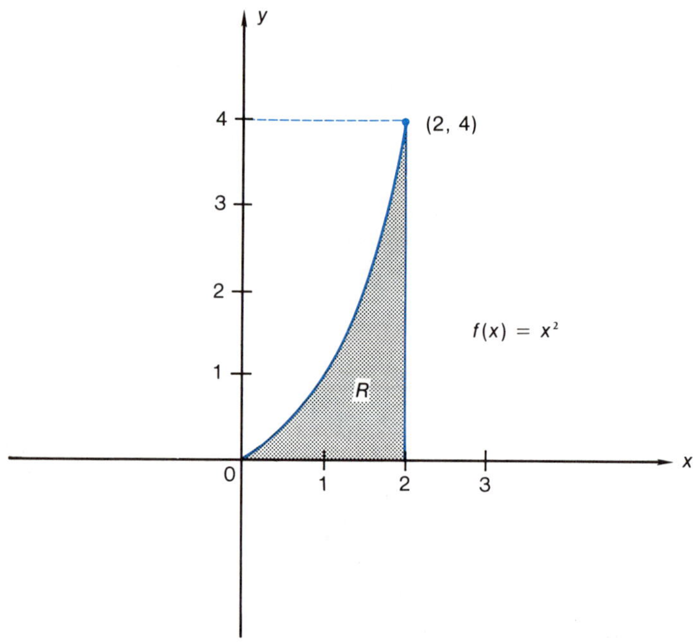

Figure 4

Suppose that we divide the interval [0, 2] into two parts and consider the point (1, 1) on the graph of f. We draw two sets of rectangles as shown in Figure 5. Obviously,

$$A(R_1) < A(R) < A(R_2) + A(R_3)$$

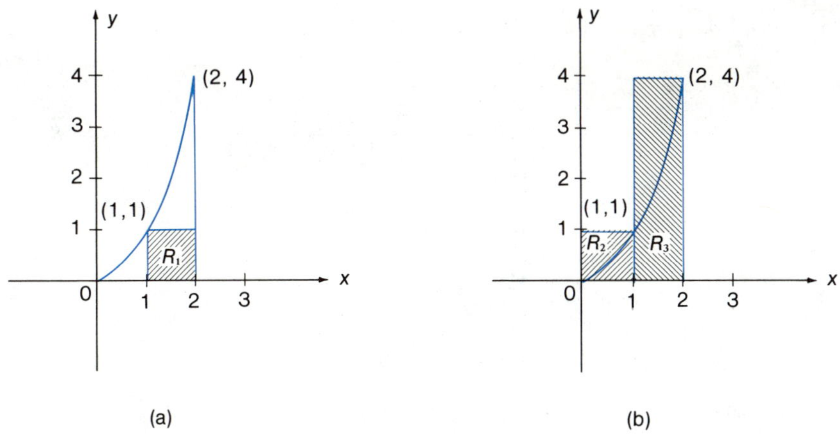

Figure 5

where

$$A(R_1) = (1)(1) = 1$$

$$A(R_2) = (1)(1) = 1$$

$$A(R_3) = (1)(4) = 4$$

Thus,

$$1 < A(R) < 1 + 4 \quad \text{or} \quad 1 < A(R) < 5$$

Next, we divide the interval [0, 2] into four parts and consider the points $(\frac{1}{2}, \frac{1}{4})$, (1, 1), and $(\frac{3}{2}, \frac{9}{4})$ on the graph of f. We draw two sets of rectangles as shown in Figure 6. Here we find that

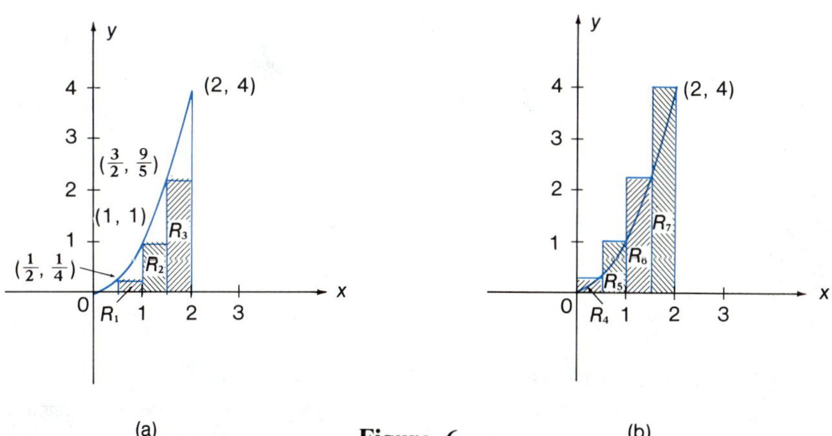

Figure 6

$$A(R_1) + A(R_2) + A(R_3) < A(R) < A(R_4) + A(R_5) + A(R_6) + A(R_7)$$

where

$$\begin{aligned}
A(R_1) &= \left(\tfrac{1}{2}\right)\left(\tfrac{1}{4}\right) = \tfrac{1}{8} \\
A(R_2) &= \left(\tfrac{1}{2}\right)(1) = \tfrac{1}{2} \\
A(R_3) &= \left(\tfrac{1}{2}\right)\left(\tfrac{9}{4}\right) = \tfrac{9}{8} \\
A(R_4) &= \left(\tfrac{1}{2}\right)\left(\tfrac{1}{4}\right) = \tfrac{1}{8} \\
A(R_5) &= \left(\tfrac{1}{2}\right)(1) = \tfrac{1}{2} \\
A(R_6) &= \left(\tfrac{1}{2}\right)\left(\tfrac{9}{4}\right) = \tfrac{9}{8} \\
A(R_7) &= \left(\tfrac{1}{2}\right)(4) = 2
\end{aligned}$$

Hence,

$$\tfrac{1}{8} + \tfrac{1}{2} + \tfrac{9}{8} < A(R) < \tfrac{1}{8} + \tfrac{1}{2} + \tfrac{9}{8} + 2$$

or

$$\tfrac{14}{8} < A(R) < \tfrac{30}{8}$$

It appears that by increasing the number of subintervals of [0, 2], the areas of the corresponding rectangles yield a better approximation for A(R).

We now leave the special function $f(x) = x^2$ and look at an arbitrary function $y = f(x)$ defined on a closed interval $[a, b]$. We assume that the function $f(x)$ is continuous on $[a, b]$. Our object is to define the *definite integral.*

We divide the interval $a \le x \le b$ into n equal parts[1] and label the parts of the subdivision by $x_0, x_1, x_2, \ldots, x_n$. Since each subinterval has length $\dfrac{b-a}{n}$, we have

$$x_0 = a,\ x_1 = a + \frac{b-a}{n},\ x_2 = a + 2\left(\frac{b-a}{n}\right),\ x_3 = a + 3\left(\frac{b-a}{n}\right), \ldots,$$

and

$$x_n = a + n\left(\frac{b-a}{n}\right) = b$$

Corresponding to $x_1, x_2, \ldots, x_n$, we find $f(x_1), f(x_2), \ldots, f(x_n)$ and form the sum (see Figure 7).

[1] The general definition of the integral is far more complicated. For simplicity, we take the liberty of giving a simplified version. However, for continuous functions, the simple definition and the complicated one are equivalent.

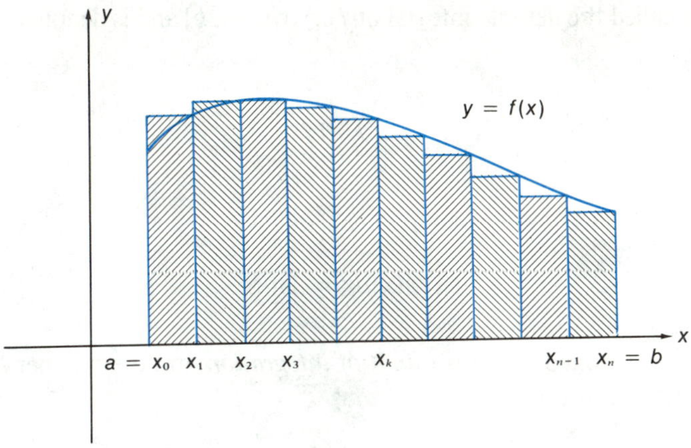

Figure 7

$$S_n = f(x_1)\left(\frac{b-a}{n}\right) + f(x_2)\left(\frac{b-a}{n}\right) + \cdots + f(x_k)\left(\frac{b-a}{n}\right) + \cdots + f(x_n)\left(\frac{b-a}{n}\right)$$

This sum S_n is called a Riemann[2] sum. We can factor out $\frac{b-a}{n}$ in S_n and write

$$S_n = \frac{b-a}{n}[f(x_1) + f(x_2) + \cdots + f(x_k) + \cdots + f(x_n)] \tag{1}$$

---

[2] GEORG FRIEDRICH BERNHARD RIEMANN (1826-1866) was born in Breselenz, Germany. His father was a Lutheran pastor; and the family, which included six children, lived in poverty. Riemann, a timid child with a horror of speaking in public, received his early education from his father and later at the Gymnasium at Hanover and Lüneburg. Riemann's interest and ability in the theory of numbers were demonstrated at Lüneburg. Schmalfuss, director of the Gymnasium, gave young Riemann Legendre's *Théorie des Nombres*, and Riemann apparently mastered the 859-page book in six days.

At the age of 19, Riemann entered the University of Göttingen, where he first studied philology and theology. A year later he transferred to the University of Berlin, where he learned much from the great mathematicians Dirichlet, Jacobi, Steiner, and Eisenstein. After two years in Berlin, he returned to the University of Göttingen to obtain his doctorate. Riemann's doctoral dissertation, "Foundations for a general theory of functions of a complex variable," impressed even Gauss. Riemann joined the faculty at the University of Göttingen as an unpaid lecturer, and in 1859 he succeeded Dirichlet as full professor.

Riemann was honored by many of the learned societies including the French Academy of Sciences and the Royal Society of London. His methods and ideas have greatly influenced research in mathematics and mathematical physics. He died of tuberculosis at the age of 40.

Now, as $n \to \infty$, it can be shown that S_n approaches a fixed number A as the limit. The limit A is called the definite integral of $f(x)$ over $[a, b]$ and is denoted by the symbol

$$\int_a^b f(x)dx$$

that is,

$$\int_a^b f(x)dx = \lim_{n\to\infty} S_n$$

The number a is called the *lower limit of integration* and the number b is called the *upper limit of integration.*

Remark 1. In general the definite integral $\int_a^b f(x)\,dx$ is a number (positive, negative, or zero). However, if $f(x) \geq 0$ on $[a, b]$, then the definite integral $\int_a^b f(x)dx$ defines the area $A(R)$ of the region R bounded by the curve $y = f(x)$, the x-axis, and the lines $x = a$ and $x = b$.

Remark 2. The similarity between $\int_a^b f(x)dx$ and the antiderivative symbol is intentional. A remarkable relationship exists between the definite integral and the antiderivative, and this relationship is stated in the *Fundamental Theorem of Calculus* (see Section 3).

EXAMPLE 2. Evaluate $\int_0^2 x^2\,dx$.

SOLUTION. Since $f(x) = x^2 \geq 0$ in the interval $[0, 2]$, the definite integral will give us the area of the region R bounded by $f(x) = x^2$, the x-axis, $x = 0$, and $x = 2$ (see Figure 4). To find the definite integral (or $A(R)$), we shall compute the limit indicated in (1).

Step 1. Divide the interval $[0, 2]$ into n equal parts. Hence, each subinterval is of length

$$\frac{b-a}{n} = \frac{2-0}{n} = \frac{2}{n}$$

Step 2. The subdivision points are:

$$x_0 = 0,\ x_1 = 0 + \frac{2}{n} = \frac{2}{n},\ x_2 = \frac{4}{n},\ \cdots,\ x_k = \frac{2k}{n},\ \cdots,\ x_n = \frac{2n}{n} = 2$$

Step 3. Calculate S_n:

$$\begin{aligned} S_n &= \frac{2}{n}[f(x_1) + f(x_2) + \cdots + f(x_n)] \\ &= \frac{2}{n}\left[\left(\frac{2}{n}\right)^2 + \left(\frac{4}{n}\right)^2 + \cdots + \left(\frac{2k}{n}\right)^2 + \cdots + \left(\frac{2n}{n}\right)^2\right] \\ &= \frac{8}{n^3}[1^2 + 2^2 + \cdots + k^2 + \cdots + n^2] \end{aligned}$$

But (see Appendix 4)

$$1^2 + 2^2 + 3^2 + \cdots + n^2 = \frac{n(n+1)(2n+1)}{6}$$

Hence,

$$S_n = \frac{8}{n^3}\left[\frac{n(n+1)(2n+1)}{6}\right] = \frac{8}{3} + \frac{4}{n} + \frac{4}{3n^2}$$

Step 4.

$$\int_0^2 x^2\,dx = \lim_{n\to\infty} S_n = \lim_{n\to\infty}\left[\frac{8}{3} + \frac{4}{n} + \frac{4}{3n^2}\right] = \frac{8}{3}$$

Thus, the area of the region R in Figure 4 is $\frac{8}{3}$.

EXAMPLE 3. Compute the definite integral

$$\int_1^5 (10 - 4x)\,dx$$

and find a meaning for the number obtained.

SOLUTION. The region bounded by $y = 10 - 4x$, the x-axis, and the lines $x = 1$ and $x = 5$ is shaded in Figure 8.

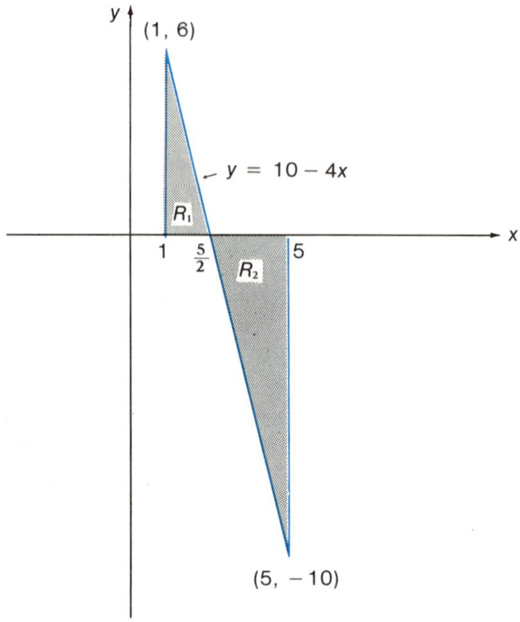

Figure 8

Step 1. Divide the interval [1, 5] into n equal parts. Hence, the length of each subinterval is

$$= \frac{b-a}{n} = \frac{5-1}{n} = \frac{4}{n}$$

Step 2. The subdivision points are:

$$x_0 = 1,\ x_1 = 1+\frac{4}{n},\ x_2 = 1+2\left(\frac{4}{n}\right),\ x_3 = 1+3\left(\frac{4}{n}\right),\ \cdots,\ x_n = 1+n\left(\frac{4}{n}\right) = 5$$

Step 3. Calculate S_n:

$$S_n = \frac{4}{n}[f(x_1)+f(x_2)+f(x_3)+\cdots+f(x_n)]$$

$$=\frac{4}{n}\{[10-4(1+\tfrac{4}{n})]+[10-4(1+\tfrac{8}{n})]+\cdots+[10-4(1+n(\tfrac{4}{n}))]\}$$

$$=\frac{4}{n}\{(10+10+\cdots+10)-4(1+1+\cdots+1)-\tfrac{16}{n}(1+2+\cdots+n)\}$$

Using (see Appendix 4) $1+2+3+\cdots+n = \dfrac{n(n+1)}{2}$, we have

$$S_n = \frac{4}{n}\left[10n - 4n - \frac{16}{n}\cdot\frac{n(n+1)}{2}\right]$$

$$= \frac{4}{n}[6n - 8(n+1)]$$

$$= \frac{4}{n}[-2n - 8]$$

$$= -8 - \frac{32}{n}$$

Step 4.

$$\int_1^5 (10 - 4x)dx = \lim_{n\to\infty} S_n = \lim_{n\to\infty}\left[-8 - \frac{32}{n}\right] = -8$$

Since the answer is negative, it certainly does not seem to be an area. From Figure 8 we see that R_1 and R_2 are triangles, and

$$\text{Area of } R_1 = \frac{1}{2}\left(\frac{3}{2}\right)(6) = \frac{9}{2}$$

$$\text{Area of } R_2 = \frac{1}{2}\left(\frac{5}{2}\right)(10) = \frac{25}{2}$$

The region R_1 is above the x-axis, and region R_2 is below the x-axis. Hence,

$$\text{Area of } R_1 - \text{Area of } R_2 = \frac{9}{2} - \frac{25}{2} = -8$$

and this is just the number obtained for the value of the definite integral. Thus, we conclude that the definite integral counts the area of the region below the x-axis with a negative sign.

If we insist that "area" is a nonnegative quantity, then we will use the phrase *geometric area*, or *true area*, or just *area*. Thus, the true area of the shaded region in Figure 8 is

$$\frac{9}{2} + \frac{25}{2} = 17$$

If we attach a sign (±) to the "area," using a negative number for regions that lie below the x-axis and a positive number for regions that lie above the x-axis, then we will use the phrase *algebraic area* or *signed area*.

Thus, $\int_a^b f(x)dx$ gives us the algebraic or signed area bounded by the curve $y = f(x)$ and the x-axis between the lines $x = a$ and $x = b$.

EXERCISES 5.2

In Problems 1 through 6, divide the interval $[a, b]$ into eight equal parts and evaluate S_8 (use a calculator):

1. $\int_1^2 (1 + x)dx$

2. $\int_1^3 \frac{x}{2} dx$

3. $\int_0^4 x^2 dx$

4. $\int_{-1}^1 x^2 dx$

5. $\int_1^2 \frac{1}{x} dx$

6. $\int_1^4 \sqrt{x + 5}\, dx$

In Problems 7 through 10, interpret and use plane geometry to compute:

7. $\int_1^4 5dx$

8. $\int_1^7 -3dx$

9. $\int_{-4}^5 (2x + 3)dx$

10. $\int_{-1}^2 (3 - 4x)dx$

In Problems 11 through 20, evaluate the indicated definite integral. Sketch the region involved.

11. $\int_0^2 x\, dx$

12. $\int_1^2 (2 + x)dx$

13. $\int_0^1 x^2 dx$

14. $\int_1^3 (1 + x^2)dx$

15. $\int_2^5 (x - 1)dx$

16. $\int_{-1}^3 (4 - x)dx$

17. $\int_{-1}^2 (1 + 2x^2)dx$

18. $\int_{-1}^1 (2 - x)^2 dx$

19. $\int_1^4 (2 - x)^2 dx$

20. $\int_0^1 x^3 dx$

21. Show that $\int_a^b x\, dx = \frac{b^2 - a^2}{2}$

***22.** Show that $\int_a^b x^2 dx = \frac{b^3 - a^3}{2}$

5.3 THE FUNDAMENTAL THEOREM OF CALCULUS

Obviously, the evaluation of a definite integral by the method of Section 2 is time-consuming and tedious. Fortunately, there is a technique available to help. It is given in the Fundamental Theorem of Calculus, which relates the definite integral to the antiderivative of a given function. This provides a simple method of evaluating the definite integral whenever an antiderivative of the function is known.

Suppose f is a nonnegative continuous function defined on $[a, b]$. We wish to compute the area of the region bounded above by the curve $y = f(x)$, below by the x-axis, and on the two sides by the vertical lines drawn at $x = a$ and $x = b$. Let $A(x)$ denote the area under the graph of f above the interval $[a, b]$ (see Figure 1). That is,

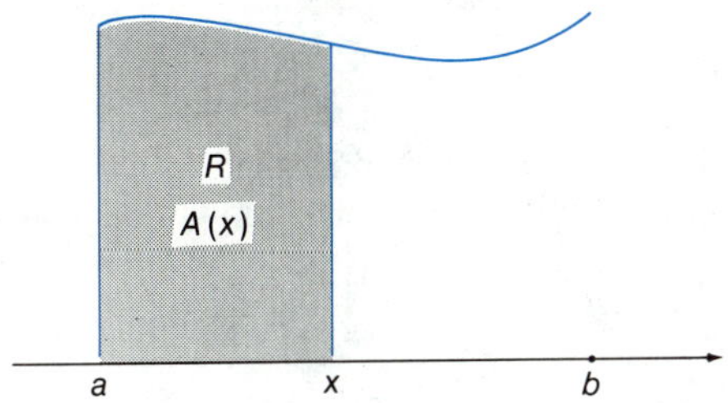

Figure 1

$$A(x) = \int_a^x f(t)dt \tag{1}$$

We call A the area function associated with f on $[a, b]$. If the variable x moves to the left until $x = a$, then the region R (see Figure 1) shrinks to a vertical line segment at $x = a$, and the area of such a segment is zero. Thus,

$$A(a) = \int_a^a f(t)dt = 0$$

When x moves to b, then the area under the graph of f from a to b is

$$A(b) = \int_a^b f(t)dt$$

A look at Figure 2 tells us that as x increases by an amount h, the region is increased by adding to R a strip that is almost a rectangle with thickness h and height close to $f(x)$. Thus,

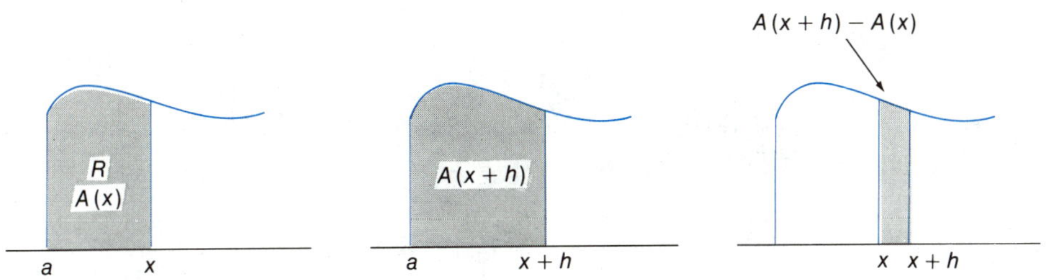

Figure 2

$$A(x+h) - A(x) \approx f(x) \cdot h$$

or

$$\frac{A(x+h) - A(x)}{h} \approx f(x) \tag{2}$$

Now take the limit in (2) as $h \to 0$. As the width of the strip shrinks to zero, the approximation in (2) gets better and better. Hence, we have

$$A'(x) = f(x) \tag{3}$$

Of course, in our discussion we have purposely omitted quite a few details. From (1) and (3) we see that, *if we differentiate the integral of f from a to x, we get f itself evaluated at x.* In other words, the area function A is an antiderivate of f.

This result is true regardless of whether or not f is a positive function. If f is not positive, the function A must be interpreted as a signed area.

EXAMPLE 1.

(a) If $F(x) = \int_2^x (t^2 + 3t)\,dt$, then $F'(x) = x^2 + 3x$

(b) If $G(x) = \int_{-5}^x (t^3 + \sqrt{t+1}\,)\,dt$, then $G'(x) = x^3 + \sqrt{x+1}$

(c) If $H(x) = \int_0^x [(t-1)^{3/2} + \sqrt{t^2+1}\,]\,dt$, then $H'(x) = (x-1)^{3/2} + \sqrt{x^2+1}$

Our result tells us that if

$$A(x) = \int_a^x f(t)\,dt$$

then

$$A'(x) = f(x)$$

That is, A is an antiderivative of f. Now, suppose F is any other antiderivative of f. Then, from Section 1, we know that for some constant C

$$A(x) = F(x) + C$$

Since $A(a) = 0$, we have

$$0 = F(a) + C$$

or

$$C = -F(a)$$

Hence,

$$F(x) - F(a) = A(x) = \int_a^x f(t)\,dt$$

Setting $x = b$, we have

$$F(b) - F(a) = \int_a^b f(t)\,dt$$

This result is known as the Fundamental Theorem of Calculus.

THEOREM 1. (*Fundamental Theorem of Calculus*). If f is a function continuous on $[a, b]$ and F is an antiderivative of f, then

$$\int_a^b f(x)\,dx = F(b) - F(a)$$

This basic relationship between the area under the curve $y = f(x)$ and the antiderivative of f was noted first by Newton's teacher, Isaac Barrow (1630-1677).

Another notation commonly used is

$$\int_a^b f(x)\,dx = F(x)\Big|_a^b = F(b) - F(a)$$

The symbol $\int_a^b f(x)\,dx$ reminds us that the definite integral is the limit of a sequence of sums. We note that $F(b) - F(a)$ does not involve the variable x, so we may write

$$\int_a^b f(x)\,dx = \int_a^b f(r)\,dr = \int_a^b f(t)\,dt$$

where the letters x, r, and t are called *dummy variables.*

Therefore, for a given function f, we can compute the definite integral of f over $[a, b]$ by finding *any* antiderivative of f, evaluating this antiderivative at b and at a, and then subtracting the second value from the first.

EXAMPLE 2. Evaluate $\int_{-1}^{2} 2x^3\,dx$

SOLUTION. Here $f(x) = 2x^3$ and an antiderivative of f is

$$F(x) = \frac{x^4}{2}$$

Evaluating F at $x = 2$ and at $x = -1$ and subtracting, we obtain

$$F(2) - F(-1) = \frac{x^4}{2}\bigg|_{-1}^{2} = \frac{(2)^4}{2} - \frac{(-1)^4}{2} = \frac{15}{2}$$

Therefore,

$$\int_{-1}^{2} 2x^3\,dx = \frac{15}{2}$$

EXAMPLE 3. Evaluate $\int_1^4 \left(x^2 - \sqrt{x} + \frac{1}{x^2}\right) dx$

SOLUTION. An antiderivative of f, where $f(x) = x^2 - \sqrt{x} + (1/x^2)$, is

$$F(x) = \frac{x^3}{3} - \frac{2}{3}x^{3/2} - \frac{1}{x}$$

Therefore,

$$\int_1^4 \left(x^2 - \sqrt{x} + \frac{1}{x^2}\right) dx = \left(\frac{x^3}{3} - \frac{2}{3}x^{3/2} - \frac{1}{x}\right)\bigg|_1^4$$

$$= \left[\frac{(4)^3}{3} - \frac{2}{3}(4)^{3/2} - \frac{1}{4}\right] - \left[\frac{(1)^3}{3} - \frac{2}{3}(1)^{3/2} - \frac{1}{1}\right]$$

$$= \frac{205}{12}$$

Suppose F is an antiderivative of f and $a < c < b$. Then, by Theorem 1,

$$\int_a^c f(x)dx = F(c) - F(a)$$

and

$$\int_c^b f(x)dx = F(b) - F(c)$$

Thus,

$$\int_a^c f(x)dx + \int_c^b f(x)dx = [F(c) - F(a)] + [F(b) - F(c)]$$

$$= F(b) - F(a)$$

$$= \int_a^b f(x)dx$$

Therefore, if $a < c < b$, we have the property

$$\int_a^b f(x)dx = \int_a^c f(x)dx + \int_c^b f(x)dx \tag{4}$$

In a similar manner, the following properties of the definite integral may be established.

$$\int_a^a f(x)dx = 0 \tag{5}$$

$$\int_a^b [Af(x) + Bg(x)]dx = A\int_a^b f(x)dx + B\int_a^b g(x)dx \tag{6}$$

where A and B are constants.

EXAMPLE 4. Find the area of the region bounded by the graphs of functions f and g, where $f(x) = x$ and $g(x) = \frac{1}{2}x^2$.

SOLUTION.

Step 1. Sketch the graph of the region R as shown in Figure 3.

Step 2. Find the points of intersection of the graphs of f and g. Here we have to solve the equation:

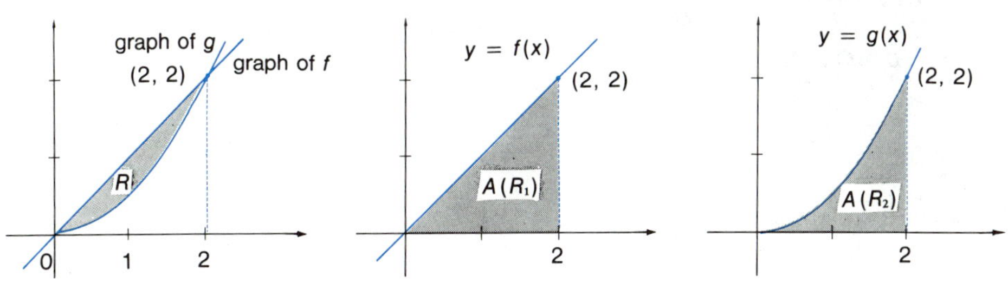

Figure 3

$$x = \tfrac{1}{2}x^2 \quad \text{or} \quad 2x - x^2 = 0$$

Thus $x = 0$ and $x = 2$, and the points of intersection are (0, 0) and (2, 2). Note that the points of intersection will define the region R.

Step 3. We observe that the area $A(R)$ of the region R is given by

$$A(R) = A(R_1) - A(R_2)$$

where $A(R_1)$ is the area of the region R_1 bounded by the graph of f, the x-axis, and the lines $x = 0$ and $x = 2$; and $A(R_2)$ is the area of the region R_2 bounded by the graph of g, the x-axis, and the lines $x = 0$ and $x = 2$. Now

$$A(R_1) = \int_0^2 f(x)dx = \int_0^2 x\,dx = \left.\frac{x^2}{2}\right|_0^2 = \frac{(2)^2}{2} - 0 = 2$$

and

$$A(R_2) = \int_0^2 g(x)dx = \int_0^2 \frac{1}{2}x^2 dx = \left.\frac{1}{2}\frac{x^3}{3}\right|_0^2 = \frac{(2)^3}{6} - 0 = \frac{4}{3}$$

Therefore, the area of the region in question is

$$A(R) = \int_0^2 x\,dx - \int_0^2 \frac{1}{2}x^2 dx = 2 - \frac{4}{3} = \frac{2}{3}$$

In the previous section we indicated that

$$A(R) = \int_a^b f(x)dx$$

is the area of the region R bounded by the graph of f, the x-axis, and the lines $x = a$ and $x = b$ provided that f is nonnegative on $[a, b]$. Now we generalize this definition.

If f and g are continuous functions on $[a, b]$ and $f(x) \geq g(x)$ for each point on $[a, b]$, then the area of the region R bounded by the graphs of f and g and the lines $x = a$ and $x = b$ is defined by

$$A(R) = \int_a^b [f(x) - g(x)]dx$$

Such a region is illustrated in Figure 4.

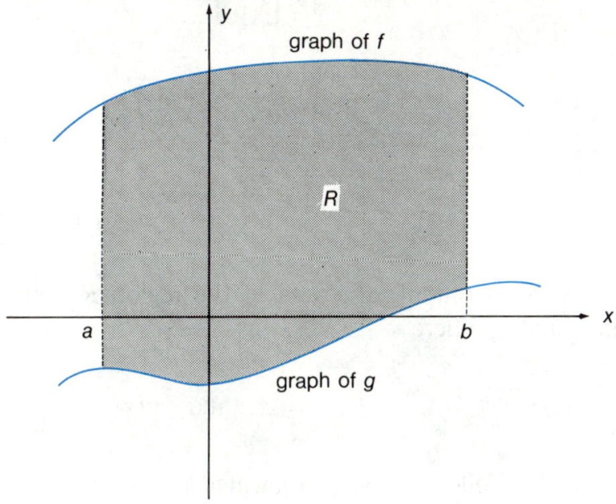

Figure 4

EXAMPLE 5. Compute the area of the region for which $x \in [-1, 2]$ and which is bounded by the graphs of the function f and g, where

$$f(x) = -\tfrac{1}{4}x^2 + 1 \quad \text{and} \quad g(x) = \tfrac{1}{2}x^2 - 3$$

SOLUTION. The region R is illustrated in Figure 5. Since $f(x) \geq g(x)$ for each x on $[-1, 2]$, we have

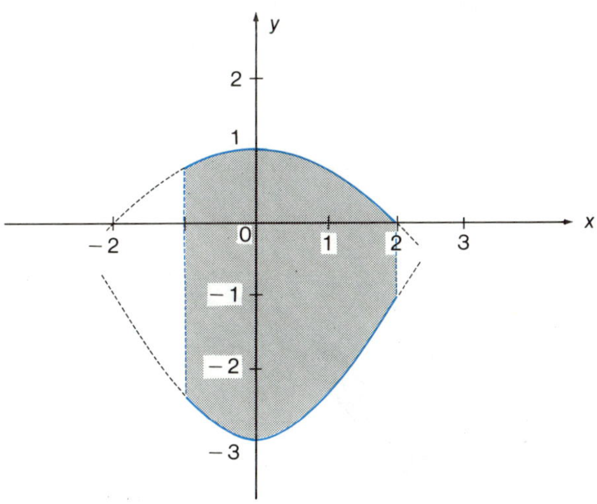

Figure 5

$$A(R) = \int_{-1}^{2} [f(x) - g(x)]dx = \int_{-1}^{2} \left[\left(-\frac{1}{4}x^2 + 1\right) - \left(\frac{1}{2}x^2 - 3\right)\right] dx$$

$$= \int_{-1}^{2} \left(-\frac{3}{4}x^2 + 4\right) dx = \left(-\frac{3}{4}\frac{x^3}{3} + 4x\right)\Bigg|_{-1}^{2}$$

$$= \left[-\frac{(2)^3}{4} + 4(2)\right] - \left[-\frac{(-1)^3}{4} + 4(-1)\right] = \frac{39}{4}$$

EXAMPLE 6. Compute the area of the region R bounded by the graphs of the functions f and g, where

$$f(x) = x \quad \text{and} \quad g(x) = x^3$$

SOLUTION. We follow the steps indicated in Example 4. First we sketch a graph of the region R as shown in Figure 6. We find the points of intersection of the graphs of f and g by solving the equation

$$x = x^3$$

from which we obtain $x = 0$, $x = 1$, and $x = -1$. Thus, the points of intersection are $(0, 0)$, $(1, 1)$, and $(-1, -1)$.

Here we see that there are two regions to be considered:

1. In R_1, $x \in [0, 1]$ and $f(x) \geq g(x)$ for each x.
2. In R_2, $x \in [-1, 0]$ and $g(x) \geq f(x)$ for each x.

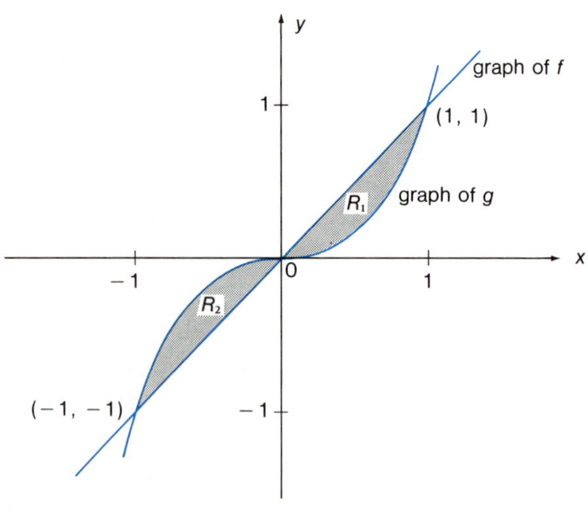

Figure 6

Hence, we treat the problem in two parts. First, we find $A(R_1)$:

$$A(R_1) = \int_0^1 [f(x) - g(x)]dx = \int_0^1 (x - x^3)dx = \left(\frac{x^2}{2} - \frac{x^4}{4}\right)\Big|_0^1$$

$$= \left[\frac{(1)^2}{2} - \frac{(1)^4}{4}\right] - \left[\frac{0}{2} - \frac{0}{4}\right] = \frac{1}{4}$$

Next, we find $A(R_2)$:

$$A(R_2) = \int_{-1}^0 [g(x) - f(x)]dx = \int_{-1}^0 (x^3 - x)dx = \left(\frac{x^4}{4} - \frac{x^2}{2}\right)\Big|_{-1}^0$$

$$= \left[\frac{0}{4} - \frac{0}{2}\right] = \left[\frac{(-1)^4}{4} - \frac{(-1)^2}{2}\right] = \frac{1}{4}$$

The area $A(R) = A(R_1) + A(R_2)$. Therefore,

$$A(R) = \frac{1}{4} + \frac{1}{4} = \frac{1}{2}$$

EXAMPLE 7. Find the area of the region bounded by the graph of $f(x) = 9 - x^2$, the x-axis, and the lines $x = -1$ and $x = 4$.

SOLUTION. We first sketch the graph of the region R (see Figure 7):

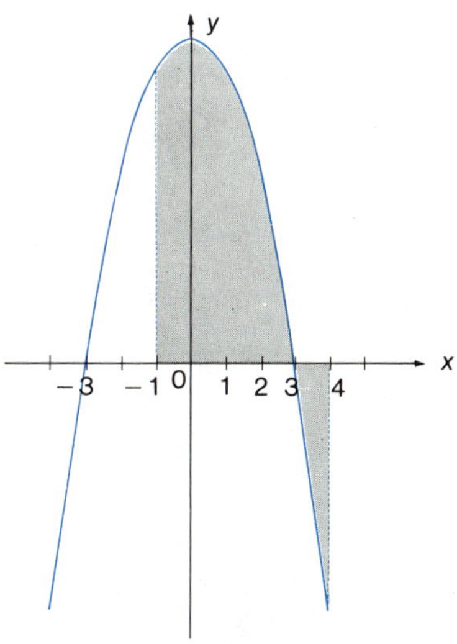

Figure 7

Since the values of f are not all nonnegative, we will get the wrong answer by just evaluating $\int_{-1}^{4} f(x)\,dx$.

However, if we think of the x-axis as the graph of $g(x) = 0$, then we can proceed as in Example 6.

The graph of f crosses the x-axis when $x = 3$. Thus, $f(x) \geq g(x)$ $[g(x) = 0]$ in $[-1, 3]$ and $g(x) = 0 \geq f(x)$ in $[3, 4]$. Hence, the desired area is given by

$$\begin{aligned} A &= \int_{-1}^{3} [(9 - x^2) - 0]dx + \int_{3}^{4} [0 - (9 - x^2)]dx \\ &= \int_{-1}^{3} (9 - x^2)dx + \int_{3}^{4} (-9 + x^2)dx \\ &= 9x - \frac{x^3}{3}\bigg|_{-1}^{3} + \left(-9x + \frac{x^3}{3}\right)\bigg|_{3}^{4} \\ &= \left[\left(27 - \frac{27}{3}\right) - \left(-9 + \frac{1}{3}\right)\right] + \left[\left(-36 + \frac{64}{3}\right) - (-27 + 9)\right] \\ &= 30 \end{aligned}$$

EXERCISES 5.3

In Problems 1 through 12, evaluate the given definite integrals.

1. $\int_{1}^{3} (x^2 - 3)dx$
2. $\int_{-1}^{2} (2x^2 + 4x)dx$
3. $\int_{-1}^{2} (r + 1)^2 dr$
4. $\int_{1}^{2} (x + 2)(x - 2)dx$
5. $\int_{4}^{9} 3\sqrt{t}\, dt$
6. $\int_{1}^{3} \frac{2}{s^2}\, ds$
7. $\int_{-2}^{4} x(x + 1)dx$
8. $\int_{2}^{5} \frac{1}{\sqrt{u}}\, du$
9. $\int_{-1}^{1} (2x^2 + x^{1/3} - 6)dx$
10. $\int_{0}^{2} \frac{1 + 2v^{1/3}}{v^{2/3}}\, dv$
11. $\int_{1}^{3} \left(\frac{1}{u}\right)^4 du$
12. $\int_{0}^{16} 2x^{1/4} dx$

13. Establish formulas (5) and (6).

In Problems 14 through 22, find the area of the region in the plane bounded by the given curves. Sketch a graph of the region.

14. $y = 2x + 1$, $x = 1$, $x = 3$, and the x-axis.

15. $y = 9 - x^2$ and the x-axis.

16. $y = x^2 + 1,\ x = 2,\ x = -3$, and the x-axis.

17. $y = x^2$ and $y = 2x$

18. $y = x^2$ and $y = x + 2$

19. $y = \sqrt{x}$ and $y = x^2$

20. $y = x^2$ and $y = 2 - x^2$

21. $y = 4x - x^2$ and $y = 3$

22. $y = 4x - x^2$ and $y = -5$

In Problems 23 through 30, determine $F'(x)$.

23. $F(x) = \int_1^x t\,dt$

24. $F(x) = \int_{-2}^x u^2\,du$

25. $F(x) = \int_0^x (r - 1)^2\,dr$

26. $F(x) = \int_1^x 2\sqrt{t}\,dt$

27. $F(x) = \int_x^2 u^2\,du$

28. $F(x) = \int_x^{3x} r^2\,dr$

29. $F(x) = \int_x^{x+2} r^2\,dr$

30. $F(x) = \int_x^{x+2} r^2\,dr + \int_{x+2}^x r^2\,dr$

In Problems 31 through 34, determine the values for x which satisfy the given equation.

31. $\int_0^x 3t^2\,dt = 8$

32. $\int_0^x (4t - 5)\,dt = -2$

33. $\int_{-x}^x \frac{u^6}{2}\,du = \frac{128}{7}$

34. $\int_x^{2x^2} dt = 6$

5.4 THE TECHNIQUES OF INTEGRATION

The reader must have certainly noticed that all of the integrals treated so far have been very elementary. Suppose that we wish to find the indefinite integrals:

$$\int (2x^3 + 5)^{7/2} 6x^2\,dx \tag{1}$$

$$\int x^2 e^x\,dx \tag{2}$$

$$\int \frac{10x - 4}{x^3 - 4x}\,dx \tag{3}$$

$$\int \sqrt{1 + x^3}\,dx \tag{4}$$

A moment's reflection will show that we cannot find any of these indefinite integrals with the methods used so far. In this chapter we will learn three new methods: (1) integration by substitution, (2) integration by parts, and (3) integration by partial fractions. These methods will suffice to compute the integrals (1), (2), and (3); but the integral in (4) is quite different. No matter how many fancy tricks we may try, we cannot express this integral with a finite number of combinations of elementary functions.

INTEGRATION BY SUBSTITUTION.

Integration by substitution is used to compute indefinite integrals of the form

$$\int kf(g(x))g'(x)dx \tag{5}$$

where k is any fixed real number. To compute (5) we make a change of variable, or substitution. We let

$$u = g(x) \tag{6}$$

then

$$\frac{du}{dx} = g'(x)$$

or

$$du = g'(x)dx \tag{7}$$

Making these substitutions (6) and (7) in (5), we obtain the integral

$$\int kf(u)du \tag{8}$$

So the success of the method of substitution hinges on our clever (sometimes lucky) choice of $g(x)$ so that the resulting integral (8) is in recognizable form (Formulas (1) through (9) of section 1).

We illustrate this method in the following examples.

EXAMPLE 1. Evaluate $\int (2x^3 + 5)^{7/2}6x^2dx$.

SOLUTION. We let $u = 2x^3 + 5$. Then

$$\frac{du}{dx} = 6x^2 \qquad \text{or} \qquad du = 6x^2dx$$

Using these substitutions, we obtain

$$\int (2x^3 + 5)^{7/2} 6x^2 dx = \int u^{7/2} du$$

$$= \frac{u^{(7/2)+1}}{\frac{7}{2} + 1} + C$$

$$= \frac{2}{9} u^{9/2} + C$$

$$= \frac{2}{9} (2x^3 + 5)^{9/2} + C$$

EXAMPLE 2. Evaluate $\int (2x^3 + 5)^{7/2} x^2 dx$.

SOLUTION. Again, letting $u = 2x^3 + 5$, we have

$$du = 6x^2 dx$$

But the integral does not contain the constant 6, which is needed for du. To take care of this, we solve $du = 6x^2 dx$ for dx to obtain

$$dx = \frac{1}{6x^2} du$$

Substituting u for $2x^3 + 5$ and $\frac{1}{6x^2} du$ for dx in the integral, we obtain

$$\int (2x^3 + 5)^{7/2} x^2 dx = \int u^{7/2} \cdot x^2 \cdot \frac{1}{6x^2} du$$

$$= \int \frac{1}{6} u^{7/2} du$$

$$= \frac{1}{6} \int u^{7/2} du$$

$$= \frac{1}{6} \frac{u^{9/2}}{\frac{9}{2}} + C$$

$$= \frac{1}{27} u^{9/2} + C$$

$$= \frac{1}{27} (2x^3 + 5)^{9/2} + C$$

EXAMPLE 3. Evaluate $\int (2x^3 + 5)^{7/2} dx$.

SOLUTION. Again, we let $u = 2x^3 + 5$; then $du = 6x^2 dx$. Proceeding as in Example 2, we would obtain

$$\int (2x^3 + 5)^{7/2} dx = \int u^{7/2} \cdot \frac{1}{6x^2} du$$

The factor $1/6x^2$ makes it impossible to evaluate the integral on the right side. The method of substitution of variables does not work in this case.

EXAMPLE 4. Find $\int e^{3x} dx$.

SOLUTION. Let $u = 3x$; then $du = 3\,dx$. Using these substitutions, we have

$$\int e^{3x} dx = \tfrac{1}{3}\int e^{3x} \cdot 3\,dx = \tfrac{1}{3}\int e^u du = \tfrac{1}{3}(e^u + C_1)$$

$$= \tfrac{1}{3}e^u + C \qquad \text{(where } C = \tfrac{1}{3}C_1\text{)}$$

$$= \tfrac{1}{3}e^{3x} + C$$

EXAMPLE 5. Find $\int xe^{2x^2+1} dx$.

SOLUTION. Letting $u = 2x^2 + 1$, we have $du = 4x\,dx$. Hence

$$\int xe^{2x^2+1} dx = \tfrac{1}{4}\int e^{2x^2+1}(4x) dx = \tfrac{1}{4}\int e^u du = \tfrac{1}{4}e^u + C$$

$$= \tfrac{1}{4}e^{2x^2+1} + C$$

EXAMPLE 6. Evaluate $\int \frac{2x}{x^2 + 4} dx$.

SOLUTION. Let $u = x^2 + 4$. Then $du = 2x\,dx$. Since $u(x) > 0$ for all x, we have

$$\int \frac{2x}{x^2 + 4} dx = \int \frac{1}{u} du = \ln u + C = \ln(x^2 + 4) + C$$

EXAMPLE 7. Evaluate $\int \frac{2x^2 + 2x + 1}{x + 1} dx$, $x \neq -1$.

SOLUTION. We first divide the integrand to obtain

$$\int \left(\frac{2x^2 + 2x + 1}{x + 1} \right) dx = \int \left(2x + \frac{1}{x + 1} \right) dx$$

$$= \int 2x\,dx + \int \left(\frac{1}{x+1}\right) dx$$

Now,

$$\int 2x\,dx = 2\int x\,dx = 2\,\frac{x^2}{2} + C_1$$

$$= x^2 + C_1$$

and to compute the second integral, we make the substitution: $u = x + 1$; $du = dx$. Thus,

$$\int \frac{1}{x+1}\,dx = \int \frac{1}{u}\,du = \ln u + C_2$$

$$= \ln(x+1) + C_2$$

Consequently,

$$\int \frac{2x^2 + 2x + 1}{x+1}\,dx = x^2 + \ln(x+1) + C$$

where $C = C_1 + C_2$.

Sometimes a change-of-variable strategy requires that we write the initial variable x in terms of the new variable u.

EXAMPLE 8. Evaluate $I = \int (3x+2)(2x-5)^{1/2}dx$.

SOLUTION. We let $u = 2x - 5$. From this, we have

$$x = \frac{1}{2}u + \frac{5}{2}$$

and

$$dx = \frac{1}{2}du$$

Substituting for x and dx in the given integral, we have

$$I = \int \left[3\left(\frac{1}{2}u + \frac{5}{2}\right) + 2\right]u^{1/2}\left(\frac{1}{2}du\right)$$

$$= \frac{1}{2}\int\left(\frac{3}{2}u + \frac{19}{2}\right)u^{1/2}\,du$$

$$= \frac{1}{2}\int\left(\frac{3}{2}u^{3/2} + \frac{19}{2}u^{1/2}\right)du$$

$$= \frac{1}{2}\left[\frac{3}{2}\cdot\frac{2}{5}u^{5/2} + \frac{19}{2}\cdot\frac{2}{3}u^{3/2}\right] + C$$

$$= \frac{3}{10}u^{5/2} + \frac{19}{6}u^{3/2} + C$$

$$= \frac{3}{10}(2x-5)^{5/2} + \frac{19}{2}(2x-5)^{3/2} + C$$

We now consider the evaluation of definite integrals. The method involves first finding an antiderivative of the integrand and then evaluating this antiderivative at the limits of integration. There are two possible ways of doing this and we shall illustrate both methods.

EXAMPLE 9. Evaluate $\int_1^2 (1 - x^{-1})^3 x^{-2} dx$.

SOLUTION 1. We first evaluate the indefinite integral

$$\int (1 - x^{-1})^3 x^{-2} dx$$

We let

$$u = 1 - x^{-1}, \quad du = u'(x)dx = x^{-2}dx$$

Substitution yields

$$\int (1 - x^{-1})^3 x^{-2} dx = \int u^3 du = \tfrac{1}{4}u^4 + C = \tfrac{1}{4}(1 - x^{-1})^4 + C$$

so that

$$\int_1^2 (1 - x^{-1})^3 x^{-2} dx = \frac{1}{4}(1 - x^{-1})^4 \Big|_1^2$$

$$= \left[\frac{1}{4}\left(1 - \frac{1}{2}\right)^4\right] - \left[\frac{1}{4}(1 - 1)^4\right] = \frac{1}{64}$$

SOLUTION 2. Let

$$u = 1 - x^{-1}, \quad du = x^{-2}dx$$

We note that the limits of the definite integral for values of x are $x = 1$ and $x = 2$. We find the corresponding values of u:

$$\text{when} \quad x = 1, \qquad u = 1 - 1 = 0$$

$$\text{when} \quad x = 2, \qquad u = 1 - \frac{1}{2} = \frac{1}{2}$$

Therefore, by substitution, we have

$$\int_1^2 (1 - x^{-1})^3 x^{-2} dx = \int_0^{1/2} u^3 du$$

$$= \frac{1}{4}u^4 \Big|_0^{1/2} = \left[\frac{1}{4}\left(\frac{1}{2}\right)^4\right] - \left[\frac{1}{4}(0)^4\right] = \frac{1}{64}$$

EXERCISES 5.4

In Problems 1 through 32, use the method of substitution of variables to evaluate the given integral.

1. $\int (x + 4)^2 dx$
2. $\int (x - 5)^6 dx$
3. $\int 3\sqrt{3x + 4}\, dx$
4. $\int \frac{1}{\sqrt{3x + 4}} dx$
5. $\int (4 - x)^3 dx$
6. $\int \left(\frac{x - 3}{2}\right)^2 dx$
7. $\int (4x - 1)^{-2} dx$
8. $\int (2x + 3)(x^2 + 3x - 1)^{-3} dx$
9. $\int x^4 (4x^5 + 6)^{100} dx$
10. $\int (x + 1)(x^2 + 2x - 5)^3 dx$
11. $\int x\sqrt{x^2 - 4}\, dx$
12. $\int x\sqrt{4 - x^2}\, dx$
13. $\int x^{-2}\sqrt{1 - x^{-1}}\, dx$
14. $\int x^{2/3}\sqrt{2 + x^{5/3}}\, dx$
15. $\int e^{7x} dx$
16. $\int e^{-3x+4} dx$
17. $\int xe^{x^2} dx$
18. $\int 3xe^{5x^2} dx$
19. $\int (e^x + e^{-x}) dx$
20. $\int \frac{2}{x^2} e^{1/x} dx$
21. $\int \frac{3}{x^2} e^{-1/x} dx$
22. $\int \frac{1}{x + 2} dx$
23. $\int \left(e^{-x} + \frac{1}{2x + 3}\right) dx$
24. $\int \frac{e^{\sqrt{x}}}{\sqrt{x}} dx$

25. $\int \frac{e^x}{1+e^x}\,dx$

26. $\int \frac{e^x - e^{-x}}{e^x + e^{-x}}\,dx$

27. $\int \frac{\ln x}{x}\,dx$

28. $\int \frac{x}{x^2+4}\,dx$

29. $\int (3x+5)\sqrt{x+1}\,dx$

30. $\int (x^2+3)\sqrt{2x-1}\,dx$

31. $\int \frac{x^2+x+1}{x+1}\,dx$

32. $\int \frac{x^3+5x^2+10x-8}{x+2}\,dx$

In Problems 33 through 42, evaluate the definite integral.

33. $\int_0^1 (3x-1)\,dx$

34. $\int_{-2}^1 \sqrt{1-x}\,dx$

35. $\int_3^7 (x-1)^{-1/2}\,dx$

36. $\int_{-1}^0 (1-3x)^{3/2}\,dx$

37. $\int_0^1 x(x^2+1)^{-2}\,dx$

38. $\int_0^1 x(1-x^2)^{-1/2}\,dx$

39. $\int_0^1 xe^{x^2}\,dx$

40. $\int_1^2 \frac{(\ln x)^2}{x}\,dx$

41. $\int_2^3 \frac{x^2-x-1}{x-1}\,dx$

42. $\int_0^{e-1} \frac{6x^2+3x-2}{x+1}\,dx$

5.5 INTEGRATION BY PARTS

Suppose f and g are differentiable functions of x. Then, by the rule of the derivative of a product, we have

$$(f\cdot g)' = f\cdot g' + f'\cdot g \tag{1}$$

Integrating both sides of Equation (1) with respect to x, we obtain

$$\int \frac{d}{dx}[f(x)\,g(x)]\,dx = \int [f(x)\cdot g'(x) + f'(x)\,g(x)]\,dx \tag{2}$$

or

$$f(x)\,g(x) = \int f(x)\,g'(x)\,dx + \int g(x)\,f'(x)\,dx \tag{3}$$

From this, we have

$$\int f(x)\,g'(x)\,dx = f(x)\,g(x) - \int g(x)\,f'(x)\,dx \tag{4}$$

Formula (4) is called *integration by parts.*

For the definite integral, we have

$$\int_a^b f(x)\,g'(x)\,dx = f(x)\,g(x)\Big|_a^b - \int_a^b g(x)f'(x)\,dx \tag{5}$$

At first glance, it might appear that we are complicating the problem. However, integration by parts facilitates the evaluation of antiderivatives of a large class of functions.

EXAMPLE 1. Evaluate $\int x\sqrt{x-1}\,dx$.

SOLUTION. There are two choices we can make for f and g'. First we let

$$f(x) = \sqrt{x-1} \quad \text{and} \quad g'(x) = x$$

then

$$f'(x) = \tfrac{1}{2}(x-1)^{-1/2} \quad \text{and} \quad g(x) = \tfrac{1}{2}x^2$$

and integration by parts yields

$$\int x\sqrt{x-1}\;dx = \tfrac{1}{2}x^2\sqrt{x-1} - \int \tfrac{1}{4}x^2(x-1)^{-1/2}\,dx$$

We note that the problem is getting more complicated. Therefore, this choice of f and g' is discarded.

A second choice is

$$f(x) = x \qquad \text{and} \qquad g'(x) = \sqrt{x-1}$$

Then

$$f'(x) = 1 \qquad \text{and} \qquad g(x) = \tfrac{2}{3}(x-1)^{3/2}$$

Substitution in formula (4) yields

$$\begin{aligned}\int x\sqrt{x-1}\;dx &= \tfrac{2}{3}x(x-1)^{3/2} - \int \tfrac{2}{3}(x-1)^{3/2}dx\\ &= \tfrac{2}{3}x(x-1)^{3/2} - \tfrac{2}{3}\cdot\tfrac{2}{5}(x-1)^{5/2} + C\\ &= \tfrac{2}{3}x(x-1)^{3/2} - \tfrac{4}{15}(x-1)^{5/2} + C\end{aligned}$$

EXAMPLE 2. Evaluate $\int_0^1 x^3\sqrt{1+x^2}\;dx$.

SOLUTION. Let

$$f(x) = x^2 \qquad \text{and} \qquad g'(x) = x\sqrt{1+x^2}$$

Then

$$f'(x) = 2x \qquad \text{and} \qquad g(x) = \frac{1}{3}(1 + x^2)^{3/2}$$

Therefore,

$$\begin{aligned}
\int_0^1 x^3\sqrt{1+x^2}\,dx &= \int_0^1 x^2(x\sqrt{1+x^2}\,)dx \\
&= \frac{1}{3}x^2(1+x^2)^{3/2}\Big|_0^1 - \int_0^1 \frac{2}{3}x(1+x^2)^{3/2}dx \\
&= \frac{1}{3}x^2(1+x^2)^{3/2}\Big|_0^1 - \frac{1}{3}\cdot\frac{2}{5}(1+x^2)^{5/2}\Big|_0^1 \\
&= \left[\frac{2\sqrt{2}}{3} - 0\right] - \frac{2}{15}[4\sqrt{2} - 1] = \frac{2}{15}(\sqrt{2} + 1)
\end{aligned}$$

We have used intergration by substitution whenever needed.

EXAMPLE 3. Find $\int x^2e^x\,dx$.

SOLUTION. We use integration by parts. Set

$$f(x) = x^2, \qquad g'(x) = e^x, \qquad \text{and} \qquad f'(x) = 2x, \qquad g(x) = e^x$$

Using formula (4), we have

$$\int x^2e^x\,dx = x^2e^x - \int e^x 2x\,dx = x^2e^x - 2\int xe^x\,dx$$

We find $\int xe^x\,dx$ by parts. Again, setting

$$f(x) = x, \qquad g'(x) = e^x \qquad \text{and} \qquad f'(x) = 1, \qquad g(x) = e^x$$

and using (4), we have

$$\int xe^x\,dx = xe^x - \int e^x\,dx = xe^x - e^x + C_1$$

Consequently,

$$\begin{aligned}
\int x^2e^x\,dx &= x^2e^x - 2\int xe^x\,dx = x^2e^x - 2(xe^x - e^x) + C \\
&= x^2e^x - 2xe^x + 2e^x + C
\end{aligned}$$

EXAMPLE 4. Find $\int \ln x \, dx$.

SOLUTION. Setting

$$f(x) = \ln x \quad \text{and} \quad g'(x) = 1, \quad \text{then} \quad f'(x) = \frac{1}{x} \quad \text{and} \quad g(x) = x$$

Using (4), we have

$$\int \ln x \, dx = x \ln x - \int x \cdot \frac{1}{x} \, dx = x \ln x - \int dx = x \ln x - x + C$$

EXERCISES 5.5

In Problems 1 through 20, use integration by parts to find the integral.

1. $\int x(2x-1)\,dx$
2. $\int x\sqrt{1+x}\,dx$
3. $\int x^3(1+x^2)^{3/2}\,dx$
4. $\int_0^8 x^2(1+x)^{-1/2}\,dx$
5. $\int xe^x\,dx$
6. $\int x^3e^x\,dx$
7. $\int x\ln x\,dx$
8. $\int x^2\ln x\,dx$
9. $\int_2^4 5x^4\ln x\,dx$
10. $\int \sqrt{x}\ln x\,dx$
11. $\int (1-x)e^x\,dx$
12. $\int (2+x^2)e^x\,dx$
13. $\int xe^{2x}\,dx$
14. $\int 3xe^{-2x+4}\,dx$
15. $\int \ln(2x+3)\,dx$
16. $\int 2x\ln(4x-3)\,dx$
17. $\int (1+x)e^{-x}\,dx$
18. $\int (1-x^2)e^{-x}\,dx$
19. $\int \frac{\ln(\ln x)}{x}\,dx$
20. $\int e^x\ln e^{-x}\,dx$
21. Find the area of the region bounded by $y = xe^x$, the x-axis, and the lines $x = 0$ and $x = 3$.
22. Find the area of the region bounded by $y = x\ln x$, the x-axis, and the lines $x = 1$ and $x = 5$.

5.6 INTEGRATION BY PARTIAL FRACTIONS

In this section we present a method for finding the antiderivative of a function of the form

$$\int \frac{P(x)}{Q(x)}\,dx$$

where P and Q are polynomial functions, and P is of a degree less[1] than the degree of Q. The method is called *integration by partial fractions*, and the procedure depends on Q being factorable. We shall consider the cases in which $Q(x)$ can be expressed as a product of linear and/or quadratic equations.

CASE 1. The function Q can be written as a product of linear functions

$$Q(x) = a(x - a_1)(x - a_2)\ldots(x - a_n)$$

in which $a_i \neq a_j$ for $i \neq j$. In this case we can write

$$\frac{P(x)}{Q(x)} = \frac{A_1}{x - a_1} + \frac{A_2}{x - a_2} + \cdots + \frac{A_n}{x - a_n} \tag{1}$$

where A_i $(i = 1, 2, \ldots, n)$ are constants that can be determined.

EXAMPLE 1. Expresss the following function as a sum of partial fractions as in (1) and find its antiderivative:

$$\frac{10x - 4}{x^3 - 4x}, \quad x \neq 0, \pm 2$$

SOLUTION. We write

$$\frac{10x - 4}{x^3 - 4x} = \frac{10x - 4}{x(x - 2)(x + 2)}$$

$$= \frac{A}{x} + \frac{B}{x - 2} + \frac{C}{x + 2} \tag{2}$$

Multiplying equation (2) by $x(x - 2)(x + 2)$ we obtain

$$10x - 4 = A(x - 2)(x + 2) + Bx(x + 2) + Cx(x - 2)$$

---

[1] Remark: In case P has degree greater than or equal to the degree of Q, one simply divides P by Q. Thus, there is no loss of generality in assuming that the degree of P is less than the degree of Q.

or

$$10x - 4 = (A + B + C)x^2 + 2(B - C)x - 4A \tag{3}$$

The polynomial on the left of equation (3) is identical to the polynomial on the right if and only if every coefficient on the left is equal to the corresponding coefficient on the right. Thus, we have

$$A + B + C = 0$$

$$2(B - C) = 10$$

$$-4A = -4$$

These equations can be solved for A, B, and C. An alternate method for finding A, B, and C is as follows. We evaluate equation (3) at the roots of $Q(x)$:

For $x = 0$, $\quad -4 = -4A \quad$ or $\quad A = 1$

For $x = 2$, $\quad 16 = 8B \quad$ or $\quad B = 2$

For $x = -2$, $\quad -24 = 8C \quad$ or $\quad C = -3$

Hence, (2) may be written as

$$\frac{10x - 4}{x^3 - 4x} = \frac{1}{x} + \frac{2}{x - 2} - \frac{3}{x + 2} \tag{4}$$

Therefore, integration with respect to x on both sides of equation (4) yields

$$\begin{aligned}\int \frac{10x - 4}{x^3 - 4x}\,dx &= \int\left(\frac{1}{x} + \frac{2}{x - 2} - \frac{3}{x + 2}\right)dx \\ &= \int \frac{1}{x}\,dx + 2\int \frac{1}{x - 2}\,dx - 3\int \frac{1}{x + 2}\,dx \\ &= \ln|x| + 2\ln|x - 2| - 3\ln|x + 2| + C\end{aligned}$$

CASE 2. The function Q can be written as a product of linear functions

$$Q(x) = a(x - a_1)(x - a_2)\cdots(x - a_n)$$

in which two or more a_j's are equal. The following are examples involving case 2:

(1) $$\frac{P(x)}{(x-a)^3} = \frac{A}{x-a} + \frac{B}{(x-a)^2} + \frac{C}{(x-a)^3}$$

(2) $$\frac{P(x)}{(x-b)(x-a)^4} = \frac{A}{x-b} + \frac{B}{x-a} + \frac{C}{(x-a)^2} + \frac{D}{(x-a)^3} + \frac{E}{(x-a)^4}$$

(3) $$\frac{P(x)}{(x-a)^2(x-b)^3} = \frac{A}{x-a} + \frac{B}{(x-a)^2} + \frac{C}{x-b} + \frac{D}{(x-b)^2} + \frac{E}{(x-b)^3}$$

EXAMPLE 2. Write the following expression as a sum of partial fractions and find its antiderivative:

$$\frac{x+4}{x^3+x^2-5x+3}, \qquad x \neq 1, -3$$

SOLUTION. In this case we have

$$\frac{x+4}{x^3+x^2-5x+3} = \frac{x+4}{(x+3)(x-1)^2}$$

$$= \frac{A}{x+3} + \frac{B}{x-1} + \frac{C}{(x-1)^2} \tag{5}$$

Multiplying equation (5) by $(x+3)(x-1)^2$, we obtain

$$x+4 = A(x-1)^2 + B(x+3)(x-1) + C(x+3) \tag{6}$$

Setting $x = 1$ we have

$$5 = 4C \qquad \text{or} \qquad C = \frac{5}{4}$$

setting $x = -3$ we have

$$1 = 16A \qquad \text{or} \qquad A = \frac{1}{16}$$

We still need to evaluate B. Since A and C have been determined, choose any number for x and solve for B. To simplify the evaluation, we set $x = 0$ in equation (6) and use the determined values for A and C to obtain

$$4 = \frac{1}{16} - 3B + \frac{15}{4} \qquad \text{or} \qquad B = -\frac{1}{16}$$

Hence, equation (5) may be written in the form

$$\frac{x+4}{x^3+x^2-5x+3} = \frac{1}{16(x+3)} - \frac{1}{16(x-1)} + \frac{5}{4(x-1)^2} \tag{7}$$

Therefore,

$$\begin{aligned}\int \frac{x+4}{x^3+x^2-5x+3}\,dx &= \int \left[\frac{1}{16(x+3)} - \frac{1}{16(x-1)} + \frac{5}{4(x-1)^2}\right]dx \\ &= \frac{1}{16}\left[\int \frac{1}{x+3}\,dx - \int \frac{1}{x-1}\,dx + 20\int \frac{1}{(x-1)^2}\,dx\right] \\ &= \frac{1}{16}\left[\ln|x+3| - \ln|x-1| - 20\frac{1}{x-1}\right] + C\end{aligned}$$

CASE 3. The function Q can be written as a product of linear and quadratic functions (with no real roots)

$$Q(x) = (x-a)\cdot(x^2+bx+c)$$

In this case we write:

$$\frac{P(x)}{(x-a)(x^2+bx+c)} = \frac{A}{x-a} + \frac{Bx+C}{x^2+bx+c}$$

EXAMPLE 3. Write the following expression as a sum of partial fractions and find its antiderivative:

$$\frac{3x^2-8x+1}{(x-4)(x^2+1)}, \qquad x \neq 4$$

SOLUTION. We have

$$\frac{3x^2-8x+1}{(x-4)(x^2+1)} = \frac{A}{x-4} + \frac{Bx+C}{x^2+1} \tag{8}$$

Multiplying Equation (7) by $(x-4)(x^2+1)$, we obtain

$$3x^2-8x+1 = A(x^2+1) + (Bx+C)(x-4)$$

or

$$3x^2-8x+1 = A(x^2+1) + Bx(x-4) + C(x-4) \tag{9}$$

Setting $x = 4$ we obtain

$$17 = 17A \qquad \text{or} \qquad A = 1$$

setting $x = 0$ we obtain, $1 = A - 4C$. Since, $A = 1$, we have

$$1 = 1 - 4C \qquad \text{or} \qquad C = 0$$

setting $x = 1$ we obtain, $-4 = 2A - 3B - 3C$. Since $A = 1$, $C = 0$, we have

$$-4 = 2 - 3B \qquad \text{or} \qquad B = 2$$

Hence,

$$\frac{3x^2 - 8x + 1}{(x-4)(x^2+1)} = \frac{1}{x-4} + \frac{2x}{x^2+1}$$

Intergration yields

$$\int \frac{3x^2 - 8x + 1}{(x-4)(x^2+1)}\,dx = \int \left[\frac{1}{x-4} + \frac{2x}{x^2+1}\right]dx$$

$$= \int \frac{1}{x-4}\,dx + \int \frac{2x}{x^2+1}\,dx$$

$$= \ln|x-4| + \ln(x^2+1) + C$$

EXERCISES 5.6

In Problems 1 through 14, evaluate:

1. $\displaystyle\int \frac{5x-3}{x^2-2x-3}\,dx$

2. $\displaystyle\int \frac{7x-13}{(x+1)(x-3)}\,dx$

3. $\displaystyle\int \frac{6x^2-11x-14}{(x-3)(x-2)(x+4)}\,dx$

4. $\displaystyle\int \frac{2x-3}{x^2-8x+16}\,dx$

5. $\displaystyle\int \frac{x^2+2x+5}{x(x+1)(x-1)}\,dx$

6. $\displaystyle\int \frac{x+3}{(x-1)^2(x+2)}\,dx$

7. $\displaystyle\int \frac{4-2x}{(x^2-1)(x-1)}\,dx$

8. $\displaystyle\int \frac{x^4+1}{x^3-x}\,dx$

9. $\displaystyle\int \frac{x+4}{(x+1)^2(x-2)}\,dx$

10. $\displaystyle\int \frac{2x-3}{(x+2)(x-1)^3}\,dx$

11. $\displaystyle\int \frac{x^3+5}{(x^2-4)^2}\,dx$

12. $\displaystyle\int \frac{2x^2}{x^4-5x^2+4}\,dx$

13. $\int \frac{3x^2 - 1}{(x-1)^2(x+3)}dx$

14. $\int \frac{x^3 + x^2 - 2x - 5}{(x+1)^2(x-2)^2}dx$

5.7 TABLES OF INTEGRALS

When we are confronted with an indefinite integral such as

$$I_1 = \int \frac{dx}{x\sqrt{x^2+5}}, \tag{1}$$

there is little hope of finding the antiderivative with the techniques we have learned so far. Certainly this integral does not fit any of our standard formulas, and a few tries with integration by parts will soon leave us discouraged.

Of course, there are other tricks that have not been covered in this text; but with or without these complicated tricks, we can always look in a table of integrals (if one is available). In Table E in the appendix, we give a very brief table of integrals; and more extensive tables are available in any reasonably good mathematics library. The student should be warned that some integrals will lead to trigonometric functions and inverse trigonometric functions; and since these functions are not treated in this book, the corresponding integrals are not listed in Table E.

To use a table of integrals with any degree of efficiency, we must know where to look. However, the rich variety of functions makes it impossible to list them all in a systematic way (we can't put them in alphabetical order). If we have a small table, we can make some order out of chaos. In Table E, the elementary formulas are given first. Then the classification of the other integrals follows the pattern in Table 1.

Table 1

	Integrand involves	Use formulas
1.	$a + bu$	11—18
2.	$\sqrt{a+bu}$	19—25
3.	$\sqrt{a^2-u^2}$	26—29
4.	$\sqrt{u^2 \pm a^2}$	30—38
5.	Exponential and natural logarithm functions	39—46

EXAMPLE 1. Find I_1 [see Equation (1)].

SOLUTION. Since the integral involves $\sqrt{x^2+5}$, this is item 4 in Table 1. Searching among formulas 30 through 38, we soon match our integral with formula number 34 in Table E, with $a^2 = 5$ or $a = \sqrt{5}$. Hence,

$$I_1 = \int \frac{dx}{x\sqrt{x^2+5}} = -\frac{1}{\sqrt{5}} \ln \frac{\sqrt{5}+\sqrt{x^2+5}}{x} + C \tag{2}$$

EXAMPLE 2. Find $I_2 = \int \frac{15u^2du}{(3+2u)^2}$.

SOLUTION. Except for the factor 15, this fits formula 13 in Table E, with $a = 3$ and $b = 2$. Therefore,

$$I_2 = \frac{15}{8}\left(3 + 2u - \frac{9}{3+2u} - 6\ln(3+2u)\right) + C$$

EXAMPLE 3. Find $I_3 = \int \frac{dx}{(x+2)^2\sqrt{x^2+4x+29}}$.

SOLUTION. If we set $x + 2 = u$, then $dx = du$, and $x^2 + 4x + 29 = x^2 + 4x + 4 + 25 = (x+2)^2 + 5^2 = u^2 + 5^2$. Then (by formula 35 of Table E)

$$I_3 = \int \frac{du}{u^2\sqrt{u^2+5^2}} = -\frac{\sqrt{u^2+5^2}}{25u} + C$$

$$= -\frac{\sqrt{(x+2)^2+25}}{25(x+2)} + C$$

In each of these problems, it is easy to check the work. We merely differentiate the answer, and, if no errors have been made, then the derivative will be the integrand. Similarly, we can check any indefinite integral formula that may look suspicious in Table E. If we differentiate the right side, we should obtain the integrand on the left side of the formula.

EXERCISES 5.7

In each of the following problems use the proper formula from Table E (or any other table of integrals) to find the indefinite integral.

1. $\int \frac{x\,dx}{3+8x}$
2. $\int \frac{y\,dy}{2+9y}$
3. $\int \frac{8\,dz}{z(2+15z)}$
4. $\int \frac{x\,dx}{(x+4)^2}$
5. $\int \frac{y\,dy}{(y-6)^3}$
6. $\int \frac{r^2\,dr}{(5r+3)^2}$
7. $\int \frac{ds}{(9-s^2)^{3/2}}$
8. $\int \frac{dx}{x\sqrt{1+5x}}$
9. $\int \frac{dx}{\sqrt{x^2+6x+13}}$
10. $\int \frac{dx}{\sqrt{x^2+6x+5}}$
11. $\int \frac{\sqrt{x^2+4x+5}}{(x+2)^2}\,dx$
12. $\int \frac{dx}{x^2(3+2x)}$

*13. $\int 48ue^{6u}\,du$

*14. $\int \frac{dv}{v\ln v}$

5.8 NUMERICAL INTEGRATION

In Section 4, we indicated that we cannot find an antiderivative of $\sqrt{1+x^3}$. There are many functions for which we cannot find an antiderivative. Thus, for such functions, we are unable to evaluate the corresponding definite integrals. When we cannot find the exact value, we opt for an approximate value. Many methods of approximating definite integrals are feasible now because of the availability of calculators and computers. A discussion of three of these methods follows.

1. DEFINITION.

We recall that if f is continuous on $[a, b]$, then

$$\int_a^b f(x)dx = \lim_{n\to\infty} S_n$$

where S_n is the Riemann Sum (see Section 2). We can then write

$$\int_a^b f(x)dx \approx h[f(a+h)+f(a+2h)+\cdots+f(a+nh)] \tag{1}$$

where $h = (b-a)/n$ and n is a positive integer to be chosen. We note that the larger the n chosen, the better the approximation we obtain.

EXAMPLE 1. Use $n = 10$ to approximate $\int_5^{10} \frac{1}{x} dx$.

SOLUTION. We have $a = 5$, $b = 10$, $n = 10$, so that $h = \dfrac{10-5}{10} = 0.5$.

We make the following table:

i	ih	$a+ih$	$f(a+ih) = \dfrac{1}{a+ih}$	$f(a+ih)\cdot h$
1	0.5	5.5	0.181818	0.090909
2	1.0	6.0	0.166666	0.083333
3	1.5	6.5	0.153846	0.076923
4	2.0	7.0	0.142857	0.071428
5	2.5	7.5	0.133333	0.066666
6	3.0	8.0	0.125	0.0625
7	3.5	8.5	0.117647	0.058823
8	4.0	9.0	0.111111	0.055555
9	4.5	9.5	0.105263	0.052631
10	5.0	10.0	0.1	0.05

By adding all the entries of the last column, we have

$$\int_5^{10} \frac{1}{x} dx \approx 0.668868$$

2. TRAPEZOIDAL RULE

The idea behind the approximation of the definite integral by definition was to approximate the area under the graph of f from $x = a$ to $x = b$ by the area of rectangles of equal width and adding up these areas. Now, instead of rectangles, we use trapezoids. Recall that the area of a trapezoid is equal to one-half the product of the sum of the bases and the altitude. As before, if we divide the interval $[a, b]$ into n equal parts, then $h = (b - a)/n$ and the subdivision points are:

$$x_0 = a,\ x_1 = a + h,\ x_2 = a + 2h, \ldots,\ x_n = a + nh = b$$

In Figure 1, we see that the area of the first trapezoid is $\frac{1}{2}h[f(x_0) + f(x_1)]$, with similar results for the remaining trapezoids.

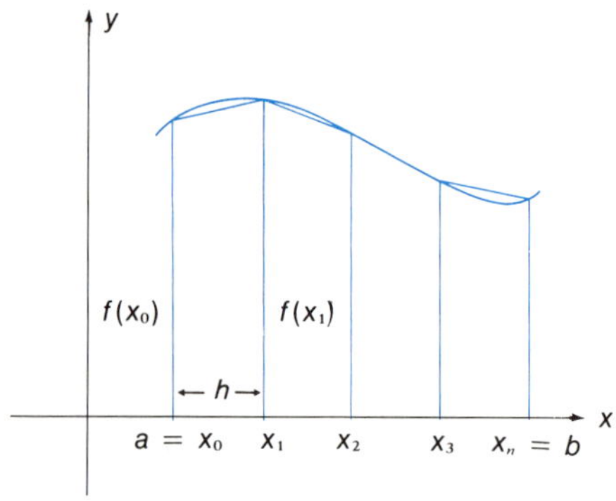

Figure 1

The total area of all the trapezoids is an approximation of the area under the curve and, hence, of the definite integral. We consequently have the *trapezoidal rule*:

$$\int_a^b f(x)dx \approx \frac{h}{2}[f(x_0) + 2f(x_1) + 2f(x_2) + \cdots + 2f(x_{n-1}) + f(x_n)]$$

where $h = \dfrac{b-a}{n}$.

EXAMPLE 2. Use $n = 10$ and the trapeziodal rule to approximate $\int_5^{10} \frac{1}{x} dx$.

SOLUTION. We have $f(x_0) = f(5) = \frac{1}{5} = 0.2$. Since $h = (10 - 5)/10 = 0.5$,

$$x_0 = 5,\ x_1 = 5.5,\ x_2 = 6.0,\ x_3 = 6.5,\ \ldots,\ x_{10} = 10$$

We can make use of the appropriate entries in the table in Example 1 and use the trapeziodal rule to obtain

$$\int_5^{10} \frac{1}{x} dx \approx \frac{0.5}{2} [f(5) + 2f(5.5) + 2f(6) + \cdots + 2f(9.5) + f(10)]$$

$$= 0.69383$$

3. SIMPSON'S RULE

Another method for numerical integration is called *Simpson's Rule*. This rule relies on the fact that given three noncollinear points, there is exactly one parabola which passes through them. We therefore approximate consecutive positions of the curve with parabolas, rather than with line segments as in the trapezoidal rule. As shown in Figure 2, the first parabola passes through the points $A\,(a, f(a))$, B, and C, and the next parabola passes through the points C, D, and E. In order to get these groupings of three points, we must have an *even* number of points on the graph in addition to the initial point $(a, f(a))$. Thus, to use Simpson's rule, we must always choose n to be even. We now state Simpson's rule:

Let n be even, and $h = \dfrac{b-a}{n}$, then

$$\int_a^b f(x)dx \approx \frac{h}{3}[f(x_0) + 4f(x_1) + 2f(x_2) + 4f(x_3) + \cdots + 2f(x_{n-2}) + 4f(x_{n-1}) + f(x_n)]$$

Note that the pattern of coefficients inside the brackets is 1, 4, 2, 4, 2, . . ., 2, 4, 1.

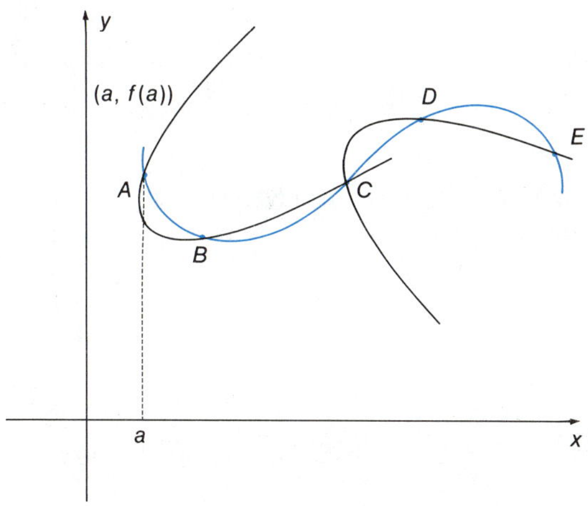

Figure 2

EXAMPLE 3. Use $n = 10$ and Simpson's rule to approximate $\int_5^{10} \frac{1}{x}\,dx$.

SOLUTION. We have $h = (10 - 5)/10 = 0.5$. The terms to be added are:

$$\begin{aligned}
f(x_0) &= \tfrac{1}{5} && = 0.2\\
4f(x_1) &= 4f(5.5) = 4(\tfrac{1}{5.5}) && = 0.72724\\
2f(x_2) &= 2f(6) && = 0.33332\\
4f(x_3) &= 4f(6.5) && = 0.61536\\
2f(x_4) &= 2f(7) && = 0.28571\\
4f(x_5) &= 4f(7.5) && = 0.53332\\
2f(x_6) &= 2f(8) && = 0.25\\
4f(x_7) &= 4f(8.5) && = 0.46980\\
2f(x_8) &= 2f(9) && = 0.22222\\
4f(x_9) &= 4f(9.5) && = 0.42104\\
f(x_{10}) &= f(10) && = 0.1\\
&\text{Sum} && = 4.15801
\end{aligned}$$

Thus, by Simpson's rule,

$$\int_5^{10} \frac{1}{x}\,dx \approx \frac{0.5}{3}(4.415801) = 0.69300$$

However, by the fundamental theorem of calculus

$$\int_5^{10} \frac{1}{x}\,dx = \ln x\Big|_5^{10} = \ln 10 - \ln 5 = \ln \frac{10}{5} = \ln 2 = 0.69315$$

EXERCISES 5.8

In Problems 1 through 10, approximate the given integral with $n = 6$ by (a) definition, (b) trapezoidal rule, and (c) Simpson's rule. Compare your results with those obtained from the fundamental theorem of calculus.

1. $\int_5^7 (x^3 + x + 2)dx$

2. $\int_5^{10} \frac{1}{x+1}\,dx$

3. $\int_2^5 xe^x dx$

4. $\int_1^{10} (\sqrt{x} + 3)dx$

5. $\int_7^{10} \frac{x+1}{x-2}\,dx$

6. $\int_3^5 2xe^{x^2} dx$

7. $\int_{10}^{20} \frac{\ln x}{x}\,dx$

8. $\int_5^7 \frac{x^2+2x+2}{x^2+2x+1}\,dx$

9. $\int_{10}^{15} \frac{\sqrt{x}+1}{\sqrt{x}}\,dx$

10. $\int_1^5 \ln x\,dx$

11. Approximate the definite integral $\int_0^1 \frac{1}{x^3+1}\,dx$ by the trapezoidal rule with $n = 8$.

12. Approximate $\int_0^1 \sqrt{1+x^3}\,dx$ by Simpson's rule with $n = 8$.

CHAPTER 5 REVIEW

Part 1 (ORAL)

Define or discuss the following:

1. Antiderivative

2. The definite integral $\int_a^b f(x)\mathrm{dx}$

3. Riemann sum

4. Fundamental theorem of calculus

5. When $f(x) \leq 0$, what is the geometric interpretation of $\int_a^b f(x)dx$?

7. State the formula for integration by parts.

8. State the trapezoidal rule.

9. State Simpson's rule.

Part 2 (WRITTEN)

In Problems 1 through 48, evaluate the given integral.

1. $\int x^5 dx$

2. $\int x^{1/2} dx$

3. $\int \frac{1}{x^3} dx$

4. $\int 6x^5\,dx$

5. $\int_1^8 x^{2/3}\,dx$

6. $\int \frac{1}{x^n}\,dx$

7. $\int (x^5 + 3x^4 + 1)dx$

8. $\int \quad 3x + 2 + \frac{5}{x} \quad dx$

9. $\int \frac{x^3 + 3x^2 + 4}{\sqrt{x}}\,dx$

10. $\int x(x^2 + 3)\,dx$

11. $\int x(x + 1)(x - 1)dx$

12. $\int \sqrt{5x}\,dx$

13. $\int (x + 2)^3 dx$

14. $\int (2x + 3)^4 dx$

15. $\int 3\sqrt{3x - 5}\,dx$

16. $\int [(x + 1)^2 + (x - 2)^3]dx$

17. $\int_2^5 \frac{1}{(x - 1)^3}\,dx$

18. $\int \left(\sqrt{x + 1} + \frac{1}{\sqrt{x + 1}}\right) dx$

19. $\int_1^2 \left(2x + \frac{1}{x}\right) dx$

20. $\int \left(2x + \frac{1}{x}\right)^2 dx$

21. $\int \left(2x + \frac{1}{x}\right)^3 dx$

22. $\int \frac{1 + x}{x}\,dx$

23. $\int_0^1 \frac{x}{1 + x}\,dx$

24. $\int \frac{x}{1 - x}\,dx$

25. $\int \frac{1 - x}{1 + x}\,dx$

26. $\int \left(\frac{1}{1 + x} + \frac{1}{1 - x}\right) dx$

27. $\int \frac{x}{(x + 2)^2}\,dx$

28. $\int_0^1 x(1 + 2x^2)^7 dx$

29. $\int \frac{x}{\sqrt{1 - x^2}}\,dx$

30. $\int x(4 + x^2)^{4/7} dx$

31. $\int 3x^2 e^{x^3+1}\,dx$

32. $\int_1^2 xe^{-x^2-4}\,dx$

33. $\int e^{-ax}\,dx$

34. $\int \frac{dx}{3x - 4}$

35. $\int \frac{dx}{3 - 7x}$

36. $\int \frac{2x}{(5x^2 + 2)^5}\,dx$

37. $\int xe^{2x}\,dx$

38. $\int 2x \ln x\,dx$

39. $\int -\ln x\,dx$

40. $\int x^2 e^{-x}\,dx$

41. $\int (1 + x^2)e^x\, dx$

42. $\int (1 + x^2)\ln x\, dx$

43. $\int \frac{1}{x^2 - 7x + 10}\, dx$

44. $\int \frac{1}{(x - 1)(x^2 + 1)}\, dx$

45. $\int \frac{1}{(x + 2)^2(x + 4)}\, dx$

46. $\int \frac{x + 1}{x^2 - 5x + 6}\, dx$

47. $\int \frac{x^3 + x + 1}{x^2 - 3x + 2}\, dx$

48. $\int \frac{x^4 + 1}{x^2 - 2x - 15}\, dx$

In Problems 49 through 54, find the area of the region whose boundaries are given.

49. The x-axis, the curve $y = x^3$, and the line $x = 3$.

50. The x-axis, the curve $y = x^2$, and the lines $x = 1$ and $x = 3$.

51. The x-axis, the curve $y = x^2 - 4$, and the lines $x = 3$ and $x = 5$.

52. The curve $y = \ln x$, the x-axis, and the lines $x = 1$ and $x = e$.

53. The curve $y = x^2 + 2x - 8$ and the line $y = 4x$.

54. The curve $y = -x^2 + 2x + 8$ and the line $y = 2x - 1$.

In Problems 55 through 58, use $n = 6$ to evaluate the given definite integral by (a) definition, (b) trapezoidal rule, and (c) Simpson's rule.

55. $\int_0^1 \frac{1}{1 + x^2}\, dx$

56. $\int_0^1 \frac{1}{\sqrt{1 + x^2}}\, dx$

57. $\int_0^1 \sqrt{1 - x^2}\, dx$

58. $\int_1^4 \sqrt{1 + x^3}\, dx$

59. Find $y = f(x)$, if $f'(x) = 2x - 3$ and the curve passes through the point $(1, 2)$.

60. If $f''(x) = 3x^2 - 6x + 1, f'(0) = 1$, and $f(2) = 0$, find $f(x)$.

Chapter Six

Applications of the Integral

In this chapter, we shall show how the process of integration can be used to solve problems arising in several diverse fields. In the previous chapter, we touched on a few problems in economics and mathematics. For example, if the marginal cost is given, then we can determine the total cost function, and we can determine the area under the curve $y = f(x)$ between $x = a$ and $x = b$. In this chapter, we shall again concentrate on such problems together with the applications of integration to determine consumer's surplus, producer's surplus, and average value of a continuous function. We shall end the chapter with a discussion of improper integrals, which are used extensively in probability.

6.1 SOME APPLICATIONS

Many problems in business, science, and engineering involve rate of change of a function. The simplest of such problems are of the form

$$y'(t) = f(t) \tag{1}$$

Such an equation is called a *differential equation*[1]. In general, a differential equation is an equation involving a variable t (or x or s, etc.), an unknown function $y(t)$, and certain derivatives of y with respect to t. The problem is usually to determine a function $y(t)$ that satisfies the given equation. Such a function is called a *solution* of the differential equation. In many applications, equations such as (1) are usually associated with additional conditions that the function $y(t)$ is to satisfy. For instance, it may be

---

[1] We shall discuss differential equations in greater detail in the next chapter.

required that

$$y(t_0) = k \tag{2}$$

at $t = t_0$. This requirement is called an *initial condition*. Equation (1) together with (2) form an *initial-value problem*. The problem is to find a function $y(t)$ that satisfies the differential equation (1) as well as the initial condition (2). The following examples illustrate the procedure for solving such problems.

EXAMPLE 1. The Goldegg Chicken Farm produces $E(t)$ dozen eggs in t days. Its *production rate* is $E'(t)$, and this can be estimated by

$$E'(t) \approx \frac{E(t+h) - E(t)}{h}$$

If the production rate (in dozens/day) is given by

$$E'(t) = 150 + \tfrac{2}{5}t \tag{3}$$

find the following:

(a) The number of eggs produced in t days, assuming that $E(0) = 0$
(b) The number of eggs produced in one year (365 days)
(c) The average daily production.

SOLUTION.

(a) We shall use two methods for solving part (a).

First method. The production E in t days is

$$E(t) = \int E'(t)\,dt = \int \left(150 + \frac{2}{5}t\right) dt = 150t + \frac{t^2}{5} + C \tag{4}$$

The student can show that $E(t)$ satisfies Equation (3). We now satisfy the initial condition. From (4), we have

$$E(0) = (150)(0) + \frac{0^2}{5} + C = 0$$

which implies that $C = 0$. Therefore, the solution to the initial-value problem is

$$E(t) = 150t + \frac{t^2}{5} \tag{5}$$

Second method. Integrating both sides of Equation (3) from 0 to t, we have[2]

---

[2] Note: (a) The lower limit of integration is the initial value for t.
(b) The dummy variable s is used to avoid confusion with the upper limit of the integration.

$$\int_0^t E'(s)\,ds = \int_0^t \left(150 + \frac{2}{5}s\right) ds \tag{6}$$

this yields

$$E(s)\Big|_0^t = \left(150s + \frac{s^2}{5}\right)\Big|_0^t$$

$$E(t) - E(0) = \left[150t + \frac{t^2}{5}\right] - \left[(150)(0) + \frac{0^2}{5}\right]$$

or

$$E(t) = 150t + \frac{t^2}{5}$$

as before. Observe that $E(0) = 0$.

(b) To obtain the production in one year, we have

$$E(365) = (150)(365) + \frac{(365)^2}{5} = 81{,}395 \text{ dozen eggs.} \tag{7}$$

(c) The average daily production is

$$\frac{E(365)}{365} = \frac{81{,}395}{365} = 223 \text{ dozens/day}$$

EXAMPLE 2. Consider an (idealized) experiment in which a colony of live bacteria is introduced into a limited food supply. Suppose that the rate of change in the number N of live bacteria with respect to time is given by

$$N'(t) = 6000t^2 - 75t^4 \tag{9}$$

Find the size $N(t)$ of the population of bacteria at time t if 1000 bacteria were introduced into the food supply initially.

SOLUTION. We have an initial-value problem involving the differential equation (9) with initial condition

$$N(0) = 1000 \tag{10}$$

Integrating both sides of Equation (9), we have

$$\int_0^t N'(s)\,ds = \int_0^t (6000s^2 - 75s^4)\,ds$$

$$N(s)\Big|_0^t = (2000s^3 - 15s^5)\Big|_0^t$$

$$N(t) - N(0) = [2000t^3 - 15t^5] - [(2000)(0)^3 - (15)(0)^5]$$

that is,

$$N(t) = 2000t^3 - 15t^5 + 1000 \tag{11}$$

Note that $N(0) = 1000$. The reader is encouraged to show that the function defined by

(11) satisfies Equations (9) and (10).

A rough sketch of the graph of N is shown in Figure 1, indicating the period of

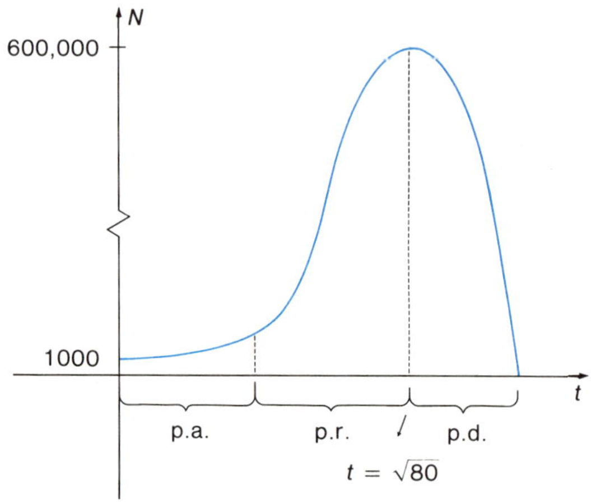

Figure 1

adaption (p.a.), period of reproduction (p.r.), and period of dying (p.d.).

We now consider the motion of an object defined by

$$s = f(t) \tag{12}$$

where s, the distance, is a function of time t. We recall that the *velocity function*, $v(t)$, is defined by

$$v(t) = \frac{ds}{dt} \tag{13}$$

which is the rate of change of the distance with respect to time. In addition, the *acceleration function*, $v(t)$, is defined by

$$a(t) = \frac{dv}{dt} = \frac{d^2s}{dt^2} \tag{14}$$

which is the rate of change of the velocity with respect to time.

EXAMPLE 3. An object on the ground is projected vertically with initial velocity of 96 ft./sec. If the acceleration $a(t) = -32$ ft./sec^2, find the following:

(a) The velocity function
(b) The distance at time t
(c) The maximum height the object will obtain
(d) The height of the object in 5 seconds

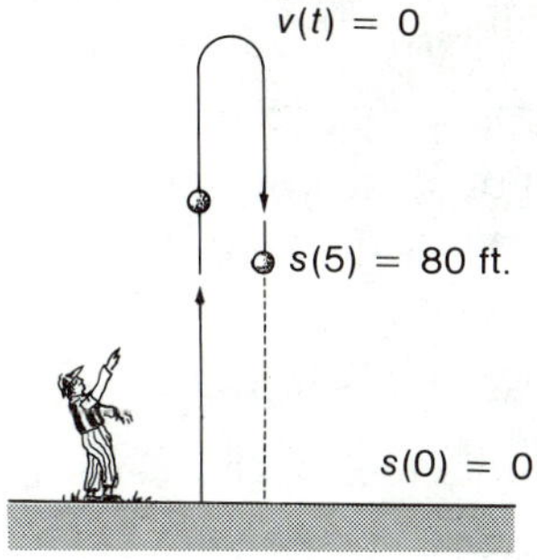

Figure 2

SOLUTION.

(a) Here we are dealing with the initial-value problem

$$v'(t) = -32 \tag{15}$$

$$v(0) = 96 \tag{16}$$

Integrating both sides of Equation (15), we obtain

$$\int_0^t v'(r)\,dr = -32\int_0^t dr \quad \text{or} \quad v(r)\Big|_0^t = -32r\Big|_0^t \tag{17}$$

that is,

$$v(t) - v(0) = -32[t-0]$$

Since $v(0) = 96$, we have

$$v(t) = -32t + 96 \tag{18}$$

(b) We observe that the object is at ground level at $t = 0$. The initial condition is $s(0) = 0$. Since

$$s'(t) = v(t) = -32t + 96$$

we have

$$\int_0^t s'(r)\,dr = \int_0^t (-32r + 96)\,dr$$

$$s(r)\Big|_0^t = (-16r^2 + 96r)\Big|_0^t$$

$$s(t) - s(0) = [-16t^2 + 96t] - [(-16)(0)^2 + (96)(0)]$$

Since $s(0) = 0$, we have

$$s(t) = -16t^2 + 96t \tag{19}$$

(c) The object will continue to move as long as $v(t) \neq 0$, and it will be at instantaneous rest when $v(t) = 0$. Therefore, from Equation (18), we have

$$v(t) = -32t + 96 = 0$$

which is satisfied when $t = \frac{96}{32} = 3$. In other words, 3 seconds after the object is projected upward, it comes to instantaneous rest and then begins its journey downward. To find the distance traveled by the object in 3 seconds, we use Equation (19) to obtain

$$s(3) = (-16)(3)^2 + (96)(3) = 144$$

that is, the object will rise to a height of 144 ft. above the ground.

(d) From Equation (19), we have

$$s(5) = (-16)(5)^2 + (96)(5) = 80$$

In 5 seconds the object will be 80 ft. above ground level. We observe that $s(6) = 0$. Hence, the object will be back on the ground in 6 seconds, and, for $t > 6$, $v(t) \equiv 0$, and $s(t) \equiv 0$. (Why?)

EXAMPLE 4. A water supply tank with a capacity of 1500 cu. ft. develops a leak. Suppose that the rate at which the water is flowing out of the tank is given by

$$V'(t) = -\tfrac{1}{30}t \tag{20}$$

where V, the volume of water, is a function of time t measured in minutes. The negative sign indicates that the volume is decreasing.

(a) Find the volume of water in the tank at time t.
(b) If the tank develops the leak at 9:05 P.M., at what time will the tank be empty?

SOLUTION.

(a) We assume that time is measured starting at 9:05 P.M. Hence, $V = 1500$ at $t = 0$, and we have the initial-value problem

$$V'(t) = -\tfrac{1}{30}t, \qquad V(0) = 1500$$

Then

$$\int_0^t V'(s)\,ds = -\frac{1}{30}\int_0^t s\,ds$$

which yields

$$V(s)\Big|_0^t = -\frac{1}{30}\frac{s^2}{2}\Big|_0^t$$

$$V(t) - V(0) = -\frac{1}{60}t^2$$

$$V(t) = -\frac{1}{60}t^2 + 1500 \tag{21}$$

(b) When the tank is empty, $V = 0$. From Equation (21), we have

$$-\tfrac{1}{30}t^2 + 1500 = 0$$

which implies that $t = \pm 300$. We disregard $t = -300$.

Therefore,

$$t = 300$$

This result means that the tank will be empty 300 minutes after the leak develops — at precisely 2:05 A.M.

It is important to note that the volume function defined by (21) has domain $\{t \mid 0 \leq t \leq 300 \text{ minutes}\}$, since we cannot have negative volume. For $t = 300$, we define $V(t) = 0$.

The definite integral can also be applied to problems that involve a sum of terms. Since the definite integral is the limit of a sum, we can often simplify computation of a sum by using integration. When we replace a sum by an integral, we make a small error, but quite often this process is sufficient to have an approximate value for the sum.

EXAMPLE 5. Use the definite integral to approximate:

$$S = \sqrt{1} + \sqrt{2} + \sqrt{3} + \sqrt{4} + \ldots + \sqrt{99} + \sqrt{100} \tag{22}$$

SOLUTION. Let us consider the function $f(x) = \sqrt{x}$. If we set $a = 0$, $b = 100$, and $n = 100$, then

$$h = \frac{b-a}{n} = \frac{100-0}{100} = 1$$

Since

$$\int_a^b f(x)\,dx \approx h\,[f(x_1) + f(x_2) + \ldots + f(x_n)]$$

we have

$$\int_0^{100} \sqrt{x}\,dx \approx 1[\sqrt{1} + \sqrt{2} + \sqrt{3} + \ldots + \sqrt{100}] \tag{23}$$

Now,

$$\int_0^{100} \sqrt{x}\,dx = \frac{2}{3}x^{3/2}\Big|_0^{100} = \frac{2}{3}(100)^{3/2} = \frac{2}{3}(1000) = \frac{2000}{3} = 666.67$$

Therefore, the sum S in (22) is approximately equal to 666.67.

EXAMPLE 6. For Mongoose Motor Co. the cost of producing the xth car is $3000 + 1000/(1 + x)$ in dollars. How much does it cost (approximately) to produce the first 400 cars? Find the cost of producing the first half of this quantity and the cost of producing the second half of this quantity.

SOLUTION. If $C(x)$ represents the total cost of producing x cars, then, as in Example 1, we can write the sum of terms involved as a definite integral. Thus,

$$C(400) = \int_0^{400} \left(3000 + \frac{1000}{1+x}\right) dx$$

$$= 3000x + 1000 \ln(1+x) \Big|_0^{400}$$

$$= (3000)(400) + 1000 \ln(401)$$

(note that $\ln 1 = 0$)

$$= 1{,}200{,}000 + 1000(5.994)$$

$$= 1{,}200{,}000 + 5994$$

$$= 1{,}205{,}994 \text{ dollars} \tag{24}$$

which is the approximate cost to produce the first 400 cars.

$$C(200) = \int_0^{200} \left(3000 + \frac{1000}{1+x}\right) dx$$

$$= 3000x + 1000 \ln(1+x) \Big|_0^{200}$$

$$= (3000)(200) + 1000 \ln(201)$$

$$= 600{,}000 + 1000 \ln(5.303)$$

$$= 600{,}000 + 5303$$

$$= 605{,}303 \text{ dollars} \tag{25}$$

the approximate cost to produce the first 200 cars.

The cost of the second 200 cars can also be computed by integration, but here it is much simpler to subtract (25) from (24):

$$C(400) - C(200) = 600{,}691 \text{ dollars.} \tag{26}$$

We note that the first 200 cars cost more to produce than the second 200 cars.

In Chapter 4, we noted that

$$A = Pe^{rt} \tag{27}$$

is the accumulated amount of a principal P dollars after t years at an annual interest rate of $100r\%$ compounded continuously.

If we solve equation (27) for P, we get

$$P = Ae^{-rt} \tag{28}$$

Equation (28) suggests that a principal P must be invested *now* at the annual interest rate of $100r\%$ compounded continuously so that we will have the desired amount A at the end of t years. We call P the *present value* of A.

The financial transactions involving equations (27) and (28) are not really the usual ones. It is not often that one deposits \$10,000 in a bank and leaves it there for 15 years at a specified rate of interest. Nor does one frequently borrow a large sum of money and return the entire amount plus interest in one payment. A far more common situation involves a large number of payments which occur at different times, for example, mortgage payments on a piece of property, life insurance, retirement funds, etc. A series of payments made at specific times over a certain period is called an *annuity*.

Suppose that you have borrowed money from a bank on your car and your remaining payments are \$200 per month for the next 30 months. You suddenly receive an inheritance (lucky you!) and decide to pay off the loan on your car. You would *not* pay $200(30) = 6000$ dollars to the bank. If money is worth 12 percent, you would pay the present value of the \$200 due at the end of the first month, plus the present value of the \$200 due at the end of the second month, plus the present value of the \$200 due at the end of the third month, and so on. The sum of these values is called the *present value of the annuity*. Just as we did in Examples 5 and 6, we can approximate this sum by a definite integral. In general, suppose a payment is made at a time t such that on an annual basis this payment is $f(t)$ from $t = 0$ to $t = T$ years, then the present value P of the annuity at a continuous interest rate of $100r\%$ is given by

$$P = \int_0^T f(t)e^{-rt}\,dt \tag{29}$$

The integral in (29) is also called the *present value of a continuous income stream*.

In our discussion above, $f(t) = 200(12) = 2400$ on an annual basis; $T = 30/12 = 2.5$ years. Thus, if your loan is at 12 percent, $r = 0.12$ compounded continuously, and you should pay the bank as follows:

$$\begin{aligned} P &= \int_0^{2.5} 2400e^{-0.12t}\,dt \\ &= \frac{2400}{-0.12}e^{0.12t}\Big|_0^{2.5} \\ &= -20{,}000e^{-0.3} + 20{,}000 \\ &= 5{,}183 \text{ dollars approximately} \end{aligned}$$

We can also look at the *future value* of an annuity. If a payment $f(t)$ is made at time t, then it earns an amount of interest for $T-t$ years. Hence, the payment $f(t)$ plus the interest it earns in $T-t$ years is the future value of $f(t)$ at the end of the period of annuity. The sum of all such future values, denoted by A, is called the *accumulated amount of a continuous annuity* and is given by:

$$A = \int_0^T f(t)e^{r(T-t)}\,dt \tag{30}$$

where the interest rate is $100r\%$ compounded continuously.

EXAMPLE 7. Suppose the payment made at time t is at the rate of $(100+2t)$ dollars per year at the interest rate of 10% compounded continuously for 10 years. Find:

(a) The present value of the continuous income stream.

(b) The accumulated amount of the continuous annuity.

SOLUTION. Here, $f(t) = 100+2t$, $r = 0.1$, $T = 10$.

(a) $$P = \int_0^{10} (100+2t)e^{-0.1t}\,dt$$

$$= 100\int_0^{10} e^{-0.1t}\,dt + 2\int_0^{10} te^{-0.1t}\,dt$$

$$= \frac{100}{-0.1}\,e^{-0.1t}\Big|_0^{10} + 2\left[\frac{te^{-0.1t}}{-0.1}\Big|_0^{10} - \int_0^{10}\frac{e^{-0.1t}}{-0.1}\,dt\right]$$

Note that for the second integral we use integration by parts. Thus,

$$P = -1000e^{-0.1t}\Big|_0^{10} + 2\left[-10te^{-0.1t}\Big|_0^{10} - \left(100e^{-0.1t}\Big|_0^{10}\right)\right]$$

$$= -1000e^{-1} + 1000 + 2[(-100e^{-1}+0) - 100e^{-1} + 100]$$

$$= 1200 - 1400e^{-1} \approx 685 \text{ dollars}$$

(b) $$A = \int_0^{10} (100+2t)e^{(10-t)(0.1)}\,dt$$

$$= \int_0^{10} 100e^{1-0.1t}\,dt + \int_0^{10} 2te^{1-0.1t}\,dt$$

$$= e\left[\int_0^{10} 100e^{-0.1t}\,dt + 2\int_0^{10} te^{-0.1t}\,dt\right]$$

$$= e(685)$$

$$\approx 1862 \text{ dollars}$$

EXERCISES 6.1

In Problems 1 through 10, find a solution of the initial-value problem.

1. $y'(t) = 1 - 3t, \quad y(0) = 4$

2. $\dfrac{dy}{dx} = x^2 + 4x^3, \quad y(1) = -2$

3. $y'(t) = t^3 + t^{-2}, \quad y(-2) = 1$

4. $y'(t) = \sqrt{t} + 4, \quad y(4) = 3$

5. $\dfrac{dy}{dx} = \dfrac{3x^2 + x}{x^5}, \quad y(1) = -1$

6. $\dfrac{dy}{dx} = 2 - x^{3/2}, \quad y(9) = 2$

7. $y' = x^{1/3}(1 + 3x^{1/3}), \quad y(-8) = -\frac{1}{4}$

8. $\sqrt{t}\,y'(t) = 2, \quad y(1) = 3$

9. $\dfrac{\sqrt{x}}{x^2 - 1}\dfrac{dy}{dx} + 1 = 0, \quad y(4) = -\frac{1}{2}$

10. $\dfrac{dy}{dx} = 3x^{1/3}(1 - x^{1/3}), \quad y(8) = -1$

11. Suppose that the marginal cost of producing x units of an item is given by

$$C'(x) = 4x - 350$$

(a) Find the total cost if there is a fixed cost of 500, i.e., $C(0) = 500$.

(b) Find the profit if each item produced sells at $10.

(c) How many units should be sold to producc maximum profit?

12. Suppose that the marginal cost of producing x gallons of Slick oil is $10\sqrt{x}$ dollars. Find the cost function $C(x)$ if there is an overhead cost of $350. (Note that the overhead cost is independent of x.)

13. Suppose that the marginal revenue of producing and selling x units of a product is $20 + (x/50)$. If the revenue for the first 500 units is $12,500, find the revenue function and the price function.

14. A snowball in the form of a perfect sphere has radius of 72 in. Suppose that it melts so that the radius decreases at the rate of 4 in./min. (Assume that the snowball retains its spherical shape until it is completely melted.)

(a) Find the time t when the diameter is 72 in.

(b) What is the diameter of the snowball after 12 minutes?

(c) At what time will the snowball be completely melted?

15. Three astronauts are marooned in a spacecraft. There are 350 units of oxygen available, and the rate of consumption in units per minute is

$$A'(t) = -0.03\sqrt{t}$$

(a) Determine the amount A of oxygen in the spacecraft as a function of time.

(b) Find the time at which the oxygen supply is exhausted.

16. A driver sees a dog crossing the highway ahead and applies the brakes when the car is 100 ft. from the dog. If the car's velocity after application of the brakes is $(70 - 26t)$ ft./sec., will the driver stop in time?

17. Find the velocity and distance functions under the following conditions:

(a) $a(t) = t + 1, \quad v(0) = 2, \quad s(0) = 1$

(b) $a(t) = t^2 - 5, \quad v(0) = -1, \quad s(0) = 4$

In Problems 18 through 20, assume $a(t) = 32$ ft./sec^2

18. An object is projected vertically from the surface of the earth with initial velocity $v(0) = 288$ ft./sec.

(a) Find the velocity and distance functions.

(b) How high will the object rise?

19. An object projected from the surface of the earth rises 228 ft. before starting to fall. Find the time it took the object to rise to this height.

20. An object is thrown from the roof of a building 64 ft. above the ground. If the initial velocity is 48 ft./sec., determine when the object will hit the ground.

In Problems 21 through 24, use the definite integral to find the estimate for the required sum.

21. Weekly production of cars at Mongoose Motors is given by $f(x) = 80\sqrt{x} - 2x$ for the xth week. Estimate the number of cars produced:

(a) During the first 25 weeks.

(b) During the first 36 week.

22. The cost of producing a widget is $20 + (8/\sqrt{x})$ for the xth widget. Find the cost of producing the first 64 widgets. Also, find the cost of producing the next 64 widgets.

23. The cost to produce a Brite TV is $300 + 200/\sqrt[3]{x}$ for the xth TV. Find the cost to produce: (a) the first 8, (b) the first 125, and (c) the first 1000 Brite TV's.

24. The Transgadget Company has a new plant at Dimville, and they expect that during the xth week of production this new plant will manfacture $500 + 8e^{-x/20}$ gadgets. Estimate the number of gadgets produced during the (a) first 10 weeks, (b) during the first 100 weeks.

In Problems 25 through 30, find the present and future values of a continuous annuity at an annual payment rate of $f(t)$ dollars at time t years, with the given interest rate of $100r\%$ compounded continuously for T years.

25. $r = 0.06, \quad T = 8, \quad f(t) = 1000$

26. $r = 0.08, \quad T = 10, \quad f(t) = 1000 + t$

27. $r = 0.10, \quad T = 15, \quad f(t) = 80 + 2t$

28. $r = 0.12, \quad T = 10, \quad f(t) = t^2$

29. $r = 0.10, \quad T = 10, \quad f(t) = 100 + e^t$

30. $r = 0.10, \quad T = 10, \quad f(t) = t + e^t$

6.2 AVERAGE VALUE OF A FUNCTION

If you score 85, 87, and 98 points in three tests, then your *average* or *mean* score is the sum of the points divided by 3. If we call this average $\overline{x}$, then

$$\overline{x} = \frac{85 + 87 + 98}{3} = 90$$

Similarly, we define the *average* value or the *mean* value $\overline{x}$ of n numbers $x_1, x_2, \ldots, x_n$ by the formula

$$\overline{x} = \frac{x_1 + x_2 + \cdots + x_n}{n} \tag{1}$$

But how should one define the average of a continuous function defined on an interal $[a, b]$? We certainly cannot define it as the sum of infinitely many values of $f(x)$ as x varies over $[a, b]$ divided by the number of terms in the sum. The clue to the definition lies in the geometric interpretation of the graph of a positive continuous function on $[a, b]$.

It appears from Figure 1 that if we choose a number c between a and b properly, then the area under the graph of $f(x)$ between $x = a$ and $x = b$ is equal to the area of the rectangle whose width is $b - a$ and whose length is $f(c)$. The existence of a number c is assumed by a theorem called the *Mean Value Theorem for Integrals*: If $f(x)$ is continuous on $[a, b]$, then there is a number c such that

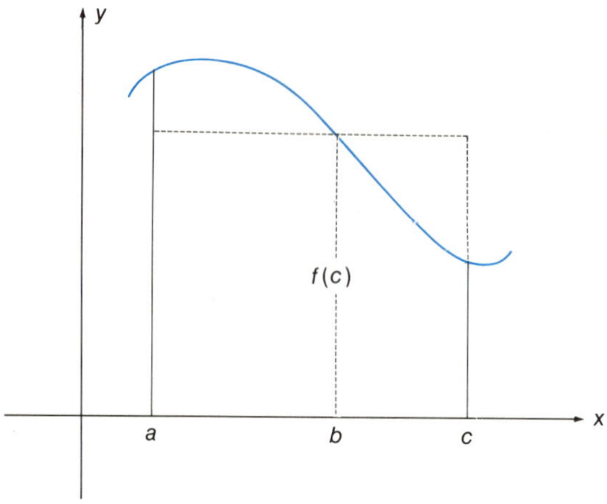

Figure 1

$$\int_a^b f(x)\,dx = f(c)(b-a) \tag{2}$$

where $a < c < b$. The result is true even if f is not positive everywhere on $[a, b]$.

If we divide both sides of (2) by $(b-a)$, we obtain

$$f(c) = \frac{1}{b-a}\int_a^b f(x)\,dx \tag{3}$$

We define the expression on the right side of (3) as the *average value* or *mean value* of $f(x)$ over the interval $[a, b]$.

EXAMPLE 1. Find the average value of $f(x) = x^2$ on $[1, 3]$.

SOLUTION. Using formula (3), we have

$$\text{Average value} = \frac{1}{3-1}\int_1^3 x^2\,dx = \frac{1}{2}\left[\frac{x^3}{3}\Big|_1^3\right] = \frac{1}{2}\left[(9)-\frac{1}{3}\right] = \frac{26}{6} = \frac{13}{3}.$$

Notice that $f(1) = 1, f(3) = 9$, and the average is not $\frac{1+9}{2} = 5$, but $\frac{13}{3}$.

The above result may be interpreted as: the average square of a number x chosen between 1 and 3 is $\frac{13}{3}$.

EXAMPLE 2. The temperature in Hotspot, Arizona, on the morning of July 8, 1980, was given by

$$T = 80 + 8t - \frac{t^2}{6}$$

where T is in degrees Fahrenheit and t is in hours ($t = 0$ in midnight and $t = 12$ is noon). Find the average temperature for that morning.

SOLUTION.

$$\text{Average Temperature } \bar{T} = \frac{1}{12-0}\int_0^{12}\left(80 + 8t - \frac{t^2}{6}\right)dt$$

$$= \frac{1}{12}\left[80t + 4t^2 - \frac{t^3}{18}\Big|_0^{12}\right]$$

$$= \frac{1}{12}\left[960 + 576 - 96\right]$$

$$= 120°\text{F}$$

EXAMPLE 3. Suppose a rock is dropped from the top of a cliff. Then as a result of gravity ($g = 32$) it falls $s(t) = 16t^2$ ft./sec., where $s(t)$ is the distance covered in time t. If it takes 4 seconds for the rock to hit the ground, find the average velocity $\bar{v}$ of the rock during the fall.

SOLUTION. Since $s(t) = 16t^2$, the velocity at any time t is given by

$$v(t) = s'(t) = 32t$$

So that, by formula (3), we have the average velocity:

$$\bar{v} = \frac{1}{4-0}\int_0^4 v(t)dt = \frac{1}{4}\int_0^4 32t\,dt$$

$$= \frac{1}{4}\,16t^2\Big|_0^4$$

$$= \frac{1}{4}[256] = 64 \text{ ft./sec.}$$

However, since $s(0) = 16(0) = 0$ and $s(4) = 16(4)^2 = 256$, then, from the usual concept of average velocity, we have:

$$\text{average velocity} = \frac{\text{distance}}{\text{time taken}} = \frac{256-0}{4-0} = 64 \text{ ft./sec.}$$

Thus, in this example the usual concept of average velocity and the definition given by formula (3) coincide. This is true not only in this example but in general. For if $s(t)$ is the distance covered in time t, then $v(t) = s'(t)$ is the instantaneous velocity. Therefore, the average value of $v(t)$ over the interval $a \le t \le b$ is given by

$$\bar{v} = \frac{1}{b-a}\int_a^b v(t)dt$$

$$= \frac{1}{b-a}\int_a^b s'(t)dt$$

$$= \frac{1}{b-a}\left[s(t)\Big|_a^b\right]$$

or

$$\bar{v} = \frac{s(b)-s(a)}{b-a} \tag{4}$$

The right side of equation (4) is indeed the usual concept of average velocity.

EXAMPLE 4. The density of a rod 9 feet long is given by $\varrho(x) = 1 + 2x - \frac{3}{2}\sqrt{x}$, where x is measured in feet from the left. What is the average density over the length of the entire rod?

SOLUTION. The symbol ϱ is the Greek letter "Rho." The average density $\bar{\varrho}$ is given by

$$\bar{\varrho} = \frac{1}{9-0}\int_0^9 (1 + 2x - \tfrac{3}{2}\sqrt{x})dx$$

$$= \frac{1}{9}[x + x^2 - x^{3/2}\Big|_0^9]$$

$$= \frac{1}{9}[9 + 81 - 27]$$

$$= 7$$

EXERCISES 6.2

In Problems 1 through 16, find the average value of the given function over the given interval.

1. $f(x) = x, \quad [5, 11]$

2. $f(x) = 2x + 3, \quad [3, 8]$

3. $f(x) = 5, \quad [1, 9]$

4. $f(x) = x^2, \quad [0, 1]$

5. $f(x) = 2x^3, \quad [0, 2]$

6. $f(x) = \sqrt{x}, \quad [0, 4]$

7. $f(x) = \dfrac{1}{\sqrt{x}}$, $[1, 4]$

8. $f(x) = xe^{x^2}$, $[0, 1]$

9. $f(x) = \dfrac{x}{\sqrt{x^2 + 16}}$, $[0, 3]$

10. $f(x) = x + \dfrac{1}{x}$, $[1, 8]$

11. $f(x) = 1 + x + x^2 + x^3$, $[0, 1]$

12. $f(x) = \sqrt[3]{x} + \dfrac{1}{\sqrt[3]{x}}$, $[1, 8]$

13. $f(x) = \ln x$, $[1, e]$

14. $f(x) = xe^x$, $[-2, 2]$

15. $f(x) = \dfrac{1}{x^2 - 3x + 2}$, $[3, 7]$

16. $f(x) = \dfrac{1}{x}\ln(\ln x)$, $[e, e^2]$

17. Find the average cube root $\bar{c}$ of a number x chosen between 1 and 8.

18. The marginal cost of producing a gadget is given by

$$\text{M.C.} = 20 + 0.008x - 0.00003x^2$$

(a) Find the additional cost if production is increased from 50 to 100 gadgets.

(b) Find the average additional cost per unit when the production is increased from (1) 50 to 100; (2) 100 to 150.

19. The marginal revenue for each additional unit sold is given by

$$\text{M.R.} = 5 - 0.001x - 0.0006x^2$$

(a) Find the additional revenue received if the sales increase from 60 to 90 units.

(b) Find the average revenue per unit sold in (a).

20. The density of a metal bar 4 feet long is given by

$$\varrho(x) = 1 + x - \frac{3}{16}\sqrt{x}$$

Find:

(a) The minimum density;

(b) The maximum density;

(c) The average density;

(d) Total mass of the rod. [*Hint*: mass $= \int_0^4 \varrho(x)\,dx$]

21. Do problem 20, if the rod is 3 feet long and $\varrho(x) = 2 + x + 6\sqrt{3 - x}$.

22. Do Example 3, if it takes 10 seconds for the rock to hit the ground.

23. An object moves in a straight line such that its distance $s(t)$ in t seconds is given by $s(t) = 3t^2 - 2t + 5$. Find the average velocity between $t = 0$ and $t = 4$.

24. The temperature in Hotsprings, Florida, on the afternoon of June 1, 1980, was given by

$$T = \frac{t^3}{3} - \frac{9}{2}t^2 + 18t + 80 \quad (t = 0, \text{noon})$$

where T is in degrees Fahrenheit and t is in hours $(0 \leq t \leq 8)$. Find:

(a) The maximum temperature that afternoon;

(b) The minumum temperature that afternoon;

(c) The average temperature between noon and 8 P.M.

6.3 CONSUMER'S SURPLUS AND PRODUCER'S SURPLUS

In Chapter 5, we showed how to calculate the area between two curves. In this section we shall see the application of this concept to a problem in economics. Before we discuss the problem, let us review the demand and supply functions.

Generally, if a producer lowers the unit price of a product, then the sales increase. Suppose p is the price per unit and x is the number of units sold at this price, then the function $p = D(x)$ is called the *demand function* (see Chapter 1). Typically, a demand function is a decreasing function (x increases as p decreases). The graph of a typical demand function is sketched in Figure 1.

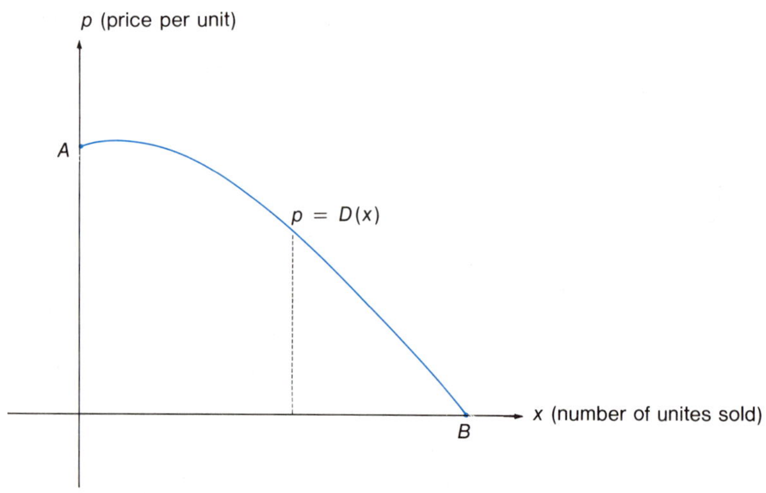

Figure 1

The points A and B on the graph pertain to the extreme situations. At the point A, we have the maximum price so that no sales occur. At the point B, the demand is exhausted, so that no more of the product can even be given away free. Generally, a producer (or seller) sells a certain number of units at a fixed price. But suppose in an ideal situation the producer charges $p = D(x)$ for the xth unit; then we say that *perfect price discrimination* has been attained. In other words, a producer practices perfect price discrimination if the consumer pays the price that the market will bear at every instant.

Of course, in a typical market situation perfect price discrimination is not practiced. That is, the prices do not change instantly with demand but rather are held fixed over a certain period of time. Let this fixed price be denote by $\overline{p}$, corresponding to $x = x_0$

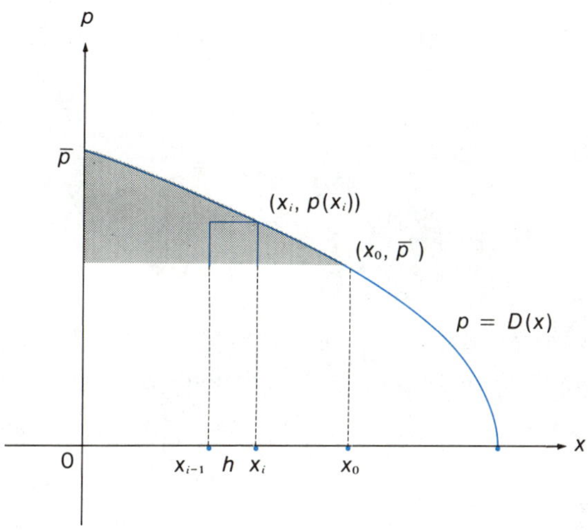

Figure 2

(see Figure 2). Suppose we choose x_i as shown in Figure 2; then according to the demand function the correct price should by $D(x_i)$. Since the price is only $\overline{p}$, the consumer is saving the amount $[D(x_i) - \overline{p}]$ for each unit purchased. If h (small) units are purchased, then the saving to the consumer is $[D(x_i) - \overline{p}\ ]h$. We now "add up" these rectangles and arrive at the area of the shaded region in Figure 2. This area represents the difference between what the consumer should pay (according to the demand function) for the product and the actual expenditure for the product and is called the *consumer's surplus*. We thus arrive at the following definition.

Definition 1. Let $p = D(x)$ be a demand function for a product. Let $\overline{p}$ be the market price per unit at demand x_0. Then the area bounded by $p = D(x)$, $x = 0$, and $p = \overline{p}$ is called the *consumer's surplus* (CS) and is given by

$$CS = \int_0^{x_0} [D(x) - \overline{p}\ ]dx = \int_0^{x_0} D(x)dx - \overline{p}\ x_0 \tag{1}$$

EXAMPLE 1. Let the demand function be given by

$$p = D(x) = 40 - \tfrac{3}{2}\sqrt{x}$$

Calculate the consumer's surplus, if the price per unit is fixed at \$16.

SOLUTION. We first find the value of x which corresponds to $p = 16$. When $p = 16$, we have

$$16 = 40 - \tfrac{3}{2}\sqrt{x}$$

or

$$\tfrac{3}{2}\sqrt{x} = 24$$

or

$$x = 256$$

The consumer's surplus is the area of the region shaded in Figure 3.

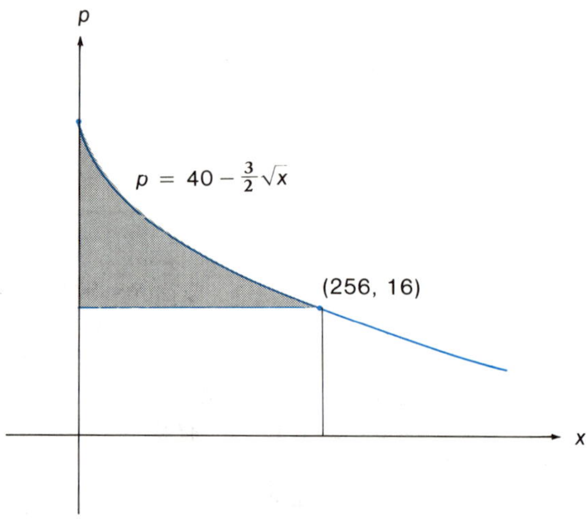

Figure 3

Hence,

$$\begin{aligned} CS &= \int_0^{256} [(40 - \tfrac{3}{2}\sqrt{x}) - 16]\, dx \\ &= \int_0^{256} (24 - \tfrac{3}{2}\sqrt{x})\, dx \end{aligned}$$

$$= (24 - x^{3/2}) \Big|_0^{256}$$

$$= 24(256) - (256)^{3/2}$$

$$= 2048 \text{ dollars}$$

In the discussion on consumer's surplus, the price $\overline{p}$ is usually not picked arbitrarily, unless we have a monopolist who insists on squeezing every penny out of us. Rather, the price, in a competitive market, depends on the law of supply and demand. Let us discuss a supply function. A supply function is a relationship between the expected price of a product and the number of units of the product the producer is willing to produce at that price. In general, if the price is higher, then the producer is willing to produce more items. Thus, a supply function is generally an increasing function. We denote a supply function by $p = S(x)$, where p is the price per unit when x units are produced. Graphs of typical supply functions are sketched in Figure 4.

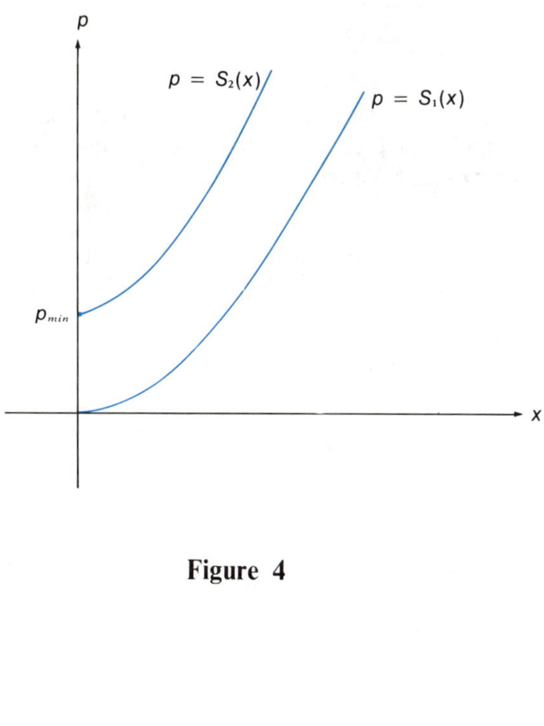

Figure 4

In Figure 4, the graph of $p = S_2(x)$ has a minimum price p_{min} that must be paid in order for the producers to manufacture any item. The graph of $p = S_1(x)$ depicts a confident producer ($p_{min} = 0$) who is willing to take his chances. But suppose the price per unit in the marketplace is $\overline{p}$ and $\overline{p} > p_{min}$. Then the producer earns more money initially than if he had sold at his expected price per unit (given by the supply function). This gain by the producer is called *producer's surplus* and is represented graphically by the area of the shaded region in Figure 5.

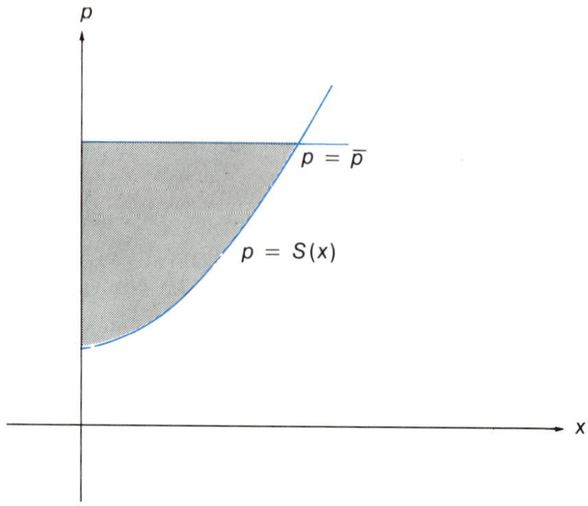

Figure 5

Definition 2. Let $p = S(x)$ be a supply function for a product. Let $\bar{p}$ be the market price per unit at supply level x_0. Then the area bounded by $p = \bar{p}$, $p = S(x)$, and $x = 0$ is called the *producer's surplus* (PS) and is given by

$$PS = \int_0^{x_0} [\bar{p} - S(x)]\, dx = \bar{p}\, x_0 - \int_0^{x_0} S(x)\, dx \tag{2}$$

EXAMPLE 2. Let the supply function be given by

$$p = S(x) = \frac{1}{256}x^2 + \frac{1}{8}x + 4$$

Find the producer's surplus if the price per unit is fixed at \$7.

SOLUTION. We first find the value of x which corresponds to $p = 7$. When $p = 7$, we have

$$\frac{1}{256}x^2 + \frac{1}{8}x + 4 = 7$$

or

$$x^2 + 32x - 768 = 0$$

or

$$(x - 16)(x + 48) = 0$$

Therefore, $x = 16$; we obviously reject $x = -48$. The producer's surplus is represented by the area of the region shaded in Figure 6.

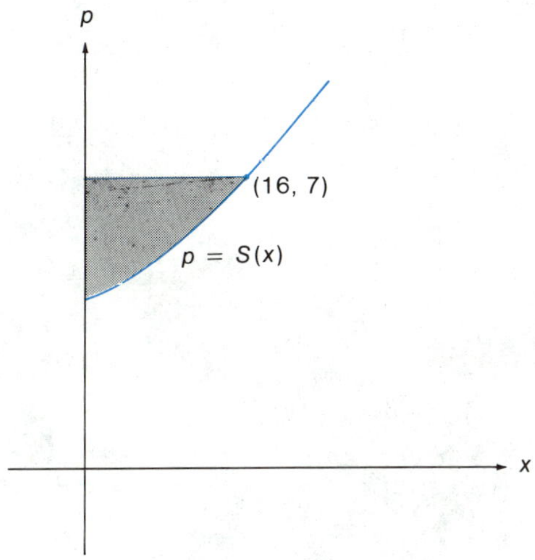

Figure 6

Hence,

$$\begin{aligned} PS &= \int_0^{16} \left[7 - \frac{1}{256}x^2 - \frac{1}{8}x - 4\right] dx \\ &= \int_0^{16} \left[3 - \frac{1}{256}x^2 - \frac{1}{8}x\right] dx \\ &= 3x - \frac{x^3}{768} - \frac{x^2}{16}\Big|_0^{16} \\ &= 48 - \frac{16}{3} - 16 \\ &= \$26.67 \text{ approx.} \end{aligned}$$

We define *pure competition* as the situation when the price is fixed so that the demand is equal to the supply. In this situation we set $S(x) = D(x), x \geq 0$, and obtain x_0, and $\bar{p}$ is obtained by either of the equations $\bar{p} = S(x_0)$ or $\bar{p} = D(x_0)$. The point $(x_0, \bar{p})$ is called the *equilibrium point*. So the equilibrium point tells us that, at the price $\bar{p}$, x_0 units are supplied and demanded; consequently, there is neither a shortage nor a surplus of the product.

EXAMPLE 3. Graph the demand and supply functions of Examples 1 and 2, respectively. Find:

(a) The equilibrium point

(b) Consumer's surplus

(c) Producer's surplus

SOLUTION.

(a) Graphs of demand and supply functions are sketched in Figure 7.

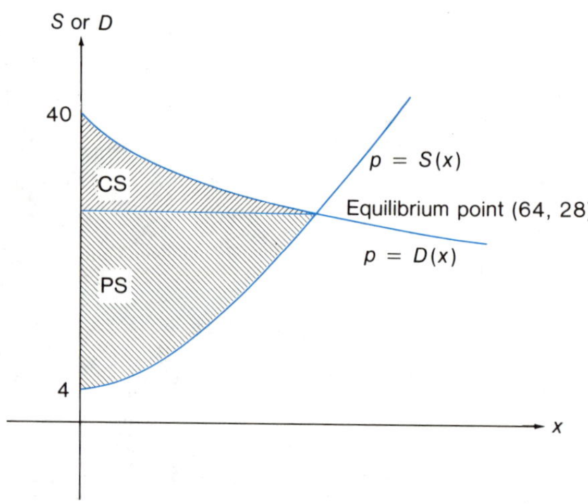

Figure 7

To find the equilibrium point, we set $S(x) = D(x)$ and solve for x to obtain x_0. Thus, we are to solve

$$\frac{1}{256}x^2 + \frac{1}{8}x + 4 = 40 - \frac{3}{2}\sqrt{x}$$

From the graph it is clear that there is only one such point ($x \geq 0$), and by trial and error we see that $x_0 = 64$. (The reader may try a systematic solution by squaring to get rid of the radical and then using the rational root theorem.) Substituting $x = 64$ into $D(x)$, we obtain

$$\bar{p} = 40 - \frac{3}{2}\sqrt{64} = 28$$

Thus, the equilibrium point is (64, 28).

(b)

$$
\begin{aligned}
CS &= \int_0^{64} (40 - \frac{3}{2}\sqrt{x} - 28)\, dx \\
&= \int_0^{64} (12 - \frac{3}{2}\sqrt{x})\, dx \\
&= 12x - x^{3/2} \Big|_0^{64} \\
&= 768 - 512 \\
&= \$256
\end{aligned}
$$

(c)

$$
\begin{aligned}
PS &= \int_0^{64} \left[28 - \left(\frac{1}{256}x^2 + \frac{1}{8}x + 4\right)\right] dx \\
&= \int_0^{64} \left(24 - \frac{1}{256}x^2 - \frac{1}{8}x\right) dx \\
&= 24x - \frac{x^3}{768} - \frac{x^2}{16} \Big|_0^{16} \\
&= \$938.67
\end{aligned}
$$

From the above example, we note that in pure competition there is a level of production and price per unit which assures profits (on paper) to the consumers as well as the producers.

EXERCISES 6.3

In Problems 1 through 6, for a given demand function and the given fixed price, calculate the consumer's surplus.

1. $p = D(x) = 8 - x, \quad \overline{p} = 2$

2. $p = D(x) = 17 - 3x, \quad \overline{p} = 2$

3. $p = D(x) = 20 - x^2, \quad \overline{p} = 4$

4. $p = D(x) = 50 - 3x^2, \quad \overline{p} = 2$

5. $p = D(x) = 27 + 5x - x^2, \quad \overline{p} = 3$

6. $p = D(x) = 85 + 27x - 2x^2, \quad \overline{p} = 5$

In Problems 7 through 12, for a given supply function and the given fixed price, calculate the producer's surplus.

7. $p = S(x) = 5 + x, \quad \overline{p} = 8$

8. $p = S(x) = 1 + \sqrt{x}, \quad \bar{p} = 6$

9. $p = S(x) = 1 + \frac{1}{2}x^2, \quad \bar{p} = 9$

10. $p = S(x) = 2 - \frac{1}{2}x + \frac{1}{8}x^2, \quad \bar{p} = 6$

11. $p = S(x) = \frac{1}{9}x^2 - \frac{8}{9}x, \quad \bar{p} = 1$

12. $p = S(x) = \frac{x^2}{200} - \frac{7x}{50} + \frac{1}{5}, \quad \bar{p} = \frac{1}{2}$

In Problems 13 through 20, for the given demand and supply functions, find: (a) the equilibrium point, (b) the consumer's surplus, and (c) the producer's surplus. Assume that pure competition is attained.

13. $D(x) = 240 - 2x, \qquad S(x) = x$

14. $D(x) = 100 - \frac{1}{2}x, \qquad S(x) = 10 + x$

15. $D(x) = 8 - x^2, \qquad S(x) = x^2$

16. $D(x) = -x^2 + 3x + 2, \qquad S(x) = x^2 + 3$

17. $D(x) = 200 - x, \qquad S(x) = 75 + x + 0.01x^2$

18. $D(x) = 1000 - 28x - x^2, \qquad S(x) = x^2 + 2x + 100$

19. $D(x) = 650 - 4x^2, \qquad S(x) = 50 + 11x + 4x^2$

20. $D(x) = 100 - \frac{x^2}{16}, \qquad S(x) = -25 + 5x$

21. Interpret the supply function of problem 20.

22. In some books on economics, the axes for the supply and demand functions are interchanged (price is horizontal, and the number of units demanded or supplied at that price is vertical). Sketch typical supply and demand functions with axes interchanged and shade the regions repesenting the consumer's surplus and the producer's surplus.

6.4 EXTENSIONS OF INTEGRALS

The definite integrals considered earlier all had the form

$$\int_a^b f(x)\,dx \tag{1}$$

where we assumed that the function $f(x)$ was continuous on the finite closed interval $[a, b]$. We shall now relax these assumptions to enlarge the class of functions for which we can compute the definite integral.

1. PIECEWISE CONTINUOUS FUNCTIONS

We know that if f is a nonnegative continuous function defined on $[a, b]$, then the integral in (1) represents the area bounded by the curve $y = f(x)$, the x-axis, and the lines $x = a$ and $x = b$. We now enlarge the class of functions for which we can compute the area under the curve to include those functions whose graphs consist of continuous pieces on $[a, b]$; these functions do not "blow up" anywhere in $[a, b]$ (see Figure 1).

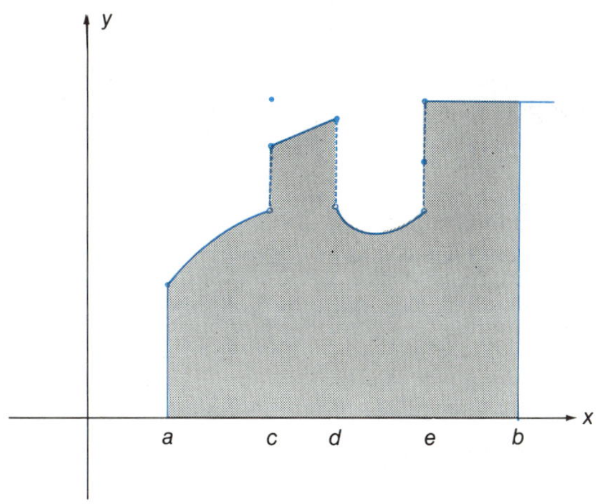

Figure 1

Definition 1. We say that a function f is *piecewise continuous* on its domain if:

(1) The function has at most a finite number of points of discontinuity in its domain.

(2) At each point c of discontinuity both $\lim_{x \to c^-} f(x)$ and $\lim_{x \to c^+} f(x)$ exist.

EXAMPLE 1. Which of the following functions are piecewise continuous?

(a) $f(x) = x^2$

(b) $f(x) = \begin{cases} x^2, & x \leq 0 \\ 2x + 1, & x > 0 \end{cases}$

(c) $f(x) = [\![x]\!]$

(d) $f(x) = \begin{cases} 2x^2 + 1, & x \geq 0 \\ \dfrac{1}{x}, & x < 0 \end{cases}$

(e) $f(x) = \begin{cases} \dfrac{x}{x-1}, & x > 0 \\ 5, & x = 0 \\ 3x, & x < 0 \end{cases}$

(f) $f(x) = \begin{cases} e^{1/x}, & x \neq 0 \\ 1, & x = 0 \end{cases}$

SOLUTION.

(a) The function $f(x) = x^2$ is continuous at each point of its domain $(-\infty, \infty)$. Thus, it has at most finitely many points (none) of discontinuity. Therefore, $f(x) = x^2$ is piecewise continuous.

In general, therefore, every continuous function is piecewise continuous on its domain (but not conversely).

(b) This function is piecewise continuous, because it is discontinuous only at $x = 0$. However,

$$\lim_{x \to 0^-} f(x) = \lim_{x \to 0^-} x^2 = 0$$

and

$$\lim_{x \to 0^+} f(x) = \lim_{x \to 0^+} (2x + 1) = 1$$

(c) This function is not piecewise continuous because it is discontinuous at an infinite number of points (at every integer).

(d) This function is not piecewise continuous. Although it has only one point of discontinuity at $x = 0$, and

$$\lim_{x \to 0^+} f(x) = \lim_{x \to 0^+} (2x^2 + 1) = 1$$

but

$$\lim_{x \to 0^-} f(x) = \lim_{x \to 0^-} \frac{1}{x} = -\infty \quad \text{(does not exist)}$$

(e) This function is not piecewise continuous, since

$$\lim_{x\to 1^+} f(x) = \lim_{x\to 1^+} \frac{x}{x-1} = \infty \quad \text{(does not exist)}$$

(f) This function is not piecewise continuous, because

$$\lim_{x\to 0^+} f(x) = \lim_{x\to 0^+} e^{1/x} = \infty \quad \text{(why?)}$$

However,

$$\lim_{x\to 0^-} f(x) = \lim_{x\to 0^-} e^{1/x} = 0 \quad \text{(why?)}$$

Before we are able to compute the area under the graph of a nonnegative piecewise continuous function and hence define the definite integral (signed area) for piecewise continuous functions, we need to clear one more obstacle.

Let us consider two functions:

$$f(x) = \begin{cases} x^2, & 1 \le x \le 3 \\ 12, & x = 3 \end{cases} \quad \text{and} \quad g(x) = x^2, \quad 1 \le x \le 3$$

and try to compute the areas under their graphs.

The graphs of f and g and the required shaded areas are sketched in Figures 2 and 3, respectively.

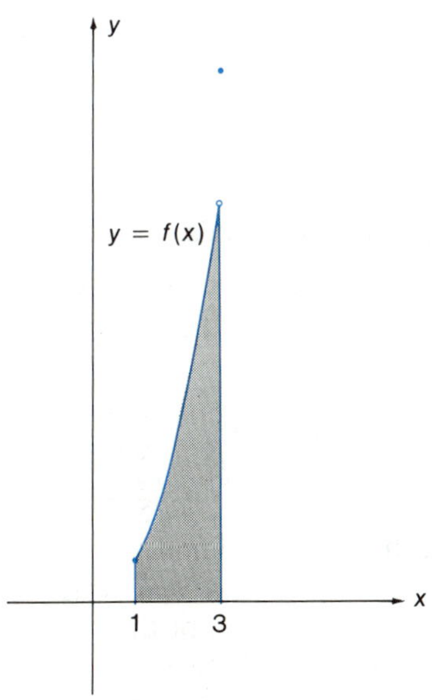

Figure 2

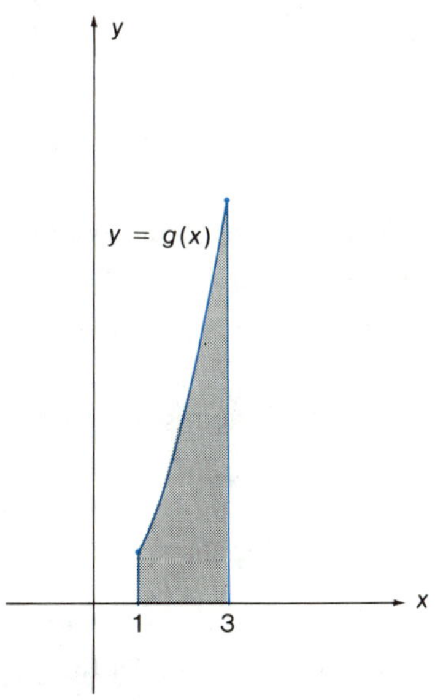

Figure 3

Although the function f is not continuous on [1, 3], the area under the graph of f is equal to the area under the graph of g. Now, since g is continuous on [1, 3], the area under its graph (and therefore under the graph of f) is given by

$$\int_1^3 g(x)\,dx = \int_1^3 x^2\,dx = \left.\frac{x^3}{3}\right|_1^3 = 9 - \frac{1}{3} = \frac{26}{3}$$

So that, even though the function f is not continuous at the endpoint 3, we may define

$$\int_1^3 f(x)\,dx = \int_1^3 g(x)\,dx = \frac{26}{3}$$

In general, suppose f and g agree (are identical) on the interval $[c, d]$, except possibly at the endpoint(s), and g is continuous on $[c, d]$, we define

$$\int_c^d f(x)\,dx = \int_c^d g(x)\,dx$$

Now, if the function f does not "blow up" at its endpoints, then the graph of g is obtained from the graph of f by simply filling in the hole(s) at the endpoint(s).

Thus, to compute the area of a region under the graph of a nonnegative piecewise continuous function f, it seems natural to compute the area under each piece and add. We illustrate in the following example.

EXAMPLE 2. For the piecewise continuous function defined by

$$f(x) = \begin{cases} x, & 1 \le x < 2 \\ 3, & x = 2 \\ 6 - x, & 2 < x \le 5 \\ 4, & x > 5 \end{cases}$$

(a) Sketch the region enclosed by the graph of f, the x-axis, and the lines $x = 1$ and $x = 6$.

(b) Compute the area of the region in (a).

SOLUTION. The region is sketched in Figure 4.

We notice that this function is discontinuous at $x = 2$ and $x = 5$. To compute the area of the region, we shall compute the area of the three subregions over the subintervals [1, 2], [2, 5], and [5, 6] and add them. The area of the subregion over [1, 2] is equal to the area under the graph of $g_1(x) = x,\ 1 \le x \le 2$; the area of the subregion over [2, 5] is equal to the area under the graph of $g_2(x) = 6 - x,\ 2 \le x \le 5$; and, finally, the area of the subregion over [5, 6] is the same as the area under the graph of $g_3(x) = 4$, $5 \le x \le 6$. The total area A is given by

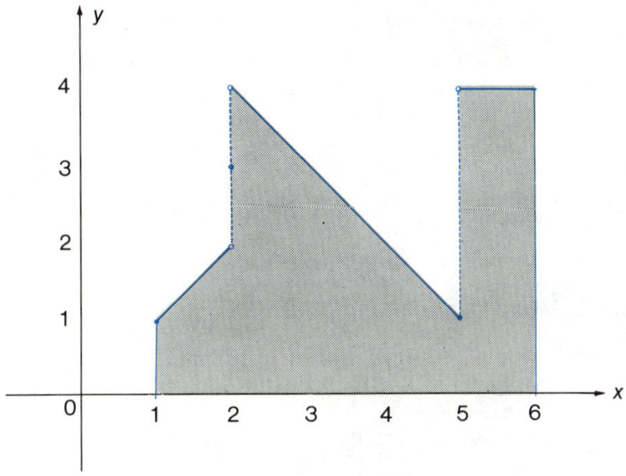

Figure 4

$$
\begin{aligned}
A &= \int_1^2 x\,dx + \int_2^5 (6-x)\,dx + \int_5^6 4\,dx \\
&= \frac{x^2}{2}\Big|_1^2 + \left(6x - \frac{x^2}{2}\right)\Big|_2^5 + 4x\Big|_5^6 \\
&= \left(2 - \frac{1}{2}\right) + \left[30 - \frac{25}{2} - (12-2)\right] + (24-20) \\
&= 13 \text{ units}
\end{aligned}
$$

If we do not mention the word area, then the procedure outlined in the above example is used to evaluate definite integrals.

EXAMPLE 3. Let $f(x) = \begin{cases} x - x^2, & x < 1 \\ 1 - 4x, & \text{otherwise} \end{cases}$

Evaluate

$$\int_0^5 f(x)\,dx$$

SOLUTION. We have

$$
\begin{aligned}
\int_0^5 f(x)\,dx &= \int_0^1 f(x)\,dx + \int_1^5 f(x)\,dx \\
&= \int_0^1 (x - x^2)\,dx + \int_1^5 (1-4x)\,dx
\end{aligned}
$$

$$= \left(\frac{x^2}{2} - \frac{x^3}{3}\right)\Big|_0^1 + (x - 2x^2)\Big|_1^5$$

$$= \left(\frac{1}{2} - \frac{1}{3}\right) + [(5 - 50) - (1 - 2)]$$

$$= \frac{1}{6} - 44$$

$$= -\frac{263}{6}$$

2. IMPROPER INTEGRALS

We now consider the integrals for which either

(1) the function "blows up" at some point in the interval $[a, b]$, or

(2) the interval of integration is infinite, such as $[a, \infty)$, $(-\infty, a]$, or $(-\infty, \infty)$.

The integrals having properties (1) or (2) are called *improper integrals*. First, we consider the case where f is defined and continuous on an interval $(a, b]$, but not on $[a, b]$. For example, let f be the function defined by

$$y = \frac{1}{\sqrt{x}}$$

and consider the integral

$$\int_0^1 \frac{1}{\sqrt{x}}\, dx \qquad (2)$$

Here f is not defined at $x = 0$ and $f(x) \to \infty$ as $x \to 0^+$ (see Figure 5). Suppose that we take $0 < a < 1$. Then

$$\int_a^1 \frac{1}{\sqrt{x}}\, dx = 2\sqrt{x}\,\Big|_a^1 = 2 - 2\sqrt{a}$$

and we have that $(2 - 2\sqrt{a}) \to 2$ as $a \to 0^+$. We make the following definition:

$$\int_0^1 \frac{1}{\sqrt{x}}\, dx = \lim_{a \to 0^+} \int_a^1 \frac{1}{\sqrt{x}}\, dx = \lim_{a \to 0^+} (2 - 2\sqrt{a}) = 2$$

Therefore, we assign 2 to the area of the region shown in Figure 5.

The integral in (2) is an example of an *improper integral* which is said to be *convergent*.

Definition 2. If f is continuous on $(a, b]$ and $\lim_{x \to a^+} f(x)$ does not exist, then

$$\int_a^b f(x)\, dx = \lim_{\varepsilon \to 0^+} \int_{a+\varepsilon}^b f(x)\, dx$$

provided this limit exists.

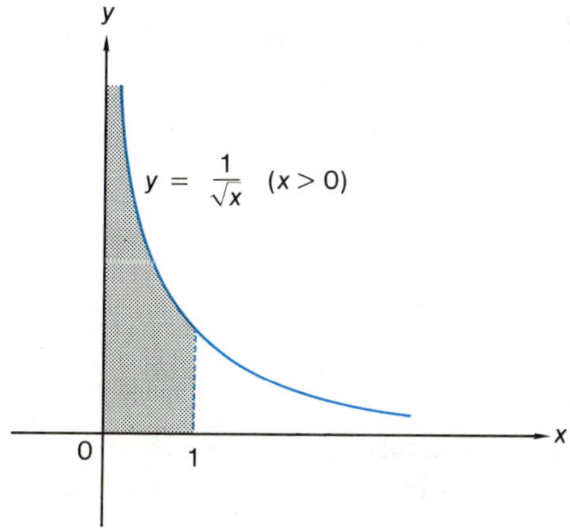

Figure 5

Another example of an improper integral is

$$\int_{-2}^{0} \frac{1}{x^2}\,dx \tag{3}$$

Here the function $y = 1/x^2$ is continuous on $[-2, 0)$ but not on $[-2, 0]$. To evaluate this integral, we compute for $-2 < b < 0$

$$\int_{-2}^{b} \frac{1}{x^2}\,dx = -\frac{1}{x}\bigg|_{-2}^{b} = -\frac{1}{b} - \frac{1}{2}$$

In this case, we find that as $b \to 0^-$, the expression $[-(1/b) - (1/2)] \to \infty$. Hence, no limit exists, and no number can be assigned to the area of the region shown if Figure 6.

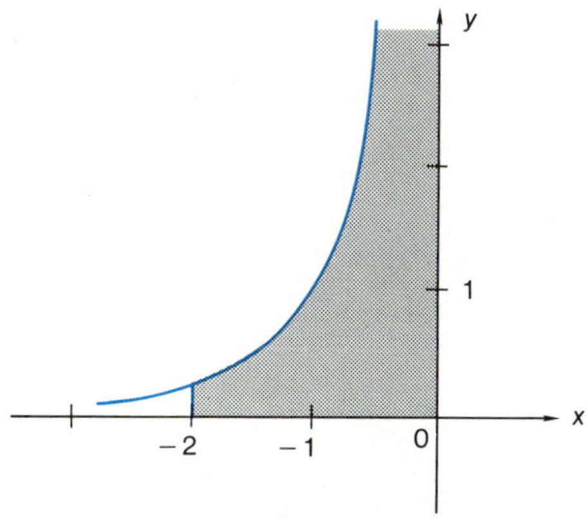

Figure 6

The integral in (3) is an example of an *improper integral* that is said to be *divergent*.

Definition 3. If f is continuous on $[a, b)$ and $\lim_{x \to b^-} f(x)$ does not exist, then

$$\int_a^b f(x)\,dx = \lim_{x \to 0^+} \int_a^{b-\varepsilon} f(x)\,dx$$

provided this limit exists.

If the integrand "blows up" at an interior point of the interval of integration, we have the following definition.

Definition 4. If f is continuous at all x in the interval $[a, b]$ except at c where $a < c < b$ and $\lim f(x)$ as $x \to c+$ or $x \to c-$ does not exist, then

$$\int_a^b f(x)\,dx = \lim_{\varepsilon_1 \to 0^+} \int_a^{c-\varepsilon_1} f(x)\,dx + \lim_{\varepsilon_2 \to 0^+} \int_{c+\varepsilon_2}^b f(x)\,dx$$

provided *both* of these limits exist.

If the limits in Definitions 2, 3, and 4 exist, we say that the improper integral is *convergent*; otherwise it is *divergent*.

EXAMPLE 4. Evaluate the following[1] improper integral if it exists:

$$\int_{-2}^{3} \frac{1}{x^2}\,dx \tag{4}$$

SOLUTION. In this case, $y = 1/x^2$ is continuous at every x in $[-2, 3]$ except at $x = 0$, and $f(0)$ does not exist. To evaluate the integral, we use Definition 4, and write

$$\begin{aligned}
\int_{-2}^{3} \frac{1}{x^2}\,dx &= \lim_{\varepsilon_1 \to 0^+} \int_{-2}^{0-\varepsilon_1} \frac{1}{x^2}\,dx + \lim_{\varepsilon_2 \to 0^+} \int_{0+\varepsilon_2}^{3} \frac{1}{x^2}\,dx \\
&= \lim_{\varepsilon_1 \to 0^+} -\frac{1}{x}\bigg|_{-2}^{-\varepsilon_1} + \lim_{\varepsilon_2 \to 0^+} -\frac{1}{x}\bigg|_{\varepsilon_2}^{3} \\
&= \lim_{\varepsilon_1 \to 0^+} \left(\frac{1}{\epsilon_1} - \frac{1}{2}\right) + \lim_{\varepsilon_2 \to 0^+} \left(-\frac{1}{3} + \frac{1}{\epsilon_2}\right)
\end{aligned}$$

Since neither of the limits exist, the improper integral (4) is divergent.

---

[1] **Note.** It is important to observe the points of discontinuity of the integrand. For instance, a careless evaluation of (4) may lead to the following erroneous answer:

$$\int_{-2}^{3} \frac{1}{x^2}\,dx = -\frac{1}{x}\bigg|_{-2}^{3} = -\frac{1}{3} + \frac{1}{2} = \frac{1}{6}$$

This example represents incorrect application of the definition of the definite integral.

EXAMPLE 5. Evaluate the following improper integral if it exists.

$$\int_{-1}^{2} \frac{1}{x^{1/3}}\, dx \tag{5}$$

SOLUTION. In this case, $y = 1/x^{1/3}$ is continuous at every x in $[-1, 2]$ except at $x = 0$, and $f(0)$ does not exist. Using Definition 4, we find that

$$\begin{aligned}
\int_{-1}^{2} \frac{1}{x^{1/3}}\, dx &= \lim_{\varepsilon_1 \to 0^+} \int_{-1}^{0-\varepsilon_1} \frac{1}{x^{1/3}}\, dx + \lim_{\varepsilon_2 \to 0^+} \int_{0+\varepsilon_2}^{2} \frac{1}{x^{1/3}}\, dx \\
&= \lim_{\varepsilon_1 \to 0^+} \tfrac{3}{2} x^{2/3} \Big|_{-1}^{-\varepsilon_1} + \lim_{\varepsilon_2 \to 0^+} \tfrac{3}{2} x^{2/3} \Big|_{\varepsilon_2}^{2} \\
&= -\tfrac{3}{2} + \tfrac{3}{2}(2)^{2/3} = \tfrac{3}{2}(\sqrt[3]{4} - 1)
\end{aligned}$$

Next we consider the case of a continuous function that is to be integrated over an infinite interval. An interesting application of such improper integrals arises in the field of economics. For example, if the number of dollars in the annual income is $f(t)$ at t years and r is the annual rate of interest compounded continuously, then the present value of all future income is given by the improper integral

$$\int_{0}^{\infty} f(t) e^{-rt}\, dt$$

Improper integrals such as

$$\int_{0}^{\infty} f(x)\, dx, \quad \int_{-\infty}^{b} f(x)\, dx, \quad \text{and} \quad \int_{-\infty}^{\infty} f(x)\, dx$$

are very common and useful in applications. Such integrals arise in the theory of probability, which provides very powerful mathematical tools for predicting events in most branches of the sciences. A biologist studying heredity or a psychologist conducting learning experiments would encounter probability theory frequently. The procedure for evaluating improper integrals of this type is similar to the procedure considered in earlier examples. We have the following definition.

Definition 5. If f is continuous for all $x \geq a$, then

$$\int_{a}^{\infty} f(x)\, dx = \lim_{b \to \infty} \int_{a}^{b} f(x)\, dx$$

provided this limit exists.

EXAMPLE 6. Evaluate the following improper integral if it exists:

$$\int_{0}^{\infty} e^{-x}\, dx \tag{6}$$

SOLUTION. By definition, we have

$$\int_0^\infty e^{-x}\,dx = \lim_{b\to\infty}\int_0^b e^{-x}\,dx = \lim_{b\to\infty} -e^{-x}\Big|_0^b = \lim_{b\to\infty}(1-e^{-b}) = 1$$

Therefore, the improper integral (6) is convergent and its value is 1.

Definition 6. If f is continuous for all $x < a$, then

$$\int_{-\infty}^a f(x)\,dx = \lim_{b\to-\infty}\int_b^a f(x)\,dx$$

provided this limit exists.

If a function f is continuous on the interval $(-\infty, \infty)$, we make the following definition.

Definition 7. If f is continuous for all values of x, then

$$\int_{-\infty}^\infty f(x)\,dx = \lim_{a\to-\infty}\int_a^0 f(x)\,dx + \lim_{b\to\infty}\int_0^b f(x)\,dx$$

provided *both* limits exist.

EXAMPLE 7. Evaluate the following improper integrals if they exist.

(a) $\displaystyle\int_{-\infty}^0 \frac{1}{(2-x)^2}\,dx$ (b) $\displaystyle\int_{-\infty}^\infty x\,dx$

SOLUTION.

(a) By definition, we have

$$\int_{-\infty}^0 \frac{1}{(2-x)^2}\,dx = \lim_{b\to-\infty}\int_b^0 \frac{1}{(2-x)^2}\,dx = \lim_{b\to-\infty} \frac{1}{2-x}\Big|_b^0$$
$$= \lim_{b\to-\infty}\left(\frac{1}{2} - \frac{1}{2-b}\right) = \frac{1}{2} + 0 = \frac{1}{2}$$

Therefore, the improper integral in (a) is convergent and has the value 1/2.

(b) In this case, we have

$$\int_{-\infty}^\infty x\,dx = \lim_{b\to-\infty}\int_b^0 x\,dx + \lim_{a\to\infty}\int_0^a x\,dx$$
$$= \lim_{b\to-\infty}\frac{x^2}{2}\Big|_b^0 + \lim_{a\to\infty}\frac{x^2}{2}\Big|_0^a = \lim_{b\to-\infty}\left(-\frac{b^2}{2}\right) + \lim_{a\to\infty}\frac{a^2}{2}$$

Since neither of these two limits exist, the improper integral in (b) is divergent.

A common mistake is to write

$$\int_{-\infty}^{\infty} f(x)\,dx = \lim_{b\to\infty} \int_{-b}^{b} f(x)\,dx$$

which is not the definition of the improper integral on the left-hand side of this equation. For example,

$$\lim_{b\to\infty} \int_{-b}^{b} x\,dx = \lim_{b\to\infty} \frac{x^2}{2}\bigg|_{-b}^{b} = \lim_{b\to\infty} \left(\frac{b^2}{2} - \frac{b^2}{2}\right) = \lim_{b\to\infty} 0 = 0$$

whereas the improper integral $\int_{-\infty}^{\infty} x\,dx$ is divergent.

EXERCISES 6.4

In Problems 1 through 6, determine which of the functions are piecewise continuous.

1. $f(x) = 3x + 5$

2. $f(x) = x + 1 + \dfrac{1}{x}, \quad x > 0$

3. $f(x) = \ln x, \quad x > 0$

4. $f(x) = \begin{cases} \dfrac{1}{x-1}, & x \neq 1 \\ 1, & x = 1 \end{cases}$

5. $f(x) = \begin{cases} xe^{-x}, & x > 0 \\ 0, & x \leq 0 \end{cases}$

6. $f(x) = \begin{cases} 2x - 3, & x > 3 \\ 1 + x, & 1 \leq x \leq 3 \\ \dfrac{2}{x-1}, & x < 1 \end{cases}$

In Problems 7 through 10, evaluate the given integral.

7. $\displaystyle\int_{-5}^{7} f(x)\,dx$, where $f(x) = \begin{cases} -x, & x < 0 \\ 1 + x, & x \geq 0 \end{cases}$

8. $\displaystyle\int_{0}^{4} f(x)\,dx$, where $f(x) = \begin{cases} xe^{x}, & x < 1 \\ 0, & \text{otherwise} \end{cases}$

9. $\displaystyle\int_{-2}^{6} f(x)\,dx$, where $f(x) = \begin{cases} x^2, & x \leq 0 \\ 2x, & 0 < x \leq 3 \\ 5, & x > 3 \end{cases}$

10. $\displaystyle\int_{-1}^{5} f(x)\,dx$, where $f(x) = [\![x]\!]$

In Problems 11 through 36, evaluate the given improper integral if it exists.

11. $\displaystyle\int_{0}^{4} \frac{1}{\sqrt{x}}\,dx$

12. $\displaystyle\int_{-1}^{0} \frac{1}{x^2}\,dx$

13. $\int_0^1 \frac{1}{x^{1/3}}\,dx$

14. $\int_0^9 \frac{1}{x^{3/2}}\,dx$

15. $\int_2^4 \frac{1}{\sqrt{x-2}}\,dx$

16. $\int_{-5}^1 \frac{1}{\sqrt{1-x}}\,dx$

17. $\int_{-2}^2 \frac{1}{x^2}dx$

18. $\int_2^{10} \frac{1}{(x-2)^{2/3}}\,dx$

19. $\int_0^2 \frac{1}{(x-1)^2}\,dx$

20. $\int_0^2 \frac{1}{(x-1)^{2/3}}\,dx$

21. $\int_0^2 \frac{x^2}{(x^3-1)^2}\,dx$

22. $\int_{-1}^2 \frac{x}{\sqrt{x^2-1}}\,dx$

23. $\int_1^\infty \frac{1}{x^2}dx$

24. $\int_1^\infty \frac{1}{x}dx$

25. $\int_0^\infty e^{-3x}dx$

26. $\int_2^\infty \frac{1}{x^2}dx$

27. $\int_1^\infty \frac{1}{x^4}dx$

28. $\int_{-\infty}^0 e^{3x}dx$

29. $\int_{-1}^\infty \frac{1}{(x+3)^2}\,dx$

30. $\int_{-\infty}^{-1} \frac{1}{x^3}dx$

31. $\int_5^\infty \frac{1}{\sqrt{x-1}}\,dx$

32. $\int_1^\infty \frac{1}{x\ln x}\,dx$

33. $\int_0^\infty \frac{1}{\sqrt{e^x}}\,dx$

34. $\int_e^\infty \frac{1}{x(\ln x)^2}\,dx$

35. $\int_{-\infty}^\infty \frac{x}{1+2x^2}\,dx$

36. $\int_{-\infty}^\infty xe^{-x^2}dx$

37. If $a > 0$, show that $\int_0^\infty e^{-ax}dx = 1/a$.

38. If $a > 0$, show that $\int_0^\infty xe^{-ax}dx = 1/a^2$ [assume that $\lim_{c\to\infty}(c/e^{ac}) = 0$].

39. If $a > 1$, show that $\int_1^\infty \frac{1}{x^a}dx = \frac{1}{a-1}$.

40. The formula for finding the present value V of all future income of D dollars per year and at an interest rate of $r\%$ compounded continuously is given by

$$V = \int_0^\infty De^{-rt}dt$$

where t is time.

(a) If the perpetual cash flow $D = \$4000$ per year and $r = 8\%$, what is the present value V?

(b) Show that for constant D, we have $V = D/r$.

6.5 PROBABILITY

One of the most fascinating and useful mathematical systems is the theory of probability. Although the earliest mathematical thought concerning probability arose from the games of chance, its methods of predictions have been used in practically every branch of science. In this section we shall take a brief look at the use of definite integrals in the theory of probability.

In any experiment the set U of all possible outcomes is called the *sample space* for the experiment. If E is a subset of U, then E is called an *event*. Intuitively, then, the *probability* of an event is a number between 0 and 1 which represents its chance of occuring. We will use the symbol $P[E]$ to denote the probability of an event E. Intuitively, then, $P[E] = 1/4$ means that the event E will occur 1 out of 4 times (25% of the time) if the experiment were repeated a large number of times.

EXAMPLE 1. When we throw a die, the number of dots showing on the top face is an outcome, and this may be any one of the numbers from 1 through 6. The sample space U in this case is $\{1, 2, 3, 4, 5, 6\}$. Throwing a four is an event. Throwing an odd number is also an event. In the latter case E is the subset $\{1, 3, 5\}$. If we have a "fair" die, then the probability assigned to the outcome of a particular face being observed face up is 1/6. Thus, $P[4] = 1/6$ and $P[1, 3, 5] = 3/6 = 1/2$. (Why?) Now suppose the die is loaded; then how do we assign probability to each event of the sample space? In such a case, the usual procedure is to rely on observations. The experiment is performed n times (n is large), and the event E is observed k times, then k/n is accepted as a reasonable approximation for the probability of E.

EXAMPLE 2. Consider the experiment of throwing two "good" dice. The sum of the number of dots on the top faces is an outcome, and this may be any one of the numbers from 2 through 12. The sample space U in this case is $\{2, 3, 4, 5, 6, 7, 8, 9, 10, 11, 12\}$. Each outcome in this sample space is obviously not "equally likely." For example, the number 2 can be obtained in only one way $(2 = 1 + 1)$, while the number 5 can be obtained in four ways $(1 + 4, 2 + 3, 3 + 2, 4 + 1)$. We assign the probabilities for each of these outcomes in the following table:

x	2	3	4	5	6	7	8	9	10	11	12
$P[x]$	$\frac{1}{36}$	$\frac{2}{36} = \frac{3}{18}$	$\frac{3}{36} = \frac{1}{12}$	$\frac{4}{36} = \frac{1}{9}$	$\frac{5}{36}$	$\frac{6}{36} = \frac{1}{6}$	$\frac{5}{36}$	$\frac{4}{36} = \frac{1}{9}$	$\frac{3}{36} = \frac{1}{12}$	$\frac{2}{36} = \frac{1}{18}$	$\frac{1}{36}$

The function $P(x)$ is called a *probability density function*. An alternative method of recording these data is with a graph (see Figure 1).

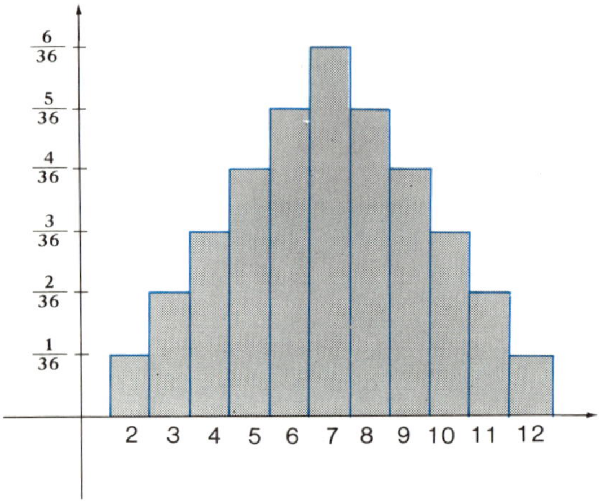

Figure 1

This type of probability density function is called *discrete* because its domain U contains a finite number of points. We notice that the rectangles of the probability "historgram" (Figure 1) are of equal width; thus, their areas are proportional to their heights. Furthermore, the sum of the areas of all the rectangles is equal to 1. Suppose we wish to find the probability of throwing the sum S, where $5 \leq S \leq 9$; when two "good" dice are thrown. All we have to do is to add the areas of the rectangles over 5, 6, 7, 8, and 9 in Figure 1. We obtain

$$P[5 \leq S \leq 9] = \frac{4}{36} + \frac{5}{36} + \frac{6}{36} + \frac{5}{36} + \frac{4}{36} = \frac{24}{36} = \frac{2}{3} = 0.666\ldots$$

In many applications the domain of the probability density function $P(x)$ is an interval (sample space has infinitely many points), or perhaps is so large that the domain is assumed to be an interval and then $P(x)$ is assumed to be a *piecewise continuous function*. For example, the life of a light bulb, in hours, the length of time in minutes that an individual talks on a long-distance telephone call, the height of a person in inches.

In certain situations it is possible to find a function P with x as an independent variable that can be used to determine the probability that x will assume a value on a given subinterval of (a, b). This may be obtained either by some mathematical reasoning associated with the problem or by fitting the function to the data on hand. Such a function P, called a *probability density function*, must satisfy the following three conditions.

(1) The domain of $P(x)$, the set of all outcomes, must be the real numbers

(2) $P(x) \geq 0$ for all $x \in [a, b]$; a and b could be $-\infty$ or ∞

(3) $\int_a^b P(x)\,dx = 1$

Furthermore, if $[c, d]$ is a subinterval of $[a, b]$, then

$$P[c \leq x \leq d] = \int_c^d P(x)\,dx$$

The statement $P[c \leq x \leq d]$ means the probability that x will lie between c and d.

EXAMPLE 3. Verify condition (3) for the following probability density functions:

(a) $f(x) = \frac{3}{26}x^2$, on $[1, 3]$

(b) $f(x) = \begin{cases} \frac{1}{63}x^2, & 3 \leq x \leq 6 \\ 0 & \text{otherwise} \end{cases}$

(c) $f(x) = \begin{cases} \frac{1}{3}e^{-x/3}, & x \geq 0 \\ 0 & \text{otherwise} \end{cases}$

SOLUTION.

(a) $\int_1^3 \frac{3}{26}x^2\,dx = \frac{3}{26}\int_1^3 x^2\,dx \qquad = \frac{3}{26}\left[\frac{x^3}{3}\Big|_1^3\right] = \frac{3}{26}\left(\frac{26}{3}\right) = 1$

(b) Here the domain of f is $(-\infty, \infty)$. Thus,

$$\begin{aligned} \int_{-\infty}^{\infty} f(x)\,dx &= \int_{-\infty}^{3} f(x)\,dx + \int_3^6 f(x)\,dx + \int_6^{\infty} f(x)\,dx \\ &= 0 + \int_3^6 \frac{1}{63}x^2\,dx + 0 \\ &= \frac{1}{63}\left[\frac{x^3}{3}\Big|_3^6\right] = \frac{1}{63}(63) = 1 \end{aligned}$$

(c)

$$\begin{aligned} \int_{-\infty}^{\infty} f(x)\,dx &= \int_{-\infty}^{0} f(x)\,dx + \int_0^{\infty} f(x)\,dx \\ &= 0 + \int_0^{\infty} \frac{1}{3}e^{-x/3}\,dx \\ &= \lim_{b\to\infty} \int_0^b \frac{1}{3}e^{-x/3}\,dx \end{aligned}$$

$$= \lim_{b \to \infty} \frac{1}{3} \left. \frac{e^{-x/3}}{-\frac{1}{3}} \right|_0^b$$

$$= \lim_{b \to \infty} [-e^{-b/3} + 1]$$

$$= 1, \text{ since } \lim_{b \to \infty} e^{-b/3} = 0$$

EXAMPLE 4. Find k (constant) such that each of the given functions is a probability density function over the given interval:

(a) $f(x) = kx^2, \quad 2 \leq x \leq 6$ (b) $f(x) = ke^{-x/6}, \quad x \geq 0$

SOLUTION. (a) We should have $\int_2^6 f(x)\,dx = 1$. Thus,

$$\int_2^6 kx^2\,dx = 1$$

or

$$k\left[\left.\frac{x^3}{3}\right|_2^6\right] = 1$$

$$k\left[\frac{208}{3}\right] = 1$$

Hence,

$$k = \frac{3}{208}$$

(b) In this case, we solve for k, the equation

$$\int_0^\infty ke^{-x/6}\,dx = 1$$

But $\int_0^\infty ke^{-x/6}\,dx = 6k$ (the reader should check this).

Therefore, $k = 1/6$.

EXAMPLE 5. The length of time in minutes that an individual talks on a long-distance telephone call has been found to be of a random nature. Let x be the length of the talk with probability density function given by

$$f(x) = ke^{(-1/4)x}, \quad x > 0$$

Find:

(a) The value of k that makes $f(x)$ a probability density function;

(b) The probability that this individual will talk for (1) less than 8 minutes, (2) between 8 and 12 minutes, and (3) more than 12 minutes.

SOLUTION. (a) We solve the equation

$$\int_0^\infty ke^{(-1/4)x}dx = 1$$

for k and obtain $k = 1/4$. Hence, the probability density function is:

$$f(x) = \tfrac{1}{4}e^{(-1/4)x}, \quad x > 0$$

(b) (1) $P[x < 8] = \int_0^8 \frac{1}{4}e^{(-1/4)x}\,dx = 1 - e^{-2}$

(2) $P[8 \leq x \leq 12] = \int_8^{12} \frac{1}{4}e^{(-1/4)x}\,dx = e^{-2} - e^{-3}$

(3) $P[x > 12] = \int_{12}^{\infty} \frac{1}{4}e^{(-1/4)x}\,dx = e^{-3}$

EXERCISES 6.5

In Problems 1 through 6, verify condition (3) of a probability density function over the given interval.

1. $f(x) = \dfrac{1}{4}$, $[7, 11]$

2. $f(x) = \dfrac{2x}{15}$, $[1, 4]$

3. $f(x) = \dfrac{3}{124}x^2$, $[1, 5]$

4. $f(x) = \ln x$, $[1, e]$

5. $f(x) = 0.1e^{-0.1x}$, $[0, \infty)$

6. $f(x) = ae^{-ax}$, $a > 0$, $[0, \infty)$

In Problems 7 through 18, find k so that each function is a probability density function over the given interval.

7. $f(x) = k$, $[2, 7]$

8. $f(x) = kx$, $[2, 6]$

9. $f(x) = kx^2$, $[3, 9]$

10. $f(x) = k(3 - x)$, $[0, 3]$

11. $f(x) = \dfrac{k}{1 + x}$, $[0, e - 1]$

12. $f(x) = ke^x$, $[0, 3]$

13. $f(x) = ke^{-4x}$, $[0, \infty)$

14. $f(x) = ke^{-0.01x}$, $[0, \infty)$

15. $f(x) = \begin{cases} kx^2, & -1 \leq x \leq 1 \\ 0 & \text{otherwise} \end{cases}$

16. $f(x) = \begin{cases} k(1 - x^2), & -1 \leq x \leq 1 \\ 0 & \text{otherwise} \end{cases}$

17. $f(x) = \begin{cases} kx, & 0 \leq x \leq 4 \\ ke^{-x}, & x > 4 \\ 0, & \text{otherwise} \end{cases}$

18. $f(x) = \begin{cases} kx^2, & 0 \leq x \leq 3 \\ ke^{-x}, & x > 3 \\ 0, & \text{otherwise} \end{cases}$

19. The Brite TV manufacturing company guarantees its TV's for two years. The time (in months) for failure after a TV is sold is given by the probability density function

$$f(x) = \begin{cases} 0.02e^{-0.2t}, & t \geq 0 \\ 0, & \text{otherwise} \end{cases}$$

Find the probability that a buyer chosen at random will have his TV fail:

(a) during warranty, (b) during the first year, (c) during the first two months.

20. The number of cars that a certain dealer sells each day is given by the probability density function:

$$f(x) = \begin{cases} k(x+1), & 0 \leq x \leq 12 \\ \dfrac{-10(6k-1)}{x^2}, & 12 < x \leq 20 \\ 0, & \text{otherwise} \end{cases}$$

(a) Find k such that $f(x)$ is a probability density function on $[0, \infty)$.

(b) Find the probability that the dealer will sell between 10 and 16 cars on a given day.

21. Suppose $P(x)$, the probability that x items of a product are demanded per week, is given by

$$P(x) = 0.03e^{-0.03x}$$

What is the probability that the demand in a given week is: (a) less than 100 items, (b) less than 500 items, and (c) more than 500 items?

22. A transportation planner finds that the average distance between two cars on a certain highway is $a = 80$ feet. If the distance x in feet between two cars on this highway has the probability density function

$$f(x) = \frac{1}{a}e^{(-1/a)x}$$

find the probability that the distance between cars is less than: (a) 50 feet, (b) 100 feet, and (c) more than 60 feet.

23. A continuous (random) variable is said to be *uniformly distributed* if it has a probability density function f given by

$$f(x) = \begin{cases} \dfrac{1}{b-a}, & a \le x \le b \\ 0, & \text{otherwise} \end{cases}$$

Show that f satisfies condition (3) of probability density functions.

24. The melting point in degrees of a specimen has a uniform probability distribution given by

$$f(x) = \begin{cases} \dfrac{1}{20}, & 100 \le x \le 120 \\ 0, & \text{otherwise} \end{cases}$$

Find the probability that such a specimen will melt between 108° and 115°.

25. The daily consumption of fuel in millions of gallons at a certain airport has a probability density function given by

$$P(x) = \begin{cases} 0.04e^{-0.04x}, & x \ge 0 \\ 0, & \text{otherwise} \end{cases}$$

Suppose the airport can store one million gallons of fuel; find the probability that this fuel supply will be inadequate on a given day.

26. The time t (in days) of a stay in a hospital due to Carcimia has a probability density function given by

$$g(t) = \begin{cases} 0.3e^{-0.3t}, & t \ge 0 \\ 0, & \text{otherwise} \end{cases}$$

Find the probability that a randomly selected person with Carcimia will stay in the hospital for less than four days.

CHAPTER 6 REVIEW

Part 1 (ORAL)

Define or discuss each of the following:

1. Differential equation
2. Initial-value problem

3. Present value

4. Annuity

5. Present value of a continuous income stream

6. Average value of a function

7. Consumer's surplus

8. Producer's surplus

9. Equilibrium point

10. Piecewise continuous function

11. Improper integral

12. Probability density function

Part 2 (WRITTEN)

In Problems 1 through 4, find a solution of the initial-value problem.

1. $y'(t) = 1 + t, \; y(0) = -1$

2. $\dfrac{dy}{dx} = e^x, \; y(1) = 2$

3. $\dfrac{dy}{dx} = xe^{-x}, \; y(0) = 3$

4. $\dfrac{dy}{dx} = \ln x, \; y(e) = -1$

5. Suppose the marginal cost of producing x items is given by

$$C'(x) = 2x - \frac{1}{18}\sqrt{x} - 100$$

   (a) Find the total cost $C(x)$ if the fixed cost is \$400.

   (b) Find the profit $P(x)$ if each item sells for \$10.

6. Suppose the marginal revenue obtained by selling x items is given by

$$R'(x) = -3x^2 + 2x + 6$$

(a) Find the total revenue if $R(0) = 0$.

(b) Find the demand function [*Hint*: $R(x) = xD(x)$].

7. Weekly production in ounces at Greeta's Gold Mine is given by $P(x) = 60\sqrt{x} - 4x$ for the xth week. Estimate (use definite integral) the number of ounces produced: (a) during the first 16 weeks; (b) during the first 64 weeks; (c) In how many weeks will the gold mine be exhausted? (d) How many ounces of gold are produced before the mine is exhausted?

8. The cost of producing a gadget is $20 + 500/\sqrt{x}$ for the xth gadget. Estimate the cost of producing: (a) the first 25 gadgets; (b) the first 100 gadgets; (c) the gadgets numbered 101 to 400. Find the average cost per gadget in each of these cases.

In Problems 9 through 12, find the present and future values of a continuous annuity at an annual payment rate of $f(t)$ dollars at time t years, with the given interest rate of $100r\%$ compounded continuously for T years.

9. $r = 0.06,\ T = 10,\ f(t) = 5000$

10. $r = 0.06,\ T = 10,\ f(t) = 5000 + 30t$

11. $r = 0.08,\ T = 8,\ f(t) = 5000 - 2t$

12. $r = 0.1,\ T = 10,\ f(t) = 5000 + e^t$

In Problems 13 through 16, find the average value of the given function over the given interval.

13. $f(x) = 60, \quad [6, 14]$

14. $f(x) = x - \dfrac{1}{x^2}, \quad [2, 4]$

15. $f(x) = x + \dfrac{1}{x}, \quad [1, e^2]$

16. $f(x) = xe^x, \quad [-1, 1]$

17. Find the average square root $\bar{c}$ of a number x chosen between 1 and 9.

18. The marginal cost of producing a widget is given by

$$C'(x) = 50 + 0.02x - 0.003x^2$$

Find the average additional cost per unit when the production is increased from 100 to 200 units.

In Problems 19 through 22 assume that pure competition is obtained. For the given supply and demand functions find: (a) the equilibrium point, (b) the consumer's surplus, and (c) the producer's surplus.

19. $S(x) = 2x, \quad D(x) = 500 - 3x$

20. $S(x) = 50 + x, \quad D(x) = 200 - 2x$

21. $S(x) = 100 - 30x + x^2, \quad D(x) = 1600 + 100x - x^2$

22. $S(x) = 500 - 70x + x^2, \quad D(x) = 3500 + 200x - 2x^2$

In Problems 23 through 28, evaluate the given improper integral if it exists.

23. $\int_0^\infty e^{(-2/3)x}\,dx$

24. $\int_1^\infty \frac{1}{\sqrt{x}}\,dx$

25. $\int_1^3 \frac{1}{(x-2)^2}\,dx$

26. $\int_0^5 \frac{1}{(3x-1)^{1/3}}\,dx$

27. $\int_{-\infty}^\infty f(x)dx$, where $f(x) = \begin{cases} 5e^{-5x}, & x \ge 0 \\ 0, & \text{otherwise} \end{cases}$

28. $\int_{-\infty}^\infty f(x)dx$, where $f(x) = \begin{cases} 2e^{2x}, & x < 0 \\ x, & 0 \le x \le 1 \\ 2e^{-3x}, & x > 1 \end{cases}$

29. Find k so that the function $f(x) = ke^{-0.02x}$ is a probability density function on the interval $[0, \infty)$.

30. Find k so that the function

$$f(x) = \begin{cases} kx^3, & -1 \le x \le 1 \\ 0, & \text{otherwise} \end{cases}$$

is a probability density function of the interval $(-\infty, \infty)$.

31. Suppose the probability that x items of a product are demanded per week is given by the probability density function

$$P(x) = 0.01x^{-0.01x}$$

What is the probability that the demand in a given week is: (a) less than 500 items, (b) between 500 and 1000 items, and (c) more than 1000 items?

32. The daily use (x gallons) of water per person has a probability density function given by

$$P(x) = 0.04e^{-0.04x}$$

Find the probability that a person chosen at random will use: (a) at least 200 gallons of water per day, (b) between 200 and 300 gallons of water per day, and (c) more than 300 gallons of water per day.

Chapter Seven

Differential Equations

The subject of differential equations[1] *is an important aspect of modern mathematics. Numerous problems encountered in the various branches of the physical and biological sciences, the social sciences, economics, and engineering lead to differential equations. Such equations arise in the study of the motion of celestial objects and atomic particles, the rate of growth of populations, the decomposition of radioactive substances, and problems involved with social diffusion. In this chapter we define and classify differential equations. We indicate in some cases how the differential equations arise and discuss simple methods of solution.*

7.1 INTRODUCTION

First, let us consider a problem involving the diffusion of some information through the inhabitants of a country.

EXAMPLE 1. A company is using the television media to advertise its latest product. Suppose the information concerning the product spreads from television commericals only. It is found that a fixed percentage (say, 5%) of the population is reached

[1] In the seventeenth century Galileo and Hooke used differential equations to study the behavior of beams under stress and strain. Such problems arose in the building of some of the famous cathedrals. Newton and others used differential equations in their study of such problems as the shape of the earth and gravitational attraction, and it was Leibniz who coined the term "differential equation."

per day and that the rate of spread of information is proportional to the number of people who have not heard of the product. Describe the situation mathematically.

SOLUTION. Let $y(t)$ denote the number of persons who have heard of the product at time t. Suppose A represents the number of persons in the country. Then $A - y(t)$ denotes the number of persons who have not yet heard of the product. Since the rate of spread of information (that is, dy/dt) is proportional to the number of persons who have not yet heard of the product, we obtain

$$\frac{dy}{dt} = k(A - y) \tag{1}$$

where k is the constant of proportionality. Initially, before airing of the commercials, the number of persons who had heard of the product was zero. (We are excluding the employees of the company.) We denote this by

$$y(0) = 0 \tag{2}$$

Thus, this situation is described by an initial-value problem. The problem involves an unknown function $y(t)$ to be determined so that Equations (1) and (2) are satisfied.

Equation (1) is called a *differential equation*. In general, an *ordinary differential equation* is an equation involving an independent variable t (or x, or u, etc.); a dependent variable y; and certain derivatives of y with respect to t. We may use the expression

$$F(t, y, y', y'', \ldots, y^{(n)}) = 0 \tag{3}$$

The relationship (3) must involve some derivative of the unknown function y. For instance,

$$y'' + ty = 0 \tag{4}$$

$$y' + ty = e^t \tag{5}$$

$$y''' + y'y = t^2 + 1 \tag{6}$$

are examples of differential equations. However, the equation

$$x^2 + ye^x + t^2 = 0$$

is *not* a differential equation because no derivatives are involved. Also, identities of the form

$$\frac{d}{dx}(e^{-x}) + e^x = 0$$

are *not* differential equations because no unknown function is involved.

A differential equation in which the highest derivative is the nth derivative is called a differential equation of *order n*. Thus, Equations (4), (5), and (6) are of order two, one, and three, respectively. Equations (1), (4), and (5) are called *linear differential equations*. In general, a linear differential equation is of the form

$$a_n(t)y^{(n)} + a_{n-1}(t)y^{(n-1)} + \cdots + a_0(t)y = f(t) \tag{7}$$

where $a_0(t), a_1(t), \ldots, a_n(t)$ and $f(t)$ are known functions of t. A differential equation that is not linear is called a *nonlinear differential equation*. Equation (6) cannot be written in the form (7). Thus, Equation (6) is a nonlinear differential equation. Other examples of nonlinear differential equations are

$$y'' + (y')^3 = x + 1 \tag{8}$$

$$\frac{dy}{dx} + xe^y = 0 \tag{9}$$

Note that Equation (6) is nonlinear because of the term $y'y$, Equation (8) is nonlinear because of the term $(y')^3$, and Equation (9) is nonlinear because of the term e^y.

Definition 1. A solution of a differential equation is a function defined on some interval such that the function and its respective derivatives on that interval satisfy the given equation.

EXAMPLE 2. Show that $y = e^t$ is a solution of the equation

$$y'' - y = 0 \tag{10}$$

SOLUTION. For $t \in R$, we find that $y'' = e^t$, and substitution into Equation (10) yields

$$y'' - y = e^t - e^t = 0$$

Thus, $y = e^t$ is a solution of the given equation for $t \in R$.

The reader can also show that $y = e^{-t}$ is a solution of Equation (10). In fact,

$$y = c_1e^t + c_2e^{-t} \tag{11}$$

where c_1 and c_2 are arbitrary constants, is also a solution of Equation (10) for $t \in R$. The function in (11) is called the *general solution* of Equation (10).

A solution of an nth-order differential equation that contains n essentially arbitrary constants[2] is called the *general solution*. The general solution of a given differential equation contains all possible solutions of the equation. A solution that is obtained from the general solution by giving particular values to one or more of the n essentially arbitrary constants is called a *particular solution* of the equation.

EXAMPLE 3. Show that $y = c_1e^t + c_2e^{3t}$ is the general solution of the equation

$$y'' - 4y' + 3y = 0 \tag{12}$$

SOLUTION. From the function

$$y = c_1e^t + c_2e^{3t}$$

we find

$$y' = c_1e^t + 3c_2e^{3t} \quad \text{and} \quad y'' = c_1e^t + 9c_2e^{3t}$$

Substituting y, y', and y'' into Equation (12), we find that for $t \in R$,

$$y'' - 4y' + 3y = c_1e^t + 9c_2e^{3t} - 4(c_1e^t + 3c_2e^{3t}) + 3(c_1e^t + c_2e^{3t}) = 0$$

Thus, $y = c_1e^t + c_2e^{3t}$ is a solution of (12) for every pair of constants c_1 and c_2. Since Equation (12) is a second-order differential equation and $y = c_1e^t + c_2e^{3t}$ contains essentially two arbitrary constants, then it is the general solution.

EXAMPLE 4. Find a particular solution of Equation (12) with the following initial conditions:

$$y(0) = 1 \quad \text{and} \quad y'(0) = 0 \tag{13}$$

SOLUTION. In this case we seek a function that is a solution of Equation (12) and satisfies the initial conditions (13). Using the general solution, we try to find values for c_1 and c_2 such that the resulting function satisfies not only Equation (12) but also (13). From the condition $y(0) = 1$ and the general solution, we find that

$$1 = c_1 + c_2 \tag{14}$$

---

[2] The function $y = (c_1 + c_2)e^t + c_3e^{-t}$ has essentially two arbitrary constants and not three since $c_1 + c_2$ may be replaced by a single arbitrary constant.

and from $y'(0) = 0$, we get

$$0 = c_1 + 3c_2 \tag{15}$$

Solving Equations (14) and (15), we obtain

$$c_1 = \tfrac{3}{2} \quad \text{and} \quad c_2 = -\tfrac{1}{2}$$

Thus, the function

$$y = \tfrac{3}{2}e^t - \tfrac{1}{2}e^{3t}$$

is a solution of (12) and (13). The reader is encouraged to verify this.

EXAMPLE 5.

(a) Show that $y = c_1e^t + c_2e^{2t}$ is the general solution of the equation

$$y'' - 3y' + 2y = 0 \tag{16}$$

(b) Find a solution satisfying the supplementary (or boundary[3]) conditions

$$y(0) = 0 \quad \text{and} \quad y(1) = 2 \tag{17}$$

SOLUTION. Direct substitution of $y = c_1e^t + c_2e^{2t}$ and its derivatives y', y'' into Equation (16) yields

$$y'' - 3y' + 2y = (c_1e^t + 4c_2e^{2t}) - 3(c_1e^t + 2c_2e^{2t}) + 2(c_1e^t + c_2e^{2t}) = 0$$

Thus, $y = c_1e^t + c_2e^{2t}$ is a solution of Equation (16). Since the differential equation is of order two and the solution contains two arbitrary constants, then it is the general solution. Now we try to solve for c_1 and c_2 so that the conditions in (17) are satisfied. The condition

$$y(0) = 0 \quad \text{implies} \quad 0 = c_1 + c_2$$

---

[3] A problem involving a differential equation associated with supplementary conditions which are related to two or more values of the independent variable is called a *boundary value problem*.

and the condition

$$y(1) = 2 \quad \text{implies} \quad 2 = c_1 e + c_2 e^2$$

Solving for c_1 and c_2, we obtain

$$c_1 = \frac{2}{e - e^2} \quad \text{and} \quad c_2 = \frac{2}{e^2 - e}$$

Thus, a particular solution of Equation (16) satisfying the boundary conditions (17) is

$$y = \frac{2}{e^2 - e}(e^{2t} - e^t) \tag{18}$$

The reader is encouraged to verify that (18) is a solution of the *boundary-value problem* (16) and (17).

EXERCISES 7.1

In Problems 1 through 12, give the order of the differential equation. State whether the equation is linear or nonlinear.

1. $y'' + 3y' + y = t$

2. $y'' + 3tyy' = 0$

3. $x\dfrac{dy}{dx} + y - y^2 = 0$

4. $(y'')^2 + 1 + y^2 = 0$

5. $y^{(4)} = t^2 + 1$

6. $\dfrac{d^3y}{dx^3} - 4\dfrac{dy}{dx} + 3y = e^x$

7. $x^2\dfrac{dy}{dx} + y^2 = 0$

8. $\dfrac{d^3y}{dx^3} + 2\left(\dfrac{dy}{dx}\right)^4 + 5y = 0$

9. $y'' + ye^x = 0$

10. $y'' + xe^y = 0$

11. $\left(\dfrac{dy}{dx}\right)^3 = \dfrac{d^2y}{dx^2} + 1$

12. $(1 + t^2)y'' + ty' = 5$

In Problems 13 through 20, show that each of the functions f is a solution of the given differential equation for some interval on the real line.

13. $f(x) = 2e^{-x} + x, \quad \dfrac{dy}{dx} + y = 1 + x$

14. $f(x) = 3e^{-x} - 5e^x, \quad \dfrac{d^2y}{dx^2} - y = 0$

15. $f(x) = 2e^x + e^{2x} + \frac{1}{6}e^{-x}, \quad \dfrac{d^3y}{dx^3} - \dfrac{d^2y}{dx^2} - 4\dfrac{dy}{dx} + 4y = e^{-x}$

16. $f(x) = 1 + 2e^x - 3e^{-x}, \quad \dfrac{d^3y}{dx^3} - \dfrac{dy}{dx} = 0$

17. $f(x) = e^x + \frac{1}{2}xe^x, \quad \dfrac{d^2y}{dx^2} - y = e^x$

18. $f(x) = 2x - xe^{1/x}, \quad x^3\dfrac{d^2y}{dx^2} + x\dfrac{dy}{dx} - y = 0$

19. $f(x) = \dfrac{1}{x^2}(1 - \ln x), \quad x^2\dfrac{d^2y}{dx^2} + 5x\dfrac{dy}{dx} + 4y = 0$

20. $f(x) = \dfrac{1}{2x+1}, \quad (2x+1)^2\dfrac{d^2y}{dx^2} - (4x+2)\dfrac{dy}{dx} - 12y = 0$

In Problems 21 through 28, show that the given function f is the general solution of the given differential equation.

21. $f(t) = -16t^2 + c_1t + c_2, \qquad y'' = -32$

22. $f(x) = c_1e^{3x} + c_2e^{-3x}, \quad \dfrac{d^2y}{dx^2} - 9y = 0$

23. $f(t) = c_1e^t + c_2te^t, \quad y'' - 2y' + y = 0$

24. $f(t) = c_1e^t + c_2e^{-2t} + t^2, \quad y'' + y' - 2y = 2(1 + t - t^2)$

25. $f(x) = c_1x^2 + c_2x^{-3}, \quad x^2\dfrac{d^2y}{dx^2} + 2x\dfrac{dy}{dx} - 6y = 0$

26. $f(t) = c_1e^t + c_2e^{-t} + \frac{1}{2}te^t, \quad y'' - y = e^t$

*27. $f(t) = c_1 + c_2e^t + c_3e^{-t} + \frac{1}{2}(t^2 - 3t)e^t, \quad y''' - y' = 2te^t$

*28. $f(t) = c_1 + c_2e^{-2t} + c_3te^{-2t} + c_4t^2e^{-2t}, \quad y^{(4)} + 6y''' + 12y'' + 8y' = 0$

29. (a) Show that $y = ce^{-2t}$, where c is an arbitrary constant, is the general solution of the differential equation $y' + 2y = 0$.
(b) Find a particular solution satisfying the initial conditions $y(0) = -3$.

30. Consider the equation $y' = -ky$.
(a) Show that $y = ce^{-kt}$ is a solution of the equation.
(b) Find c given that $y(0) = 2$.

(c) Find k given that $y(1) = 1$.
(d) What is $y(2)$?

31. (a) Show that $y = c_1e^{2x} + c_2e^{3x}$, where c_1 and c_2 are arbitrary constants, is the general solution of the differential equation

$$\frac{d^2y}{dx^2} - 5\frac{dy}{dx} + 6y = 0$$

(b) Find a particular solution satisfying the initial conditions $y(0) = 3$ and $y'(0) = 0$.

32. (a) Show that $y = c_1e^{-x} + c_2e^{2x}$ is the general solution of the equation

$$\frac{d^2y}{dx^2} - \frac{dy}{dx} - 2y = 0$$

(b) Find a particular solution satisfying the boundary conditions $y(0) = 1$ and $y(1) = 2$.

33. Show that $y = 1/(1 - x^2)$ is a solution of the initial-value problem

$$\frac{dy}{dx} - 2xy^2 = 0, \quad y(0) = 1$$

34. (a) Show that $y = 1/(\sqrt{2x + c}\,)$ is a solution of the equation

$$\frac{dy}{dx} + y^3 = 0$$

(b) Find the solution satisfying the initial condition $y(1) = 2$.

7.2 SOME METHODS OF SOLUTION

In this section we shall develop methods for solving two classes of first-order differential equations. First, we shall consider first-order differential equations in which the variables can be separated and then we shall consider the linear first-order differential equations for which the variables cannot be separated.

One of the earliest procedures used for solving differential equations is *separation of variables*. In certain cases a first-order differential equation may be written in the form

$$f(x) + g(y)\frac{dy}{dx} = 0 \tag{1}$$

Equation (1) can be integrated directly using the chain rule to obtain

$$\int\left[f(x) + g(y)\frac{dy}{dx}\right]dx = C, \qquad \int f(x)\,dx + \int g(y)\frac{dy}{dx}\,dx = C$$

or simply

$$\int f(x)\,dx + \int g(y)\,dy = C$$

where C is the constant of integration. We illustrate this procedure in the following examples.

EXAMPLE 1. Find the general solution of the equation:

$$2x\frac{dy}{dx} = 2x^3 - 1 \tag{2}$$

SOLUTION. For $x \neq 0$, we separate the variables as follows:

$$\frac{dy}{dx} = x^2 - \frac{1}{2x} \tag{3}$$

Equation (2) and (3) are equivalent; that is, the general solution of (3) is also the general solution of (2). Direct integration with respect to x yields

$$\int \frac{dy}{dx}\,dx = \int\left(x^2 - \frac{1}{2x}\right)dx \qquad \text{or} \qquad y = \frac{x^3}{3} - \frac{1}{2}\ln|x| + C$$

where C is an arbitrary constant. The reader can verify that y is indeed the general solution of Equation (2) on any interval of the x-axis, where $x \neq 0$.

EXAMPLE 2. Find the general solution of the given differential equation.

$$\frac{dy}{dx} = \frac{xy^2 - x}{xy^3 - y^3} = \frac{x(y^2 - 1)}{y^3(x - 1)} \tag{4}$$

SOLUTION. Separation of variables yields

$$\frac{y^3}{y^2 - 1}\frac{dy}{dx} = \frac{x}{x - 1}$$

Direct integration with respect to x yields

$$\int \frac{y^3}{y^2 - 1}\frac{dy}{dx}\,dx = \int \frac{1}{x - 1}\,dx$$

$$\int\left(y + \frac{y}{y^2 - 1}\right)dy = \int\left(1 + \frac{1}{x - 1}\right)dx$$

or

$$\frac{y^2}{2} + \frac{1}{2}\ln|y^2 - 1| = x + \ln|x - 1| + C \tag{5}$$

where C is an arbitrary constant.

In this case we are unable to write $y = f(x)$. We say that (5) defines an *implicit solution* of (4).

EXAMPLE 3. Solve the given initial-value problem.

$$\frac{dy}{dx} = \frac{3x^2 - 1}{2(y-1)}, \qquad y(0) = -2 \tag{6}$$

SOLUTION. Separation of variables and integration with respect to x yield

$$\int 2(y-1)\,dy = \int (3x^2 - 1)\,dx$$

or

$$y^2 - 2y = x^3 - x + C \tag{7}$$

where C is an arbitrary constant. To determine the particular solution satisfying the given initial condition, we set $x = 0$ and $y = -2$ in Equation (7) and obtain $C = 8$. Hence, the desired particular solution is given implicitly by

$$y^2 - 2y = x^3 - x + 8 \tag{8}$$

To obtain y in terms of x, we solve Equation (8) using the quadratic formula:

$$y = 1 \pm \sqrt{x^3 - x + 9} \tag{9}$$

Thus, Equation (9) provides two solutions of the differential equation; the one that satisfies the initial condition is

$$y = 1 - \sqrt{x^3 - x + 9} \tag{10}$$

The reader should verify that Equation (10) gives the solution to the initial-value problem.

Unfortunately, not all first-order equations can be solved by the method we have just described. For example, the following equations are not separable:

$$xy\frac{dy}{dx} = x^2 + y^2 \quad \text{and} \quad 3x^2y + y^3 + (x^2 + 3xy^2)\frac{dy}{dx} = 0$$

Methods for solving such equations are usually discussed in a first course on differential equations.

A very important class of first-order differential equations with applications in a wide variety of disciplines is the first-order linear differential equation

$$\frac{dy}{dx} + a(x)y = b(x) \tag{11}$$

Except for special cases, such equations are not separable. We shall give a method for solving linear equations.

Consider the function v, where

$$v = e^{\int a(x)\,dx} \tag{12}$$

If we multiply Equation (11) by v, we obtain

$$\frac{dy}{dx}e^{\int a(x)\,dx} + a(x)\,ye^{\int a(x)\,dx} = b(x)\,e^{\int a(x)\,dx} \tag{13}$$

from which we get

$$\frac{d}{dx}(ye^{\int a(x)\,dx}) = b(x)\,e^{\int a(x)\,dx} \tag{14}$$

Equation (14) can be integrated with respect to x and thus solved for y. The solution of Equation (14) is the solution of Equation (11). The function v defined by (12) is called an *integrating factor*. It should be noted that (12) is an integrating factor provided the differential equation is of the form (11). We illustrate this method in the next two examples.

EXAMPLE 4. Solve the given differential equation.

$$\frac{dy}{dx} + 2y = x \tag{15}$$

Note that the method separation of variables does not work in this case.

SOLUTION. First we find an integrating factor:

$$v = e^{\int 2\,dx} = e^{2x} \tag{16}$$

Multiplying Equation (15) by e^{2x}, we get

$$e^{2x}\frac{dy}{dx} + 2e^{2x}y = xe^{2x}$$

or

$$\frac{d}{dx}(ye^{2x}) = xe^{2x} \tag{17}$$

Note that when the linear equation is multiplied by the integrating factor, the left-hand side is always the derivative of y times the integrating factor. Integrating equation (17) with respect to x, we have

$$\int \frac{d}{dx}(ye^{2x})\,dx = \int xe^{2x}\,dx$$

$$ye^{2x} = \tfrac{1}{2}xe^{2x} - \tfrac{1}{4}e^{2x} + C$$

$$y = \tfrac{1}{2}x - \tfrac{1}{4} + Ce^{-2x}$$

where C is an arbitrary constant. The reader should verify that y is indeed the general solution of (15).

EXAMPLE 5. Solve the given initial-value problem.

$$x\frac{dy}{dx} + 2y = x^3, \qquad y(1) = -2 \tag{18}$$

SOLUTION. Writing the differential equation in the form (11), we have

$$\frac{dy}{dx} + \frac{2}{x}y = x^2 \tag{19}$$

and an integrating factor is

$$v = e^{\int (2/x)\,dx} = e^{2\ln x} = e^{\ln x^2} = x^2$$

Multiplying Equation (19) by x^2, we get

$$x^2\frac{dy}{dx} + 2xy = x^4 \quad \text{or} \quad \frac{d}{dx}(yx^2) = x^4$$

Integration with respect to x yields

$$\int \frac{d}{dx}(yx^2)\,dx = \int x^4\,dx, \qquad yx^2 = \frac{x^5}{5} + C$$

or

$$y = \tfrac{1}{5}x^3 + Cx^{-2} \tag{20}$$

Equation (20) gives the general solution of the differential equation. Using the initial condition, we set $x = 1$ and $y = -2$ to obtain $C = -\frac{11}{5}$. Therefore, the solution of the initial-value problem is

$$y = \tfrac{1}{5}\,(x^3 - 11x^{-2})$$

The reader is encouraged to verify this result.

EXERCISES 7.2

In problems 1 through 18, use the method of separation of variables to find the general solution of the given differential equation.

1. $y' = y$

2. $y' + 2y = 0$

3. $y' + \sqrt{3}y = 0$

4. $2y' + 3y = 0$

5. $\dfrac{dy}{dx} = \dfrac{4y}{x}$

6. $\dfrac{dy}{dx} = 3y$

7. $(x+1)\dfrac{dy}{dx} = y^2$

8. $(y-4)\dfrac{dy}{dx} = x+3$

9. $x^2y\dfrac{dy}{dx} = e^{-y}$

10. $e^{2x}\dfrac{dy}{dx} = 1$

11. $\dfrac{dy}{dx} = e^{x+y}$ [*Hint:* $e^{x+y} = e^xe^y$]

12. $2\dfrac{dy}{dx} = e^{-2x+y}$

13. $\dfrac{dy}{dx} = -3y+2$

14. $\dfrac{dy}{dx} = \dfrac{1}{-4x+3}$

15. $\sqrt{1-x^2}\,\dfrac{dy}{dx} = -x$

16. $\dfrac{dy}{dt} = -4yt$

17. $e^x\dfrac{dy}{dx} = -xy^2$

18. $\dfrac{dy}{dx} = xe^{y-x^2}$

In problems 19 through 26, solve the linear differential equation.

19. $y' + 2y = 3$

20. $y' - y = 10$

21. $\dfrac{dy}{dx} + 3y = x$

22. $\dfrac{dy}{dx} + \dfrac{3y}{x} = 3x^2$

23. $x\dfrac{dy}{dx} + 6y = x$

24. $y' = t^3 - 2ty$

25. $x\dfrac{dy}{dx} + (2x + 1)\,y = xe^{-2x}$

26. $x\dfrac{dy}{dx} = 8x^2 - 2y$

In problems 27 through 31, solve the initial-value problem.

27. $x\dfrac{dy}{dx} - 2y = x^4, \quad y(1) = 3$

28. $y' + y = e^{-t}, \quad y(0) = 2$

29. $yy' = t^3, \quad y(1) = 10$

30. $\dfrac{dy}{dx} = 5 - \dfrac{2y}{10 - x}, \quad y(0) = 5$

31. $(2 + y)\dfrac{dy}{dx} = x\sqrt{y}, \quad y(0) = 1$

32. Suppose that f_1 and f_2 are each solutions of the equation

$$\frac{dy}{dx} + a(x)y = 0$$

on an interval I. Show that $f_1 + f_2$ is also a solution on I.

33. Suppose that f_1 and f_2 are solutions of the equation

$$\frac{dy}{dx} + a(x)y = b(x)$$

on an interval I, where $b(x) \neq 0$ on I. Show that $f_1 + f_2$ is not a solution of the equation on I.

34. Suppose that f_1 is a solution of

$$\frac{dy}{dx} + a(x)y = b_1(x)$$

and f_2 is a solution of

$$\frac{dy}{dx} + a(x)y = b_2(x)$$

where a, b_1, and b_2 are all defined on the same interval I. Show that $f_1 + f_2$ is a solution of

$$\frac{dy}{dx} + a(x)y = b_1(x) + b_2(x)$$

on I.

35. (a) Show that there exist only two values of m such that $y = e^{mx}$ is a solution of the equation

$$\frac{d^2y}{dx^2} - 4\frac{dy}{dx} + 3y = 0$$

(b) If m_1 and m_2 are the two values, show that

$$y = c_1e^{m_1x} + c_2e^{m_2x}$$

is the general solution, where c_1 and c_2 are arbitrary constants.

In Problems 36 through 40, repeat Problem 35 for the given equation.

36. $y'' + 3y' + 2y = 0$

37. $y'' - 4y = 0$

38. $y'' - 2y = 0$

39. $y'' - 2y' = 0$

40. $2y'' - 2y' - 3y = 0$

7.3 SOME APPLICATIONS

In this section we illustrate some applications of differential equations. First we return to Example 1 of Section 1.

EXAMPLE 1. In Example 1 of Section 1, suppose that the advertising of the product reaches 5% of the population per day. Find the number of persons who would have heard of the product at time t.

SOLUTION. We are asked to solve the initial-value problem

$$\frac{dy}{dt} = k(A - y) \tag{1}$$

$$y(0) = 0 \tag{2}$$

Also, we are given that at the end of the first day 5% of the population have been reached; that is,

$$y(1) = \tfrac{5}{100}A = \tfrac{1}{20}A \tag{3}$$

Thus, we seek a function $y = f(t)$ that satisfies Equations (1), (2), and (3). From the differential equation (1), we find by separation of variables

$$\int \left(\frac{1}{A - y}\frac{dy}{dt}\right) dt = \int k\,dt, \qquad -\ln(A - y) = kt + C_1$$

or

$$A - y = e^{-kt-C_1} = Ce^{-kt}$$

where $C = e^{-C_1}$. Hence,

$$y = A - Ce^{-kt} \tag{4}$$

Using condition (2), we set $t = 0$ and $y = 0$ in (4) to find that $C = A$; and equation (4) becomes

$$y = A(1 - e^{-kt}) \tag{5}$$

Using condition (3), we set $t = 1$ and $y = \frac{1}{20}A$ in (5) to obtain $k = -\ln\frac{19}{20}$. Therefore, the number of persons who have been reached by advertising at any time t is given by

$$y = A(1 - e^{t\,\ln(19/20)})$$

Next, we illustrate the use of differential equations in problems involving rates of change in the temperature of an object.

Suppose an object A with temperature T_A is immersed in a medium B with constant temperature T_B. It is found experimentially that the rate of change of temperature of A is proportional to the difference in temperature between A and B. Mathematically, this situation is expressed by

$$\frac{dT_A}{dt} = k(T_B - T_A) \tag{6}$$

where $k > 0$ is a constant of proportionality. Consider the following example.

EXAMPLE 2. An object at a temperature of 180° is allowed to cool in a medium having a constant temperature of 60°. If in 1 minute the temperature of the object is 120°, in how many minutes will its temperature be 90°?

SOLUTION. Here $T_B = 60°$. If T represents the temperature of the object at any time t, then

$$T(0) = 180° \tag{7}$$

and

$$T(1) = 120° \tag{8}$$

Using the differential equation (6), we have

$$\frac{dT}{dt} = k(60 - T) \tag{9}$$

Equations (7), (8), and (9) form a boundary-value problem. We solve to obtain (observe that the initial value of t [i.e., $t = 0$] is used in the lower limit of integration.)

$$\int_0^t \frac{1}{60 - T(s)} \frac{dT}{ds}\, ds = k \int_0^t ds$$

$$-\ln|60 - T(s)|\,\Big|_0^t = ks\Big|_0^t$$

$$-\ln|60 - T(t)| + \ln|60 - 180| = kt$$

$$\ln \frac{120}{T(t) - 60} = kt$$

or

$$\frac{120}{T(t) - 60} = e^{kt}$$

and thus

$$T(t) = 60(1 + 2e^{-kt}) \tag{10}$$

In (10) the number k is still unknown. To evaluate k, we use (8) and (10). Thus,

$$120 = 60(1 + 2e^{-k})$$

$$\tfrac{1}{2} = e^{-k}$$

and

$$\ln \tfrac{1}{2} = -k$$

Equation (10) becomes

$$T(t) = 60(1 + 2e^{[\ln(1/2)]t}) \tag{11}$$

To find t when $T = 90°$, we use Equation (11) and obtain

$$90 = 60(1 + 2e^{[\ln(1/2)]t})$$

$$\tfrac{1}{4} = e^{[\ln(1/2)]t}$$

and

$$\ln \tfrac{1}{4} = (\ln \tfrac{1}{2})t$$

Hence,

$$t = \frac{\ln\frac{1}{4}}{\ln\frac{1}{2}} = \frac{-\ln 4}{-\ln 2} = \frac{2\ln 2}{\ln 2} = 2$$

that is, in 2 minutes the temperature of the body decreases to 90°.

Substances that decrease in quantity continuously at a rate proportional to the amount present are said to *decay* exponentially. In mathematical symbols, we write

$$\frac{dQ}{dt} = -kQ \tag{12}$$

where $k > 0$ and Q is the amount of substance at time t.

EXAMPLE 3. Suppose a radioactive substance decays according to the law in (12). If 20 lbs. are stored now, 18 lbs. would be left at the end of one century. Find the amount present at time t.

SOLUTION. We have the initial-value problem

$$\frac{dQ}{dt} = -kQ \tag{13}$$

$$Q(0) = 20 \tag{14}$$

Integration yields

$$\int_0^t \frac{1}{Q}\frac{dQ}{ds}\,ds = -\int_0^t k\,ds$$

$$\ln Q(s)\Big|_0^t = -ks\Big|_0^t$$

$$\ln Q(t) - \ln Q(0) = -kt$$

$$\ln\frac{Q(t)}{20} = -kt$$

and

$$Q(t) = 20e^{-kt} \tag{15}$$

Since $Q = 18$ at $t = 1$, we have

$$18 = 20e^{-k}$$

$$0.9 = e^{-k}$$

and

$$-k = \ln 0.9$$

Therefore, the desired function is defined by

$$Q(t) = 20e^{(\ln 0.9)t}$$

Carbon 14 (^{14}C) is a radioactive substance. The date at which an organism lived may be estimated by analyzing the remains of carbon 14 in the organism, since all living organisms contain carbon 12 and carbon 14. From the moment the organism dies, carbon 14 continues to decay and is no longer replaced, while carbon 12 remains unchanged. The change in the amount of carbon 14 relative to carbon 12 makes it possible to calculate the (approximate) time at which the organism lived.

EXAMPLE 4. An archaeological expedition working in the Gobi Desert came across some ancient objects. It was determined that about 30% of the original amount of ^{14}C was still present in the objects. If ^{14}C decays continuously, find the approximate age of the objects (assume that the half-life[4] of ^{14}C is 5600 years).

SOLUTION. Let $A(t)$ be the amount of ^{14}C at time t. If $A(t)$ satisfies the differential equation (12), then

$$A(t) = Ce^{-kt} \tag{16}$$

where C is a constant. At $t = 0$ the amount present is A_0. Hence,

$$A(0) = Ce^0 \qquad C = A_0$$

and

$$A(t) = A_0e^{-kt} \tag{17}$$

[4] The half-life of a substance is the time required for one-half the amount of the substance to decay.

After 5600 years the amount of ^{14}C present is $\frac{1}{2}A_0$. Thus,

$$A(5600) = A_0 e^{-5600k} = \tfrac{1}{2}A_0$$

$$\tfrac{1}{2} = e^{-5600k}$$

or

$$\ln\tfrac{1}{2} = \ln e^{-5600k}$$

$$k = \tfrac{1}{5600}\ln 2$$

Thus,

$$A(t) = A_0 e^{(-\ln 2/5600)t}$$

In this case 30% of the ^{14}C is present. Therefore,

$$\tfrac{30}{100}A_0 = A_0 e^{(-\ln 2/5600)t}$$

$$\tfrac{3}{10} = e^{(-\ln 2/5600)t}$$

and

$$\ln\tfrac{3}{10} = -\left(\frac{\ln 2}{5600}\right)t$$

Hence,

$$t = -5600\frac{\ln\frac{3}{10}}{\ln 2} = -5600\frac{\ln 3 - \ln 10}{\ln 2}$$

$$= -5600\,\frac{1.0986 - 2.3026}{0.6931}$$

or

$$t \approx 9700 \text{ years}$$

In other words the objects found were about 9700 years old.

EXAMPLE 5. Suppose in an autocatalytic reaction 1 unit of a substance is present initially. Let q denote the amount of the product. If the rate of formation of q is given by

$$\frac{dq}{dt} = kq(1-q) \tag{18}$$

where k is a constant,

(a) Find the amount of the product at time t;

(b) Find $q(t)$ given that at $t = 0$ the amount of the product is $\frac{1}{2}$ units.

SOLUTION. (a) We write equation (18) in the form

$$\frac{1}{q(1-q)}\frac{dq}{dt} = k \tag{19}$$

Integrating both sides of equation (19) with respect to t we obtain

$$\int \frac{1}{q(1-q)}\frac{dq}{dt}\,dt = \int k\,dt \tag{20}$$

We note that

$$\frac{1}{q(1-q)} = \frac{A}{q} + \frac{B}{1-q}$$

Solving for A and B we find

$$1 = A(1-q) + Bq$$

Setting $q = 0$ we have $A = 1$; setting $q = 1$ we have $B = 1$. Hence, equation (20) may be written as

$$\int \left(\frac{1}{q} + \frac{1}{1-q}\right) dq = \int k\,dt$$

$$\ln q - \ln(1-q) = kt + C$$

$$\ln \frac{q}{1-q} = kt + C \tag{21}$$

We find the exponential of each side of equation (21):

$$\frac{q}{1-q} = e^{kt+C}$$

or

$$\frac{q}{1-q} = Me^{kt}$$

Solving for q we obtain

$$q(t) = \frac{Me^{kt}}{1 + Me^{kt}} \tag{22}$$

(b) At $t = 0, q = \frac{1}{2}$. Hence, from equation (22) we have

$$\frac{1}{2} = \frac{Me^0}{1 + Me^0}$$

Solving for M we have $M = 1$. Therefore,

$$q(t) = \frac{e^{kt}}{1 + e^{kt}} \tag{23}$$

EXERCISES 7.3

1. The president of a university announces that no X-rated movies will be shown on campus. Suppose that 20% of the student body of 20,000 heard the announcement by the end of the first hour and transmitted this information by word of mouth. Further suppose that the rate of spread of information (per hour) is proportional to the number of students who have not heard the news. How many students would have heard the news at time t? How long will it take the news to spread through one-half the student body?

In Problems 2 through 5, the rate of change of temperature is proportional to the difference of temperatures between two mediums.

2. At a room temperature of 25°, the temperature of a liquid is observed to be 75°. In 15 minutes the temperature of the liquid is observed to be 65°. Find the temperature of the liquid 90 minutes after the first observation.

3. Rosita prepares an apple pie for her friends Bob, Manuel, and Arlene. The pie is removed from the oven at 420° and left at a room temperature of 70°. In one-half hour the pie is at 210°. Eager to have a piece of the pie, Bob calculates the time when the pie will be at 100°. How long will the friends have to wait?

4. A thermometer reading 75° is placed outdoors. In 4 minutes the reading is 30°. If the outside temperature is fixed at 20°, find the following:

(a) The thermometer reading 8 minutes after it was placed outdoors;

(b) The time it takes for the reading to drop to 21°.

5. An object is placed in a medium having a constant temperature. Initially, the temperature of the object is 40° higher than the medium. In 8 minutes the temperature of the object is 20° higher than the medium.

(a) Determine k, the constant of proportionality.

(b) In how many minutes will the difference in temperature be 10°?

6. The population P of a country increases continuously at 4% per year. Assume

$$\frac{dP}{dt} = kP, \quad k > 0$$

(a) If the present population is 200 million, what will it be in 10 years? 20 years?

(b) When will the population double? triple?

7. The birth rate of a population of ants is 3.5% per week. If there are no deaths among the ants, what is the population in 20 weeks when it is known that the population $P = 10^3$ at $t = 0$?

8. According to the 1940 census the population of Nuroville was 64,000 and in 1960 it was 80,000. If the rate of increase of population is proportional to the population, find the year in which the population will be 160,000. (Use $2\ln 2 = \ln 4 = 1.38628$, $\ln 5 = 1.60944$.)

9. A population P of mosquitoes increases by 3000 in 1 day. Given that P satisfies the initial-value problem

$$\frac{dP}{dt} = kP, \quad P(0) = 1200$$

(a) Find k.

(b) Find the size of the population after 12 hours.

10. Suppose the half-life of a substance that decays continuously is 1200 years. What is the amount of the substance left in 750 years? 500 years?

11. Suppose the rate of change in the amount of oil in this country is proportional to the amount present. If the constant of proportionality is 5% per year and the amount present in 1800 was 10^{12} barrels, find the amount that was present in 1900; in 1970.

12. A chemical plant discharges fumes into the atmosphere. The level h is 5 feet above the stacks at 6 A.M. and 20 feet at noon. If the rate of change in the level is proportional to h, what is the level at 5 P.M.?

13. Find the point on the curve $y = 2\ln(x - 3)$ at which the tangent to the curve has slope $\frac{1}{2}$.

14. A particle moves according to the law

$$s = ae^{kt} + be^{-kt}$$

where a, b, and k are constants. Show that its acceleration is proportional to its distance from the origin.

15. Let E be the electromotive force, R the resistance, L the inductance of the circuit; then the current i at time t satisfies the differential equation

$$L\frac{di}{dt} + Ri = E$$

If $i = 0$ when $t = 0$, show that the solution of this differential equation is given by

$$i = \frac{E}{R}[1 - e^{-Rt/L}]$$

16. Show that the solution of the equation

$$\frac{dy}{dx} = abe^{-bx}, \quad y(0) = 0$$

is given by the curve $y = a - ae^{-bx}$. Such a curve is also called a *learning curve.* For $a > 0, b > 0$, graph the learning curve. The learning curve derives its name from the fact that learning increases rapidly at first but gradually increases at a slower and slower rate and finally approaches an upper limit.

17. Suppose glucose is infused to the blood stream at a constant rate. The concentration $C(t)$ of glucose at time t is given by

$$\frac{dC}{dt} = a - kt$$

where a and k are constants. If $C(0) = 0$, show that the solution of this differential equation is also a learning curve.

18. In Example 5, if $q = \frac{1}{4}$ at $t = 1$, find $q(t)$.

19. Two chemicals react to form a new substance x. The rate of formation of x is given by

$$\frac{dx}{dt} = k(1 - x)(2 - x)$$

where k is a constant.

(a) Find $x(t)$ given that $x(0) = 0$.

(b) Find k if $x(1) = \frac{1}{2}$.

20. Let P_0 be the population of a city struck by an epidemic. Suppose the rate of spreading of the disease is given by

$$\frac{dP}{dt} = P(P_0 - P)$$

where P is the population that has contracted the disease.

(a) Determine $P(t)$ if $P(0) = 0$.

(b) If t is measured in days, how long will it take before half the population is affected by the disease.

21. The model depicted by

$$\frac{dP}{dt} = kP(L - P)$$

is called the "model of inhibited growth." Show that the differential equation has solution

$$P(t) = \frac{CLe^{kLt}}{1 + Ce^{kLt}}$$

where k and L are known constants and C is an arbitrary constant.

22. In problem 21, if $P(0) = P_0$, show that

$$P(t) = \frac{P_0Le^{kLt}}{P_0e^{kLt} + (L - P_0)}$$

Show that the point of inflection of $y = P(t)$ occurs at that value of t where $P = \frac{1}{2}L$.

CHAPTER 7 REVIEW

Part 1 (ORAL)

Define or discuss each of the following.

1. Differential equation
2. Order of a differential equation
3. Linear differential equation
4. Nonlinear differential equation
5. Solution of a differential equation
6. General solution
7. Particular solution
8. Initial condition
9. Boundary-value problem

10. Method of separation of variable

11. Implicit solution

12. Integrating factor

13. Exponential decay

14. Half-life of a radioactive substance

Part 2 (WRITTEN)

1. Find the general solution of the differential equation

$$\frac{dy}{dx} = -3x^4 + 4$$

2. Find the general solution of the differential equation

$$\frac{d^2y}{dx^2} = 3 + 8x$$

3. Find a particular solution of the initial-value problem

$$\frac{dy}{dx} + 2y = 3x, \quad y(0) = 2$$

4. Find a particular solution of the boundary-value problem

$$\frac{d^2y}{dx^2} = 2x + 1, \qquad y(0) = 2 \quad \text{and} \quad y(-2) = 3$$

5. Find a particular solution of the boundary-value problem

$$\frac{d^2y}{dx^2} + e^x = 2, \qquad y(0) = -2 \quad \text{and} \quad y(1) = \frac{3}{2}$$

6. Suppose that a rumor in a town spreads at a rate given by

$$y' = kt$$

where y is the number of people who have heard the rumor and t is the time in hours.

(a) Find k given that $y(0) = 0$ and $y(2) = 100$.

(b) Find y when $t = 3, 6,$ and 10.

7. A radioactive substance decays at a rate given by

$$y' = -\frac{5}{100}y$$

where y represents the amount (in grams, g) present at time t (in months).

(a) Find the general solution of the equation.

(b) Find a particular solution if $y(0) = 100$.

(c) Find the amount of substance left 10 months later.

8. Let y denote the demand for a certain product and x denote the price per item. Suppose the rate of change of demand for the product is given by

$$\frac{dy}{dx} = -2x + 20, \qquad y(0) = 10$$

Find the demand at each of the following price levels: $x = 5, x = 10, x = 15$.

9. If the marginal cost is given by

$$\frac{dC}{dx} = 40x + 1000$$

find the total cost function. If the fixed cost is 2500, find the cost of producing 100 units.

10. If the marginal profit is given by

$$\frac{dP}{dx} = -4x + 20$$

find the total profit function. If $P(0) = -20$, at what level of sales will maximum profit be realized? What is the maximum profit?

11. Let p denote the production and x the capitalization (in hundred thousand dollars). Suppose the rate at which production changes for a unit change in capitalization, in the *marginal productivity*, for a certain company is given by

$$p'(x) = 4x^2 - 15$$

(a) If the company's present capitalization is $200,000 while producing 150 units of an item per month, find the production $p(x)$.

(b) If the company's capitalization is increased to $400,000 how much would the production increase?

12. Do problem 1 of Exercises 7.3 if the rate of spread of information is given by

$$\frac{dS}{dt} = kS(T - S)$$

where S is the number of students who have heard the announcement after t hours and $T = 20{,}000$.

In Problems 13 through 20, find the general solution of the given differential equation.

13. $y\dfrac{dy}{dx} = x^2 - 2$

14. $y\dfrac{dy}{dx} = x^2 + 2x$

15. $\dfrac{dy}{dx} + 2xy = 0$

16. $\dfrac{dy}{dx} = x^2y$

17. $\dfrac{dy}{dx} + xy = 2x^2y$

18. $\dfrac{dy}{dx} + 2xy = x^2$

19. $x\dfrac{dy}{dx} + 2y = x$

20. $\dfrac{dy}{dx} + y = e^{-x}$

Chapter Eight

Functions of Several Variables

8.1 Functions of Two Variables

In this chapter we shall study functions of two variables. The theory for the functions of three or more variables can be developed along similar lines. Functions of several (two or more) variables arise in many disciplines. For example, the area A of a rectangle is given by (see Figure 1)

$$A = xy$$

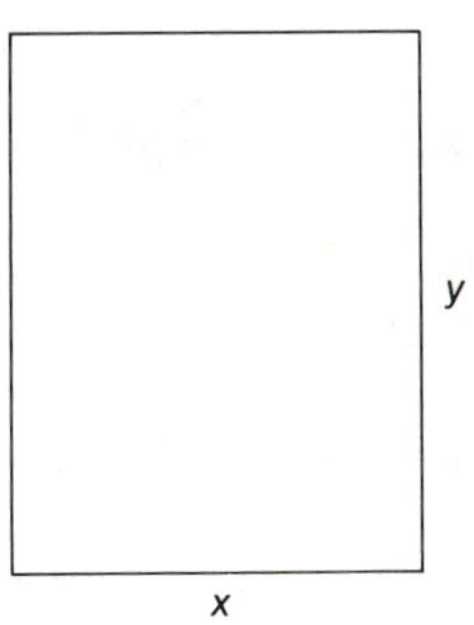

Figure 1

Thus, the area A is a function of two independent variables. Similarly, the selling price P of an anitbiotic depends upon its cost, its quality, and the competitor's selling price. Thus, P is a function of several variables.

Definition 1. A real-valued function of two variables is a function f that has a set of ordered pairs of real numbers as its domain and a subset of real numbers as its range.

In the case of functions of one variable, we write $f(x)$ to denote the value of the function at x. Similarly, if f is a function of two variables, we write $f(x, y)$ to denote the value of f at the ordered pair (x, y).

EXAMPLE 1. Consider the relation

$$z = f(x, y) = \sqrt{1 - x^2 - y^2} \tag{1}$$

The relation (1) between x, y, and z determines a value for z or $f(x, y)$ corresponding to every ordered pair of numbers (x, y) such that $x^2 + y^2 \leq 1$.

Denoting the pair of numbers (x, y) geometrically by a point on the plane, we see that the points (x, y) for which $x^2 + y^2 \leq 1$ lie on or within a circle with center at the origin and radius 1. Thus the domain of f is a region in the plane given by $\{(x, y) \mid x^2 + y^2 \leq 1\}$. The range of this function is the closed interval $[0, 1]$.

EXAMPLE 2. Describe the domain and the range of the function f, defined by

$$z = f(x, y) = e^{x^2+y^2}$$

SOLUTION. Domain of $f = \{(x, y) \mid x \in R,\ y \in R\}$, range of $f = \{z \mid z \geq 1\}$

EXAMPLE 3. Describe the domain of the function f defined by

$$f(x, y) = \sqrt{(1 - x)(x - 2)} + \sqrt{(3 - y)(y - 5)}$$

SOLUTION. Since $(1 - x)(x - 2)$ is nonnegative if $1 \leq x \leq 2$ and $(3 - y)(y - 5)$ is nonnegative if $3 \leq y \leq 5$, the domain of f is $\{(x, y) \mid 1 \leq x \leq 2;\ 3 \leq y \leq 5\}$. This domain is a rectangle (see Figure 2) in the xy-plane bounded by the lines

$$x = 1, \quad x = 2, \quad y = 3, \quad y = 5$$

We have seen in Chapter 1 that the graph of a function of one variable can be represented pictorially in the plane. To graph a function of two variables, we need a three-dimensional coordinate system. We draw three mutually perpendicular coordinate axes meeting at a common point, called the *origin* (see Figure 3). The *coordinate lines* are the x-axis, y-axis, and z-axis. The y-axis and the z-axis are usually drawn in the plane of the paper (the y-axis horizontal and the z-axis vertical), and the x-axis is drawn to point toward the reader. The positive directions of the axes are indicated by arrowheads. Each pair of coordinate axes determines a plane. Thus, we have the xy-plane, yz-plane, and zx-plane. Any two of these three planes (each unlimited in extent) divide space into four equal parts, and all three of them divide space into eight equal parts known as *octants*.

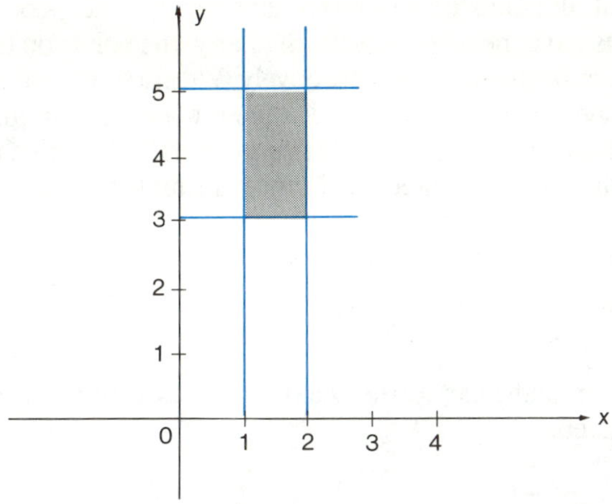

Figure 2

Let P be any point in space. The position of the point P is described by (a, b, c), where a, b, and c are directed distances from the three coordinate planes, as shown in Figure 4. For example, the point $(3, -4, 7)$ is located 3 units in front of the yz-plane, 4 units left of the zx-plane, and 7 units above the xy-plane. Thus, given an ordered triple of real numbers (a, b, c), we can find a point in space which represents this ordered triple. Conversely, given a point P in space, we construct the three planes through P which are parallel to the three coordinate planes. The plane parallel to the yz-plane intersects the x-axis at the *x-coordinate*; the plane parallel to the xz-plane intersects the y-axis at the *y-coordinate*; and the plane parallel to the xy-plane intersects the z-axis at the *z-coordinate*. Thus, there is a unique correspondence between points in three-dimensional space and ordered triples of real numbers.

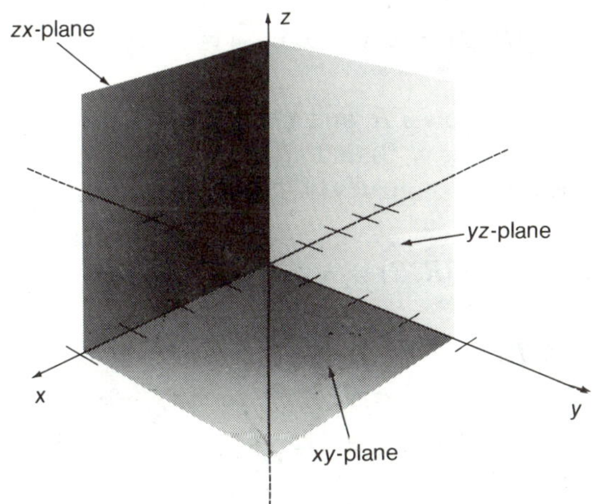

Figure 3

We observe that all points on the x-axis have their y- and z-coordinates both zero; that is, these points have the form $(x, 0, 0)$. Similarly, the points on the y-axis and z-axis have the forms $(0, y, 0)$ and $(0, 0, z)$, respectively. We also note that all points in a plane perpendicular to the z-axis have the same value for their z-coordinates. The equation $z = c$ (where c is a real constant) is satisfied at every point (x, y, c) lying in a plane perpendicular to the z-axis and directed distance of c units from the xy-plane. Similarly,

$$y = c, \qquad x = c$$

represent planes perpendicular to the y-axis and x-axis, respectively. We now prove the following theorem.

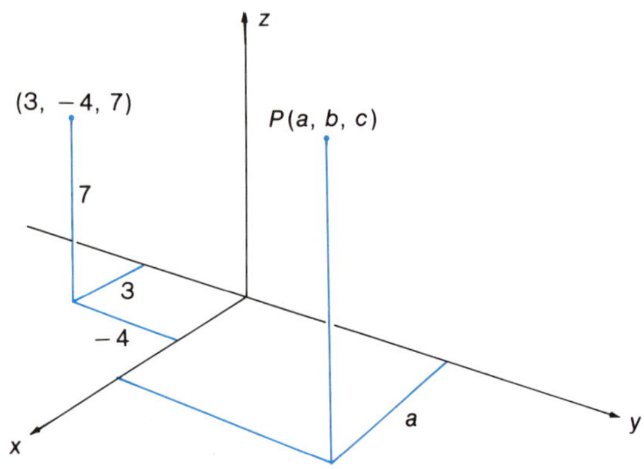

Figure 4

THEOREM 1. The distance between two points $P(x_1, y_1, z_1)$ and $Q(x_2, y_2, z_2)$ is

$$d(P, Q) = \sqrt{(x_2 - x_1)^2 + (y_2 - y_1)^2 + (z_2 - z_1)^2} \tag{2}$$

Proof. See Figure 5. Draw PR and QS perpendicular to the xy-plane. The coordinates of R and S are $(x_1, y_1, 0)$ and $(x_2, y_2, 0)$, respectively. Also, draw PT perpendicular to QS. Then the coordinates of T are (x_2, y_2, z_1). From the figure we see that

$$d(R, S) = \sqrt{(x_2 - x_1)^2 + (y_2 - y_1)^2}$$

In the rectangle $RPTS$, $d(P, T) = d(R, S)$. Furthermore,

$$d(T, Q) = |z_2 - z_1| = \sqrt{(z_2 - z_1)^2}$$

By the Pythagorean theorem,

$$d(P, Q) = \sqrt{[d(P, T)]^2 + [d(T, Q)]^2}$$

$$= \sqrt{[d(R, S)]^2 + [d(T, Q)]^2}$$

$$= \sqrt{(x_2 - x_1)^2 + (y_2 - y_1)^2 + (z_2 - z_1)^2}$$

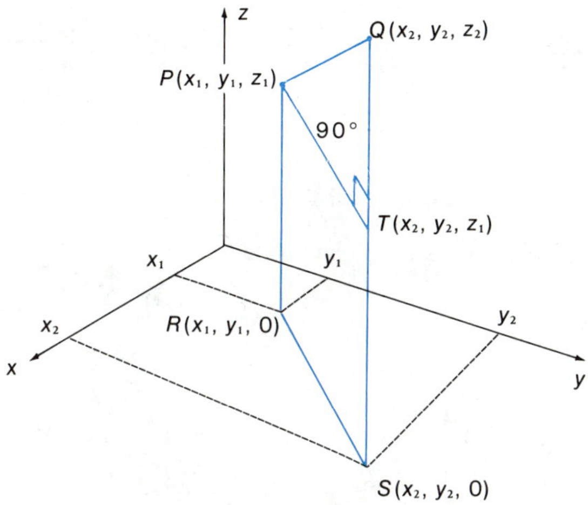

Figure 5

EXAMPLE 4. Find the distance between the points

$$P(2, 3, -1) \quad \text{and} \quad Q(-1, 0, 2)$$

SOLUTION. By direct substitution in (2), we obtain

$$d(P, Q) = \sqrt{(2-(-1))^2 + (3-0)^2 + (-1-2)^2}$$

$$= \sqrt{3^2 + 3^2 + 3^2} = \sqrt{27} = 3\sqrt{3}$$

EXERCISES 8.1

In Problems 1 through 4, find the value of the function (if defined) at the given points.

1. $f(x, y) = x^2 - xy + y$ at $(1, 0)$, $(2, -1)$ and $(-2, 3)$
2. $f(x, y) = ye^x + x^2 - 2y$ at $(0, 2)$, $(1, 0)$ and $(1, 1)$
3. $f(x, y) = 3x^2 + 4xy + y^3$ at $(0, 0)$, $(1, -1)$ and $(-1, 1)$
4. $f(x, y) = \ln xy + x \ln y$ at $(1, 1)$, $(1, e)$ and $(e, 1)$

In Problems 5 through 9, find the domain of f.

5. $f(x, y) = xy^2 + x^2y$

6. $f(x, y) = \sqrt{1 - \frac{x^2}{4} - \frac{y^2}{9}}$

7. $f(x, y) = \sqrt{(1 - x)(x - 2)} + \sqrt{(3 - y)(y - 4)}$

8. $f(x, y) = \dfrac{1}{\ln x + \ln y}$

9. $f(x, y) = \dfrac{x^3 + y^3}{x^2 + y^2}$

10. Plot the following points.

(a) $(1, 0, 0)$ (b) $(0, 2, 0)$ (c) $(0, 0, 3)$

(d) $(1, 2, 0)$ (e) $(1, 2, 3)$ (f) $(2, 1, 2)$

(g) $(-1, 2, 0)$ (h) $(-1, 0, -1)$ (i) $(2, 2, 2)$

(j) $(-1, -2, -3)$

11. Describe the set of all points (x, y, z) that satisfy the given conditions.

(a) $x = 0$ (b) $y = 0$ (c) $z = 3$

(d) $x = 0$ and $y = 0$ (e) $x = 1$ and $y = 0$ (f) $z = -2$

(g) $x = y = z$ (h) $x = 1, y = -1$, and $z = 2$

12. Calculate the distance between the points.

(a) $(0, 0, 0)$, (a, b, c) (b) $(1, 0, -1)$, $(0, 2, -3)$

(c) $(1, 2, 3)$, $(1, -2, 0)$ (d) $(3, 0, 1)$, $(-1, 0, -2)$

(e) $(0, 4, 5)$, $(0, -8, 0)$ (f) $(1, 2, 3)$, $(-4, -5, -6)$

8.2 GRAPHS

We recall from Chapter 1 that the graph of a function f of one variable is the set of all points (x, y), such that $y = f(x)$.

Definition 1. The graph of a function f of two variables is the set of all points (x, y, z), such that $z = f(x, y)$.

For a function of one variable, the graph of $y = f(x)$ is generally a curve in the plane. However, in general, the graph of a function f given by $z = f(x, y)$ is a surface in three dimensions. Obtaining the shape or even a rough sketch of the graph of a function of two variables is a difficult task. Attempting to imitate the case of a function of one variable by plotting a reasonable number of points (x, y, z) in three dimensions and connecting them by a smooth surface will not succeed. Even if we are able to plot many points in space (which itself is quite difficult), we may not be able to recognize the shape of the surface. In this section we shall discuss some of the simplest surfaces.

The proof of the following theorem is beyond the scope of this book and is therefore omitted.

THEOREM 1. The graph of the linear equation

$$ax + by + cz = d \qquad (a, b, c \text{ not all equal to zero}) \tag{1}$$

is a plane in three-dimensional space.

EXAMPLE 1. Discuss the plane

$$2x + 3y + 4z = 12 \tag{2}$$

SOLUTION. It is known that three noncollinear points in space determine a unique plane. We can find points on (2) by choosing two coordinates and solving for the third. If we let $x = 0$ and $y = 0$, we obtain $z = 3$. Consequently, the point (0, 0, 3) is on the plane given by (2). Similarly, (0, 4, 0) and (6, 0, 0) are also points on the plane. In fact, these are points where the plane intersects the axes. Figure 1 is a sketch of the plane.

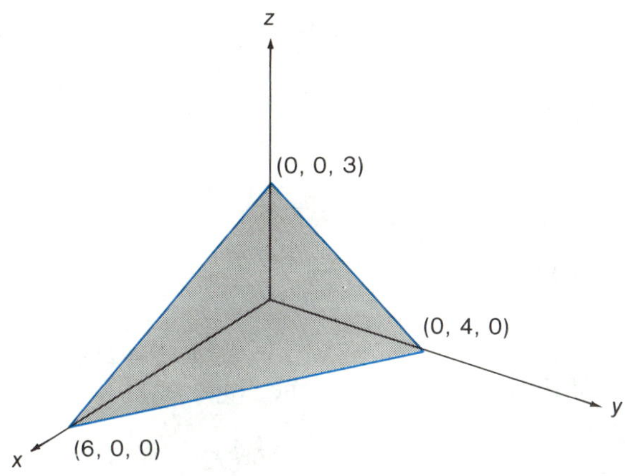

Figure 1

Suppose that we wish to graph the function given by

$$z = f(x) \tag{3}$$

in three-dimensional space. We note that one of the variables, namely y, is missing; i.e., Equation (3) is independent of y. Consequently, the point $(x_0, y_0, f(x_0))$ is on the graph if and only if the point $(x_0, 0, f(x_0))$ is on the graph. It follows that the graph of the surface given by (3) can be obtained by sliding the graph of $z = f(x)$ on the zx-plane parallel to the y-axis. In other words, we draw the curve G given by $z = f(x)$ on the zx-plane, and the graph of $z = f(x)$ in three-dimensional space is generated by moving the curve G parallel to the y-axis. Similar results hold for equations with x or z missing.

EXAMPLE 2. Sketch the graph in three-dimensional space of

$$2x + y = 4 \tag{4}$$

SOLUTION. Note that the variable z is missing in Equation (4). From the preceding discussion, we draw the graph of (4) in the xy-plane; we know that this graph is a straight line. We now slide this line parallel to the z-axis, and obtain the required graph, which is a plane, as shown in Figure 2.

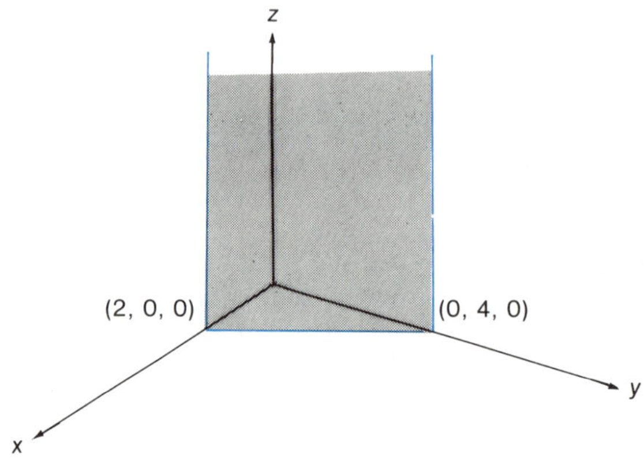

Figure 2

EXAMPLE 3. Sketch the graph of

$$z = x^2 \tag{5}$$

in three-dimensional space.

SOLUTION. We know that the graph of $z = x^2$ in the zx-plane is a parabola. Sliding this parabola parallel to the y-axis gives the graph of (5) in three-dimensional space as sketched in Figure 3.

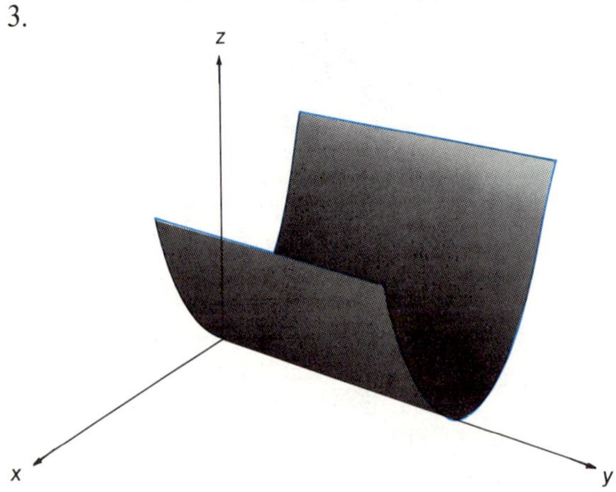

Figure 3

As we pointed out earlier, graphing a surface given by $z = f(x, y)$ is very difficult. However, if we hold y constant, say at $y = b$, then $z = f(x, b)$ is a function of one variable—namely, x. The graph of $z = f(x, b)$ lies on the plane $y = b$; in fact, it is the intersection of the graph of $z = f(x, y)$ with the plane $y = b$. Such an intersection is called a *cross section* of $z = f(x, y)$. Drawing several cross sections $z = f(x, b_1)$, $z = f(x, b_2), \ldots$ yields a fairly good idea of the shape of the surface $z = f(x, y)$. We can find cross sections by holding x constant also.

The cross sections of a given surface resulting from the intersection of the surface with the coordinate planes are called the *traces* of the surface. For example, the trace of the surface $z = x^2 - y$ in the xy-plane is $x^2 - y = 0$. The trace in the xy-plane is obtained by letting $z = 0$. Similarly, the traces in the yz- and zx-planes are obtained by letting $x = 0$ and $y = 0$, respectively, in the equation $z = f(x, y)$.

EXAMPLE 4. Sketch the graph of

$$x^2 + y^2 + z^2 = 9 \tag{6}$$

in three-dimensional space.

SOLUTION. The trace in the xy-plane is given by

$$x^2 + y^2 + 0 = 9 \quad \text{or} \quad x^2 + y^2 = 9$$

which is the equation of a circle with center at $x = 0$ and $y = 0$ and radius 3. Similarly, the traces on the yz-plane and xz-plane are, respectively,

$$y^2 + z^2 = 9 \quad \text{and} \quad x^2 + z^2 = 9$$

We also note that if we take a plane parallel to any of the coordinate planes, say $z = c$ $(-3 < c < 3)$, then the cross section of this plane with (6) is

$$x^2 + y^2 = 9 - c^2$$

which is a circle.

These traces and cross sections suggest that the graph of (6) is a sphere with center at (0, 0, 0) and radius 3 (see Figure 4).

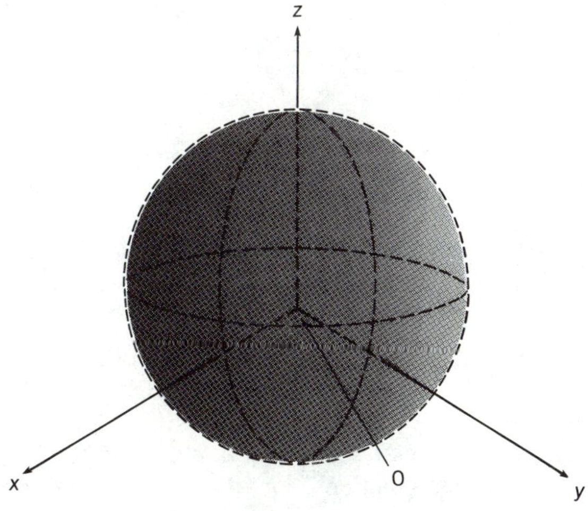

Figure 4.

EXAMPLE 5. Sketch the graph of the function

$$f(x, y) = z = x^2 + y^2 \tag{7}$$

SOLUTION. We note that the trace in the yz-plane is the parabola

$$z = 0 + y^2 = y^2$$

Similarly, the trace in the xz-plane is also the parabola

$$z = x^2 + 0 = x^2$$

The cross section of the surface (7) on a plane $z = c$ $(c > 0)$ is a circle.

$$x^2 + y^2 = c$$

Thus, the surface (7) is a cup-shaped surface called a *paraboloid* and is sketched in Figure 5.

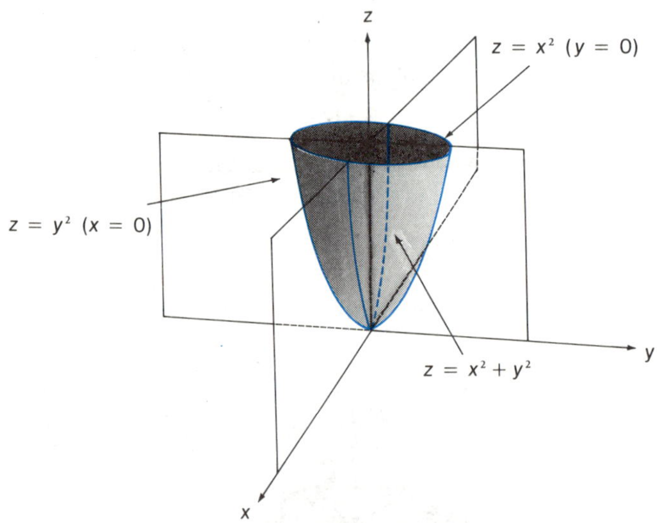

Figure 5

EXAMPLE 6. Sketch the graph of the function

$$f(x, y) = z = x^2 - y^2 \tag{8}$$

SOLUTION. The trace in the yz-plane is the parabola $z = y^2$; the trace in the xz-plane is the parabola $z = -x^2$. The cross sections on the planes $y = c$ are parabolas: $z = x^2 - c$. Similarly, the cross sections on the planes $x = k$ are parabolas: $z = k - y^2$, while the cross sections on the planes $z = h$ are "hyperbolas"; $h = x^2 - y^2$. From this information, we can draw a reasonable sketch of the surface given by (8) and obtain a saddle-shaped surface shown in Figure 6. The figure shows the point (0, 0, 0), called a *saddle point*, which is neither a relative maximum nor a relative minimum. We shall say more about relative maximum and relative minimum in Section 8.5.

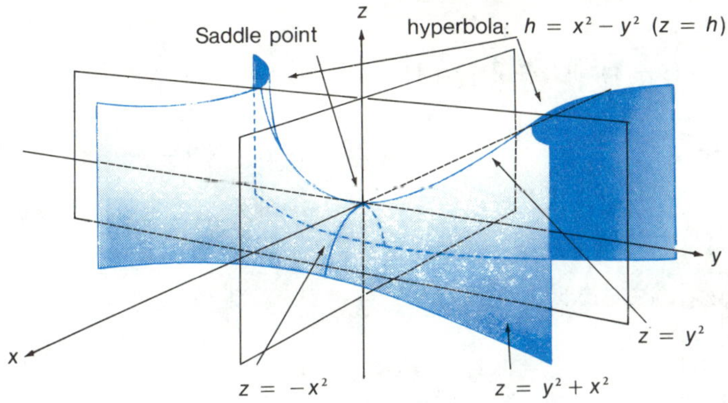

Figure 6

EXERCISES 8.2

In Problems 1 through 12, sketch the graphs of the given planes.

1. $x = 2$

2. $y = -3$

3. $z = \frac{1}{2}$

4. $x + y = 1$

5. $x + z = 3$

6. $x - y = -4$

7. $2y + z = 4$

8. $z - 3x = 6$

9. $x + y + z = 1$

10. $x - 2y + 3z = 6$

11. $2x + y + z = 4$

12. $-x - y + z = 1$

In Problems 13 through 32, sketch the following graphs in three dimensions.

13. $x^2 - y = 0$

14. $x^2 = 4y - 2$

15. $x^2 + y^2 = 4$

16. $y^2 + z^2 = 1$

17. $x^2 + z^2 = 2y$

18. $y = x^2$

19. $x = z^2$

20. $x = (y - 2)^2$

21. $x^2 + y^2 + z^2 = 1$

22. $x^2 + y^2 + z^2 = 16$

23. $(x-1)^2 + y^2 + z^2 = 9$

24. $x^2 + (y+1)^2 + (z+2)^2 = 4$

25. $(x-1)^2 + (y-2)^2 + (z-3)^2 = 16$

26. $(x+3)^2 + (y-1)^2 + (z+2)^2 = 16$

27. $z = \sqrt{4 - x^2 - y^2}$

28. $-x = \sqrt{9 - y^2 - z^2}$

29. $x = \sqrt{y - z^2}$

30. $-y = \sqrt{4z - x^2}$

31. $z = 2x^2 - y^2$

32. $z = x^2 - 2y^2$

8.3 PARTIAL DIFFERENTIATION

We know how to differentiate many kinds of functions of one independent variable and interpret the derivative dy/dx as the rate of change of y with respect to x. In particular, if $y = f(x)$, then $f'(a)$ measures the rate at which f is increasing or decreasing as we move away from $x = a$ on the x-axis. Ideally, then, if $z = f(x, y)$ is a function of two independent variables, we want to measure the rate of change of $z = f(x, y)$ as we move away from a point (a, b) in the xy-plane in some direction. This general situation is rather complicated. We can avoid the difficulty by considering only those two directions parallel to the x- and y-axes.

Let $z = f(x, y)$ be a function of two variables. If y is held constant or *fixed*, then the function $z = f(x, y)$ becomes a function of x alone, and its derivative may be found by the rules discussed in Chapter 2. This derivative is called the *partial derivative of z with respect to x* and is denoted by

$$\frac{\partial z}{\partial x}, \quad \frac{\partial f}{\partial x}, \quad \text{or} \quad f_x$$

The partial derivative with respect to y has a similar meaning.

When taking a partial derivative with respect to x (or y), we are in essence considering the behavior (rate of change) of a cross section of the surface on a plane parallel to the zx- (or yz-) plane.

Formally, we have the following definition.

Definition 1. Let $z = f(x, y)$ be a function of two variables. The partial derivatives of z with respect to x and with respect to y are defined by

$$\frac{\partial z}{\partial x} = \frac{\partial f}{\partial x} = f_x = \lim_{h \to 0} \frac{f(x+h, y) - f(x, y)}{h} \tag{1}$$

$$\frac{\partial z}{\partial y} = \frac{\partial f}{\partial y} = f_y = \lim_{k \to 0} \frac{f(x, y+k) - f(x, y)}{h} \tag{2}$$

provided that the limits in (1) and (2) exist.

Notations such as $(\partial z/\partial x)_{(a,\,b)}$ and $f_y(a, b)$ are used to indicate the value of the partial derivative at the point (a, b).

Definition 1 can be extended to functions of more than two variables. For example, if $w = f(x, y, u, v)$, then

$$\frac{\partial w}{\partial u} = \lim_{h\to 0} \frac{f(x, y, u+h, v) - f(x, y, u, v)}{h}$$

provided that the limit exists.

EXAMPLE 1. Calculate $\partial z/\partial x$ and $\partial z/\partial y$, where

$$z = x^2y + xy^{-3}$$

Also find the value of each partial derivative at (1, 2).

SOLUTION. Treating y as a constant, we have

$$\frac{\partial z}{\partial x} = 2xy + y^{-3}$$

Thus,

$$\left(\frac{\partial z}{\partial x}\right)_{(1,2)} = 2(1)(2) + 2^{-3} = 4 + \tfrac{1}{8} = \tfrac{33}{8}$$

Similarly, treating x as a constant, we obtain

$$\frac{\partial z}{\partial y} = x^2 - 3xy^{-4}$$

and

$$\left(\frac{\partial z}{\partial y}\right)_{(1,2)} = (1)^2 - 3(1)(2)^{-4} = 1 - \tfrac{3}{16} = \tfrac{13}{16}$$

The partial derivatives themselves are functions of two variables. We can, therefore, calculate the partial derivatives of $\partial z/\partial x$ and $\partial z/\partial y$, just as we calculated the partial derivatives of z. Thus, we have four *second partial derivatives* (with different notations).

(1) $$\frac{\partial}{\partial x}\left(\frac{\partial z}{\partial x}\right) = \frac{\partial^2 z}{\partial x^2} = \frac{\partial^2 f}{\partial x^2} = f_{xx}$$

(2) $$\frac{\partial}{\partial y}\left(\frac{\partial z}{\partial x}\right) = \frac{\partial^2 z}{\partial y\,\partial x} = \frac{\partial^2 f}{\partial y\,\partial x} = f_{xy}$$

(3) $$\frac{\partial}{\partial x}\left(\frac{\partial z}{\partial y}\right) = \frac{\partial^2 z}{\partial x\,\partial y} = \frac{\partial^2 f}{\partial x\,\partial y} = f_{yx}$$

(4) $$\frac{\partial}{\partial y}\left(\frac{\partial z}{\partial y}\right) = \frac{\partial^2 z}{\partial y^2} = \frac{\partial^2 f}{\partial y^2} = f_{yy}$$

Note that in the last equation of (2) and (3) the order of the symbols is reversed:

$$\frac{\partial^2 f}{\partial y\,\partial x} = f_{xy}$$

since $\frac{\partial^2 f}{\partial y\,\partial x}$ stands for $\frac{\partial}{\partial y}\left(\frac{\partial f}{\partial x}\right)$, and f_{xy} stand for $(f_x)_y$. Both symbols indicate that the partial derivative is first calculated with respect to x and then with respect to y.

The derivatives f_{xx} and f_{yy} are called the second partial derivatives with respect to x and y, respectively, while the derivatives f_{xy} and f_{yx} are called the *mixed partial derivatives.*

The mixed partial derivatives f_{xy} and f_{yx} are distinguished by the order in which f is successively differentiated with respect to x and y.

There are functions for which $f_{xy}(a, b) \neq f_{yx}(a, b)$. However, in most cases of interest $f_{xy} = f_{yx}$.

EXAMPLE 2. Calculate all the second partial derivatives of

$$f(x, y) = x^3y^4 + x^2e^y$$

SOLUTION.

$$f_x = 3x^2y^4 + 2xe^y, \qquad f_{xx} = 6xy^4 + 2e^y, \qquad f_{xy} = 12x^2y^3 + 2xe^y$$

$$f_y = 4x^3y^3 + x^2e^y, \qquad f_{yy} = 12x^3y^2 + x^2e^y, \qquad f_{yx} = 12x^2y^3 + 2xe^y$$

Note that $f_{xy} = f_{yx}$.

EXERCISES 8.3

In problems 1 through 16, find f_x and f_y.

1. $f(x, y) = 50 + 2x - 3y$
2. $f(x, y) = 2xy$
3. $f(x, y) = 40 - 5xy$
4. $f(x, y) = 2x + 5xy - 3y$
5. $f(x, y) = x^2y - xy^2$
6. $f(x, y) = x^3 + 3x^2y - 3xy^2 - y^3$
7. $f(x, y) = (xy)^3$
8. $f(x, y) = (x^2y^2 + 1)^5$
9. $f(x, y) = \frac{x}{y}$
10. $f(x, y) = (x^3 + y)(3x^2 - 5xy + y^2)$
11. $f(x, y) = \frac{x - y}{x + y}$
12. $f(x, y) = xe^y$

13. $f(x, y) = e^{x^2+2y^2}$

14. $f(x, y) = ye^{x^2} - x^2 \ln y$

15. $f(x, y) = e^{x+y} \ln (x^2 + y^2)$

16. $f(x, y) = x^3 + y^2 \ln x + xe^{xy^2+x^3y}$

17. Let $f(x, y) = x^2y^3$. Calculate all second partial derivatives at $(1, 2)$.

18. Let $g(x, y) = x \ln y$. Calculate all second partial derivatives at $(2, e)$.

19. Show that the function $f(x, y) = x^3 - 3x^2y$ satisfies the partial differential equation

$$f_{xx} + f_{yy} = 0$$

20. Show that the function of problem 19 satisfies the partial differential equation

$$f_{xy} + 6x = 0$$

21. Let $z = f(x, y, u, v) = (x^2 + y^3) \cdot e^{u+v}$. Calculate

(a) f_x, f_y, f_u, f_v (b) f_{xy}, f_{xu}, f_{uv} (c) $f_{xyu}, f_{uyx}, f_{uxy}$

22. If $u = xz^2 + yx^2 + zy^2$, prove that

$$\frac{\partial u}{\partial x} + \frac{\partial u}{\partial y} + \frac{\partial u}{\partial z} = (x + y + z)^2$$

8.4 SOME APPLICATIONS; THE TOTAL DIFFERENTIAL

One application of partial derivatives is obvious from its counterpart in functions of a single variable. The partial derivative f_x represents the rate of change of f with respect to x, when the variable y is kept fixed. A similar comment applies to f_y.

EXAMPLE 1. **Ecology.** Suppose a nuclear power plant and a chemical company are situated on a large lake. Let the pollution index P of the lake be a function of x and y, where x measures the quantity of heat entering the lake from the power plant, and y measures the amount of chemical waste pouring into the lake. We can write $P = f(x, y)$. then $(\partial P/\partial x)_{(a,b)}$ simply means that, if y is held constant at value b, and x alone is allowed to vary, then the pollution index P is changing $(\partial P/\partial x)_{(a,b)}$ units for each unit of change in x at the instant x takes the value a. Similar interpretation is given for $(\partial P/\partial y)_{(a,b)}$.

EXAMPLE 2. **Economics.** Suppose the productivity P of a company is a function of the size of the labor force x and the amount of capital invested y: $P = f(x, y)$. Then

$\partial P/\partial y$ is called the *marginal productivity of capital*, which means that for $x = a$, $y = b$, the productivity of the company increases approximately by $(\partial P/\partial y)_{(a, b)}$ (if it is positive) if the capital is increased by an additional dollar. Similarly, $\partial P/\partial x$ is called the *marginal productivity of labor*, which means that for $x = a$, $y = b$, the productivity of the company increases approximately by $(\partial P/\partial x)_{(a, b)}$ if one more person is added to the labor force.

EXAMPLE 3. **Geometry.** Let $z = f(x, y)$. We mentioned earlier that geometrically this equation represents a surface. If y is held constant, say $y = b$, then we obtain a cross section of this surface, determined by the surface $z = f(x, y)$ and the plane $y = b$. Thus, we have a curve $z = f(x, b)$ in the plane $y = b$. We note that z is now a function of the single variable x. The partial derivative $\partial z/\partial x$ is, therefore, the slope of the tangent to the curve of intersection of the surface and the plane $y = b$ (see Figure 1).

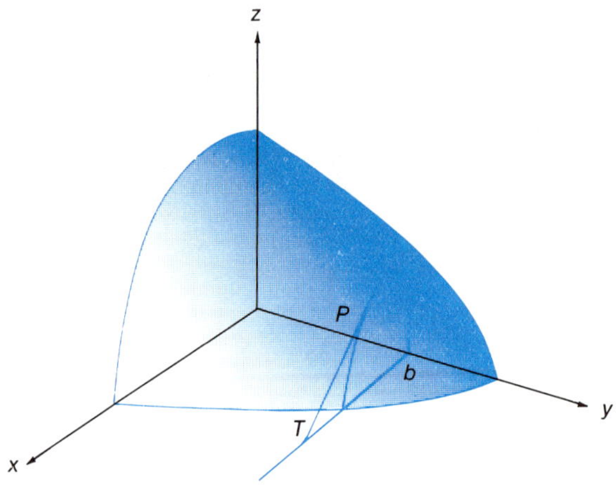

Figure 1

In Chapter 3, we defined the differential dy of the function $y = f(x)$ by

$$dy = f'(x)\,dx$$

We shall define a similar term for a function of two variables.

Let $z = f(x, y)$ be a function of two variables x and y. Let (x, y) and $(x + \Delta x, y + \Delta y)$ be any two points, so that Δx and Δy are the changes in the variables x and y, respectively. Let Δz denote the consequent change in z. We have

$$z + \Delta z = f(x + \Delta x, y + \Delta y)$$

or

$$\Delta z = f(x + \Delta x, y + \Delta y) - z = f(x + \Delta x, y + \Delta y) - f(x, y) \qquad (1)$$

Just as in the case of a single independent variable, for a function $z = f(x, y)$ we have two independent variables x and y; so we define the differentials

$$dx = \Delta x \quad \text{and} \quad dy = \Delta y$$

We define the differential dz by

$$dz = f_x \, \Delta x + f_y \, \Delta y = f_x \, dx + f_y \, dy$$

and call dz the *total differential.*

Thus, as in the case of a single variable, dz serves as an approximation of Δz. We note that the closer Δx and Δy are to zero, the closer the approximation dz is to the exact value of Δz.

EXAMPLE 4. Find the percentage error in the area of a rectangle, when an error of 1% is made in measuring the length and the width of the rectangle.

SOLUTION. Let x, y, and A denote the length, width, and area of the rectangle, respectively. Since $A = xy$, the partial derivatives are

$$\frac{\partial A}{\partial x} = y \quad \text{and} \quad \frac{\partial A}{\partial y} = x$$

Since we are given

$$dx = \frac{x}{100} \qquad \text{(1\% error in length)}$$

and

$$dy = \frac{y}{100} \qquad \text{(1\% error in width)}$$

we have

$$dA = \frac{\partial A}{\partial x} dx + \frac{\partial A}{\partial y} dy = y \frac{x}{100} + x \frac{y}{100} = \frac{2xy}{100} = \frac{2A}{100}$$

Therefore, there is a 2% error in A.

EXERCISES 8.4

1. In Example 1, let the pollution index P be given by

$$P = 2x^2 + 4xy + xy$$

Find $(\partial P/\partial x)_{(3, 5)}$ and $(\partial P/\partial y)_{(2, 3)}$. Interpret your results.

2. Repeat Problem 1 for

$$P = 3e^{x+y+xy}$$

3. In Example 2, suppose the productivity P of the ABC Company is given by

$$P = \frac{e^{xy}}{x^2 + y^2 + y}$$

(a) Find $\dfrac{\partial P}{\partial x}, \dfrac{\partial P}{\partial y}$.

(b) Find $\left(\dfrac{\partial P}{\partial x}\right)_{(200,\,5)}$ and $\left(\dfrac{\partial P}{\partial y}\right)_{(5,\,100)}$

(c) Interpret (b).

4. Repeat Problem 3 for

$$P = (x^2 + e^y)^2$$

5. A toothpaste company finds that sales per day are a function of x, the number of times that its commercial is shown on television, and y, the number of people (in millions) who see the commercial. This function is

$$f(x, y) = 50xy + 1000$$

(a) Find f_x, f_y

(b) Interpret $f_x(2, 5)$

(c) Interpret $f_y(2, 5)$

6. Repeat Problem 5 for

$$f(x, y) = x^2 + y^2 + 20xy + 500$$

7. Find the total differential dz if

(a) $z = xy^2 + xe^y$ (b) $z = x + y \ln x$

8. Find the percentage error in the area of an ellipse when an error of 1% is made in the major and minor axes. [*Hint:* If $2a$ and $2b$ are the lengths of the major and minor axes, then area $= \pi ab$.]

9. Find the percentage error in the area of an ellipse when an error of 1% is made in measuring its major axis, and an error of 2% is made in measuring its minor axis.

10. In measuring a right circular cone, errors of 2% and 1% are made in the height and radius, respectively. Find the percentage error in the volume.

11. In an experiment with a simple pendulum, the acceleration due to gravity g is calculated from the formula

$$T = 2\pi\sqrt{l/g}$$

where T is the period, and l is the length of the pendulum. Calculate approximately the percentage error in g if l and T are measured within 1% of accuracy.

12. Find

$$3.001 \times 8.997$$

approximately. [*Hint:* Let $f(x, y) = xy$. Then $f(3, 9) = 27, dx = 0.001$, and $dy = -0.003$.]

13. Suppose that the cost C of producing two items x and y is given by $C = f(x, y)$. Then $\partial C/\partial x$ *and* $\partial C/\partial y$ are called the *partial marginal costs* with respect to x and y, respectively. Find the partial marginal costs with respect to x and with respect to y for the following functions.

(a) $C = 4x^2 + xy - 13y^2$ (b) $C = y \ln x + 3x^2y$

(c) $C = xe^y + y$ (d) $C = x \ln (16x + e^y)$

8.5 MAXIMA AND MINIMA

The partial derivatives can be used to find extreme values (largest value, smallest value) of a function of several variables. We recall (Chapter 3) that $f(a)$ is a local extremum of a function of one variable if $f(a)$ is either the largest or the smallest value of the function in some open interval that contains the point a. We now give an analogous definition for a function of two variables (the definition can be easily generalized to functions of several variables).

Definition 1. We say that $f(a, b)$ is a *relative maximum* if there is a circular region N in the domain of f with center at (a, b) ,such that

$$f(a, b) \geq f(x, y)$$

for all (x, y) in region N. Similarly, we say that $f(a, b)$ is a *relative minimum* if there exists a circular region N in the domain of f with (a, b) as the center such that

$$f(a, b) \leq f(x, y)$$

for all (x, y) in the region N.

We recall that if a function of one variable has a relative extremum at a point and is differentiable there, then its derivative at that point is zero. A similar result for a function of two variables is given as follows.

THEOREM 1. Let $f(a, b)$ be a relative extreme value of a function f. If both f_x and f_y exist at (a, b), then

$$f_x(a, b) = 0 \quad \text{and} \quad f_y(a, b) = 0 \tag{1}$$

Proof. If $f(a, b)$ is a relative extreme value of the function f of two variables x and y, then it must also be a relative extreme value of the function $f(x, b)$ of one variable x for $x = a$. Consequently (from Chapter 3), its derivative, $f_x(a, b)$ for $x = a$ must necessarily be zero. Similarly, we can show that

$$f_y(a, b) = 0$$

We remark that the aforementioned conditions are necessary but not sufficient; that is, the converse of Theorem 1 is not true. (See Example 2.)

Definition 2. A point (a, b) in the domain of a function $z = f(x, y)$ is called a *critical point* if

$$f_x(a, b) = 0 \quad \text{and} \quad f_y(a, b) = 0 \tag{2}$$

Theorem 1 tells us that we must first find all the critical points (a, b) and then check further to determine whether $f(a, b)$ is (1) a relative maximum, (2) a relative minimum, or (3) neither.

EXAMPLE 1. Find the extreme values (if there are any) for the function

$$f(x, y) = x^4 + y^4 - 4x - 4y - 9 \tag{3}$$

SOLUTION. We first look for the critial points. Thus, we set each partial derivative equal to zero:

$$f_x = 4x^3 - 4 = 0 \tag{4}$$

$$f_y = 4y^3 - 4 = 0 \tag{5}$$

If we solve these two equations simultaneously, we find that $x = 1$ and $y = 1$. Since $f(1, 1) = 1 + 1 - 4 - 4 - 9 = -15$, it appears as though -15 is a minimum value, and it occurs at the point (1, 1).

We have not proved that -15 is a minimum value for $x^4+y^4-4x-4y-9$, although we suspect that it is. Unfortunately, the problem is a little more complicated than one might suppose. We will return to this question after looking at the next example.

EXAMPLE 2. Find the extreme values (if there are any) for the function

$$f(x, y) = y^2 - x^2$$

SOLUTION. Looking for the critical points, we set each partial derivative equal to zero:

$$\frac{\partial f}{\partial x} = -2x = 0 \qquad \text{and} \qquad \frac{\partial f}{\partial y} = 2y = 0 \tag{6}$$

Hence, for a critical point we must have $x = 0$ and $y = 0$. Clearly (0, 0) is the only critical point. The graph of $z = y^2 - x^2$ in the neighborhood of the origin is shown in Figure 1.

From this figure it is clear that at (0, 0) the function $f(x, y) = y^2 - x^2$ has neither a relative maximum nor a relative minimum. Along the x-axis the surface $z = y^2 - x^2$ goes down, and along the y-axis the surface goes up. A surface that has this type of point is called a *saddle surface*, and the point (in this case (0, 0)) is called a *saddle point*. Thus, a critical point can be a saddle point.The function $f(x, y) = y^2 - x^2$ has a saddle point, but it does not have a relative maximum point, nor does it have a relative minimum point.

If the function is relatively simple, as in Example 2, then it is an easy matter to separate the saddle points from the extreme points. A systematic method is given by the following theorem.

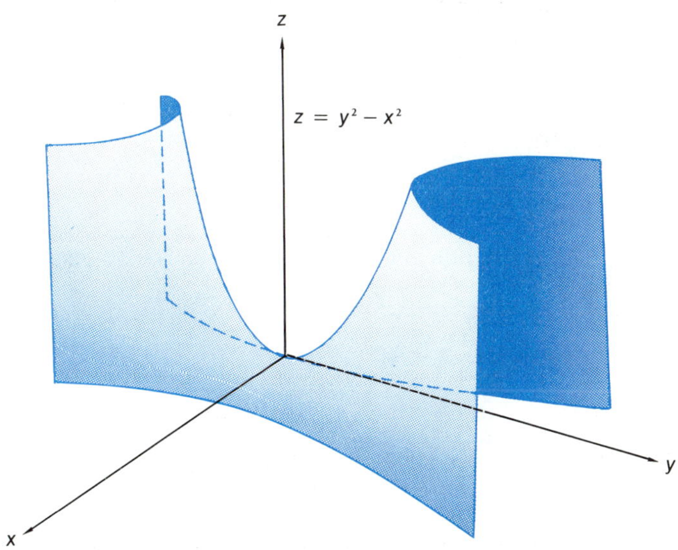

Figure 1

THEOREM 2. If:

1. $z = f(x, y)$
2. $f_x(a, b) = 0$ and $f_y(a, b) = 0$; i.e., (a, b) is a critical point.
3. All higher-order partial derivatives of f exist in some circular region containing (a, b) as center.
4. $A = f_{xx}(a, b)$, $B = f_{xy}(a, b)$, and $C = f_{yy}(a, b)$.

Then:

(1) $f(a, b)$ is a relative maximum value if $AC - B^2 > 0$ and $A < 0$.

(2) $f(a, b)$ is a relative minimum value if $AC - B^2 > 0$ and $A > 0$.

(3) f has a *saddle point* at (a, b) if $AC - B^2 < 0$.

(4) The case $AC - B^2 = 0$ is doubtful and needs further considerations.

Theorem 2 corresponds to the *second derivative test* for functions of one variable. We can now complete the solution to Example 1. From equations (4) and (5) we have

$$f_{xx} = 12x^2, \qquad f_{xy} = 0, \qquad f_{yy} = 12y^2$$

so that

$$A = f_{xx}(1, 1) = 12, \quad B = f_{xy}(1, 1) = 0, \quad \text{and} \quad C = f_{yy}(1, 1) = 12$$

Thus, $AC - B^2 = 144 > 0$ and $A = 12 > 0$. Consequently, by Theorem 2 (part (2)), the minimum value occurs at $(1, 1)$, just as we expected.

EXAMPLE 3. Use Theorem 2 to check the nature of the critical point for

$$f(x, y) = y^2 - x^2$$

SOLUTION. This is the function from Example 2. The second partial derivatives for this function are:

$$\frac{\partial^2 f}{\partial x^2} = -2, \qquad \frac{\partial^2 f}{\partial x\, \partial y} = 0, \qquad \frac{\partial^2 f}{\partial y^2} = 2$$

Then $AC - B^2 = (-2)(2) - 0 = -4 < 0$. Hence, $(0, 0)$ is a saddle point (see Figure 1).

EXAMPLE 4. If x, y, and z are positive numbers such that $x + 3y + 4z = 36$, find the maximum value of the product $P = xyz$.

SOLUTION. P involves three variables, but they are not all independent, since they are related by $x + 3y + 4z = 36$. We solve this equation for x and substitute in P. Thus, we are to maximize

$$P = (36 - 3y - 4z)yz \tag{7}$$

with y and z positive. Setting the partial derivatives equal to zero, we have

$$P_y = 36z - 6yz - 4z^2 = z(36 - 6y - 4z) = 0 \tag{8}$$

$$P_z = 36y - 3y^2 - 8yz = y(36 - 3y - 8z) = 0 \tag{9}$$

Since $y > 0$ and $z > 0$, we can drop those factors (y and z) in (8) and (9). The resulting system gives $y = 4$ and $z = 3$. Then $x = 36 - 3y - 4z$ yields $x = 12$. At this critical point, $P = (12)(4)(3) = 144$. The reader may apply Theorem 2 to prove that this is the *maximum value* of P. However, we regard this as an obvious fact from the nature of the problem.

WORKING RULES FOR LOCATING EXTREMA:

Step 1. Find f_x and f_y.

Step 2. Solve $f_x = 0$ and $f_y = 0$ simultaneously.

Step 3. Find $A = f_{xx}$, $B = f_{xy}$, $C = f_{yy}$, and evaluate at the solution set of Step 2.

Step 4. Check $AC - B^2$ for its sign. Check A for its sign. Find the part of Theorem 2 that is applicable.

EXAMPLE 5. Find the extreme values of

$$f(x, y) = x^2 + y^2 - 4x - 6y + 1$$

SOLUTION. We have

$$f_x = 2x - 4 \quad \text{and } f_y = 2y - 6$$

Setting $f_x = 0$ and $f_y = 0$, we get $x = 2$ and $y = 3$. Thus, (2, 3) is the only possible location for a relative extremum. Applying the second derivative test to the given function at (2, 3), we have

$$A = f_{xx} = 2, \quad B = f_{xy} = 0, \quad \text{and} \quad C = f_{yy} = 2$$

Therefore,

$$AC - B^2 = 2 \cdot 2 - 0^2 = 4 > 0$$

Since $A = 2 > 0$, then part (2) of Theorem 2 applies, and hence the function f has a relative minimum value at (2, 3).

EXAMPLE 6. Find the extreme values of

$$f(x, y) = xy(3 - x - y)$$

SOLUTION. We have

$$\begin{aligned} f_x &= 3y - 2xy - y^2 = y(3 - 2x - y) \\ f_y &= 3x - x^2 - 2xy = x(3 - 2y - x) \\ A &= f_{xx} = -2y \\ B &= f_{xy} = 3 - 2x - 2y \\ C &= f_{yy} = -2x \end{aligned}$$

We now solve the equations

$$\begin{cases} f_x = 0 \\ f_y = 0 \end{cases}$$

or

$$\begin{cases} y(3 - 2x - y) = 0 \\ x(3 - 2y - x) = 0 \end{cases}$$

from which we obtain four pairs of equations to consider:

(1) $y = 0, \quad x = 0$

(2) $3 - 2x - y = 0, \quad x = 0$

(3) $y = 0, \quad 3 - 2y - x = 0$

(4) $3 - 2x - y = 0, \quad 3 - 2y - x = 0$

Solving these equations, we find the following possible locations of the extreme values:

$$(0, 0), \quad (0, 3), \quad (3, 0), \quad (1, 1)$$

For (0, 0), we have $A = 0, B = 3$, and $C = 0$, so that $AC - B^2 < 0$. Hence, f has a saddle point at (0, 0). For (0, 3), we have $A = -6, B = -3$, and $C = 0$, so that $AC - B^2 < 0$. Hence, (0, 3) is another saddle point of f. Similarly, we can show that (3, 0) is also a saddle point of f. For (1, 1), we have $A = -2, B = -1$, and $C = -2$, so that $AC - B^2 > 0$. Since $A = -2 < 0$, part (1) of Theorem 2 applies. Consequently, $f(1, 1)$ is a relative maximum value.

EXERCISES 8.5

In Problems 1 through 8, find all relative extrema and saddle points.

1. $f(x, y) = x^2 - y^2$

2. $f(x, y) = x^2 + y^2 - 1$

3. $f(x, y) = x^2 - y^2 + 2x - 4y + 3$

4. $f(x, y) = x^2 - xy$

5. $f(x, y) = x^2 - xy + y^2$

6. $f(x, y) = x^3 - 4y^2$

7. $f(x, y) = y^3 + x^2 - 3x$

8. $f(x, y) = xy(x - y)$

9. Find the points on the surface

$$z = \sqrt{xy + 1}$$

whose distance from the origin is minimum. *Hint*: Let

$$u = d[(0, 0, 0), (x, y, z)]^2 = x^2 + y^2 + z^2$$

and find the minimum of u, when $z^2 = xy + 1$.

10. A rectangular tank is open at the top and holds 32 cu. ft. Find its dimensions so that the surface area of the tank is a minimum.

11. Find the dimensions of a rectangular box of maximum volume, if the box has no top and has a surface area of 108 sq. in.

12. Consider all triangles with a given perimeter, $2s$. Show that the one with the largest area is an equilateral triangle. *Hint*: If a, b, and c are the sides of a triangle, then $a + b + c = 2s$, and the area A is given by

$$A = \sqrt{s(s - a)(s - b)(s - c)}$$

13. The sum of three positive numbers is 30. Show that their product is maximum when they are equal.

14. Suppose a rectangular box of volume 64 cu. ft. is to be constructed from three different materials. The cost of the bottom is 50¢/sq.ft., of the sides, 30¢/sq.ft.; and of the top, 40¢/sq.ft. Find the dimensions of the most economical box.

15. Do Problem 14 if the bottom costs 50¢/sq.ft. and the sides and top cost 20¢/sq.ft.

16. The cost of a day's production at ABC Company is given by

$$C = x^2 + y^2 + xy - 20x - 25y + 1500$$

where x is the labor force and y (in thousands of tons) is the amount of raw material. Find the x and y that minimize the cost C.

17. Do Problem 16 if

$$C = 3x^2 + 2y^2 + 3xy - 66x - 58y + 1600$$

18. A company produces x units of product A at a selling price of $500 - 2x$ and y units of product B at a selling price of $400 - 3y$. Given that the cost of producing both products is

$$C(x, y) = x^2 + 20x + 40y + y^2$$

(a) Show that the profit function P is given by

$$P(x, y) = 480x - 3x^2 + 360y - 4y^2$$

(b) How many items of each product should be produced to maximize profits?

19. Suppose that the weekly revenue from the manufacture and sale of x units of item A is $75x - 2x^2$, and the weekly revenue from the manufacture and sale of y units of B is $100y - 3y^2$. The weekly cost to produce x and y units of A and B, respectively, is given by $C(x, y) = 100 + 6x + 7y + 3xy$. Find the weekly production of these two items that will maximize the profit.

20. Suppose that the cost of a certain construction job is

$$C(x, y) = 7x^2 - 10xy + 4y^2 - 18x + 200$$

in units of \$1000, where x is the number of skilled workers and y is the number of unskilled workers. Find the number of workers of each type that should be hired to minimize the cost of the job.

8.6 THE REGRESSION LINE

So far in our applications of calculus to real-world problems, we assumed that the functions that describe such phenomena are known and readily available. As a practical matter this is seldom the case. For example, how does a manufacturer determine functions such as the cost function or the demand function for its product? The determination of functions which describe the real-life phenomena belong to a fascinating field in statistics — called the *regression analysis*. Here we shall attempt to estimate functional relationship between two variables when they seem to be linearly related.

Suppose that in some experiment we measure two related variables x and y. For example, x might be the price of a certain item, and y could be the number of items (in thousands) that sell at that price. The pairs (x_i, y_i) of these measured values are then plotted on a coordinate plane to produce a so-called *scatter diagram*, as shown in Figure 1.

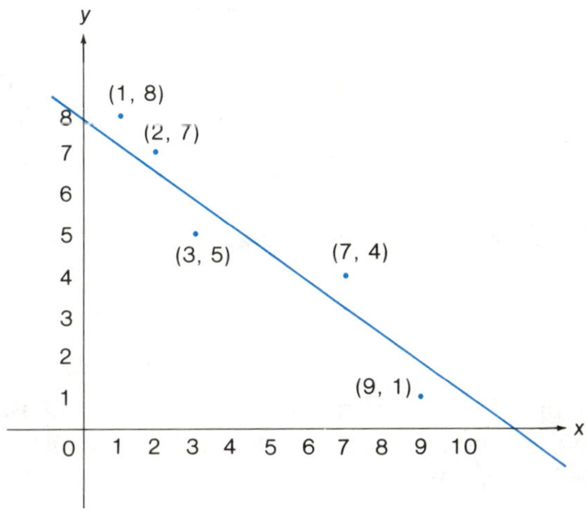

Figure 1

These points *do not* lie on a *straight line*, yet they seem to lie close to some line, as indicated in Figure 1. Our problem is to find the line that fits these points as closely as possible. Such a line is called the line of *best fit*. (We must emphasize that it is not always reasonable to approximate the observed data by a straight line.) One way to find a line of best fit is to use the so-called *theory of least squares*. According to this theory, we let $y = mx + b$ be the equation of the line we are seeking, and we wish to minimize the sum of the squares of the vertical distances of each point from this line. We clarify these ideas as follows:

Let the experimental points be $P_1, P_2, \ldots, P_n$ with coordinates $(x_1, y_1), (x_2, y_2), \ldots, (x_n, y_n)$, respectively. If $y = mx + b$ is a nonvertical line, then for each $k = 1, 2, \ldots, n$ the point on the line with the same horizontal displacement as (x_k, y_k) is $(x_k, mx_k + b)$. The vertical distance d_k between these two points is

$$d_k = |y_k - (mx_k + b)| \tag{1}$$

Hence, the square of d_k is

$$d_k^2 = |y_k - (mx_k + b)|^2 = (y_k - mx_k - b)^2 \tag{2}$$

We make the following definition:

Definition 1. (*Regression Line*). The regression line for a set of points $P_1(x_1, y_1)$, $P_2(x_2, y_2), \ldots, P_n(x_n, y_n)$ is that line that makes the sum

$$S = d_1^2 + d_2^2 + \cdots + d_n^2 \tag{3}$$

a minimum, where d_k^2 is given by (2).

We substitute the expression for each d_k^2 from (2) into (3) and obtain

$$S = (y_1 - mx_1 - b)^2 + (y_2 - mx_2 - b)^2 + \cdots + (y_n - mx_n - b)^2 \tag{4}$$

and in sigma notation (4) can be written as

$$S = \sum_{k=1}^{n} (y_k - mx_k - b)^2 \tag{5}$$

We observe that S is a function of two variables m and b (since the x_k's, y_k's, and n are known). We now find the values of m and b that minimize S by using the technique of the previous section. We shall assume without proof that S does have an absolute minimum at critical points.

We first compute the partial derivatives of S in (5) with respect to m and b (these are the variables). We find that

$$\frac{\partial S}{\partial m} = \sum_{k=1}^{n} 2(y_k - mx_k - b)(-x_k) \tag{6}$$

$$\frac{\partial S}{\partial b} = \sum_{k=1}^{n} 2(y_k - mx_k - b)(-1) \tag{7}$$

If we set these two partial derivatives equal to zero and drop the common factor -2 we will (after some algebra) arrive at

$$m\sum_{k=1}^{n} x_k^2 + b\sum_{k=1}^{n} x_k = \sum_{k=1}^{n} x_k y_k \tag{8}$$

and

$$m\sum_{k=1}^{n} x_k + bn = \sum_{k=1}^{n} y_k \tag{9}$$

Despite the complicated appearance, Equations (8) and (9) are two linear equations in two unknowns, m and b. The unique solution of this pair of equations, the only critical point, gives the absolute minimum of S.

Solving (8) and (9), after a little labor, we obtain

$$m = \frac{n\sum_{k=1}^{n} x_k y_k - \left(\sum_{k=1}^{n} x_k\right)\left(\sum_{k=1}^{n} y_k\right)}{n\sum_{k=1}^{n} x_k^2 - \left(\sum_{k=1}^{n} x_k\right)^2} \tag{10}$$

and

$$b = \frac{1}{n}\left(\sum_{k=1}^{n} y_k - m\sum_{k=1}^{n} x_k\right) \tag{11}$$

The nasty-looking Equations (10) and (11) can be written in a more attractive form by using the concept of mean or average of the set of numbers. We define

$$x = \frac{1}{n}\sum_{k=1}^{n} x_k, \qquad y = \frac{1}{n}\sum_{k=1}^{n} y_k$$

$$x^2 = \frac{1}{n}\sum_{k=1}^{n} x_k^2, \qquad xy = \frac{1}{n}\sum_{k=1}^{n} x_k y_k$$

If we divide the numerator and denominator of (10) and (11) by n^2 and use our new notation, we obtain, respectively:

$$m = \frac{xy - x\,y}{x^2 - (x)^2} \tag{12}$$

and

$$b = \bar{y} - m\bar{x} \tag{13}$$

When we obtain m and b from (12) and (13), then the equation $y = mx + b$ is the required regression line for the given data.

EXAMPLE 1. Find the regression line for the data shown in Figure 1.

SOLUTION. We arrange the data in a table and use the given points to fill in places in the table (see Table 1).

Table 1

k	x_k	y_k	$x_k y_k$	x_k^2
1	1	8	8	1
2	2	7	14	4
3	3	5	15	9
4	7	4	28	49
5	9	1	9	81
$\sum_{k=1}^{5}$	22	25	74	144
$\frac{1}{5}\sum_{k=1}^{5}$	$\bar{x} = 4.4$	$\bar{y} = 5.0$	$\overline{xy} = 14.8$	$\overline{x^2} = 28.8$

We use the numbers given in the last row of Table 1 in Equations (12) and (13), and we obtain

$$m = \frac{14.8 - (4.4)(5.0)}{28.8 - (4.4)^2} = \frac{14.8 - 22}{28.8 - 19.36} = \frac{-7.2}{9.44} \approx -0.763$$

and

$$b = 5.0 - (-0.763)(4.4) \approx 8.357$$

Therefore, the equation of the regression line is

$$y = -0.763x + 8.357$$

EXERCISES 8.6

In Problems 1 through 8, some experiment gave the values of two related variables. In each case find the equation of the regression line (use a calculator to find the coefficients to the nearest three significant digits).

1.

x	1	5	7	13
y	5	6	7	9

2.

x	3	4	6	7
y	−1	2.5	6	8.5

3.

x	0	1	2	4
y	4	1	−2	−8

4.

x	0	2	4	6
y	9	9.5	8.5	7.5

5. The ABC Chewing Gum Co. finds that at a price of x cents per piece of gum the number of pieces of gum (in millions) is given by the following table:

x	1	2	3	4	6
y	11	8	6	4	3

(a) Find the regression line for the data.

(b) Estimate the number of pieces of gum the company will sell if the price per piece of gum is 5 cents.

6. The Minor Motors dealer in Pollutville keeps a careful record x of his weekly advertising expenses in dollars on local TV and the number of cars y the dealership sells the following week. The table shows the data collected:

x	1000	3000	5000	7000
y	20	30	60	78

Use the regression line to predict the sales:(a) if he quits advertizing, and (b) if he spends \$10,000 on advertizing in a special week.

7. The following table gives the yearly gross revenue R (in millions of dollars) of the XYZ Company. It also gives profit P per share of stock in dollars. Find the regression lines for $R(t)$ and $P(t)$ using the years 1970 as the zero point on the time axis. Use these lines to predict R and P in 1986.

year	1970	1972	1974	1976	1978
R	98	200	422	729	972
P	0.29	0.49	0.94	1.70	2.72

8. Do Problem 7 for the ABC Company:

year	1970	1974	1976	1978	1980
R	50	68.2	78	210	350
P	−0.1	0.1	0.15	0.60	1.40

8.7 EXTREMAL PROBLEMS WITH CONSTRAINTS

In many maxima and minima problems, the functions defined are subject to certain restrictions (or conditions) called *constraints*. For example, suppose that we wish to find the minimum value of the function

$$z = f(x, y) = x^2 + y^2 \tag{1}$$

The solution is obvious, since $x^2 + y^2 \geq 0$ for all $x, y \in R$ and $x^2 + y^2 = 0$ if and only if $x = 0$ and $y = 0$. Consequently, the minimum value of f is 0, and it is attained at (0, 0). However, we may be asked to find the minimum value of the function f defined by (1), subject to the constraint

$$x + y - 1 = 0 \tag{2}$$

Obviously, f cannot have minimum value at (0, 0) since (0, 0) does not satisfy (2). To solve the problem in this case, we solve Equation (2) for y:

$$y = 1 - x \tag{3}$$

Substituting this value of y in (1), we have

$$z = x^2 + y^2 = x^2 + (1 - x)^2 \tag{4}$$
$$= x^2 + 1 - 2x + x^2 = 2x^2 - 2x + 1$$

Since z is a function of one variable x, we can use the method of Chapter 3. We find dz/dx and set it equal to zero:

$$\frac{dz}{dx} = 4x - 2 = 0 \Rightarrow x = \tfrac{1}{2}$$

Also, using the second derivative test we find

$$\frac{d^2z}{dx^2} = 4 > 0$$

Therefore, $x = \frac{1}{2}$ gives the minimum value of z. Substituting $x = \frac{1}{2}$ in (3), we get $y = \frac{1}{2}$. Thus, the minimum value of f is attained at $(\frac{1}{2}, \frac{1}{2})$, and

$$f(\tfrac{1}{2}, \tfrac{1}{2}) = (\tfrac{1}{2})^2 + (\tfrac{1}{2})^2 = \tfrac{1}{4} + \tfrac{1}{4} = \tfrac{1}{2}$$

The following example illustrates how to use a similar method when three variables are involved (see also Example 4, Section 8.5):

EXAMPLE 1. Find the minimum value of

$$u = x^2 + y^2 + z^2 \tag{5}$$

subject to the constraint

$$x + y + z = 30 \tag{6}$$

SOLUTION. Solving (6) for z, we obtain

$$z = 30 - x - y \tag{7}$$

We substitute this value of z in (5):

$$u = x^2 + y^2 + (30 - x - y)^2 \tag{8}$$

Then

$$u_x = 2x + 2(30 - x - y)(-1) = 4x + 2y - 60$$

$$u_y = 2y + 2(30 - x - y)(-1) = 2x + 4y - 60$$

$$u_{xx} = 4; \qquad u_{xy} = 2; \qquad u_{yy} = 4$$

Solving

$$\begin{cases} u_x = 0 \\ u_y = 0 \end{cases}$$

we find that $x = 10$, $y = 10$. Also, $u_{xx} \cdot u_{yy} > (u_{xy})^2$, and $u_{xx} > 0$. Hence, u is minimum for $x = 10$, $y = 10$, and

$$z = 30 - x - y = 30 - 10 - 10 = 10$$

and this minimum value is

$$u = 10^2 + 10^2 + 10^2 = 300$$

The previous method works, since the constraint equation (6) could be solved for one of the variables in terms of the other variables. Suppose that Equation (6) was replaced by

$$x^3y^2 + 2xy^2 + y^5z^3 + 7xy + 8yz = \sqrt{71} \tag{9}$$

It would indeed be difficult to solve (9) for any one of the three variables. There is an alternative method developed by Lagrange. We state this method in the following theorem.

THEOREM 1. Let $f, g_1, g_2, \ldots, g_m$ be differentiable functions of the variables $x_1, x_2, \ldots, x_n$. Let f be subject to the following constraints:

$$\begin{aligned} g_1(x_1, x_2, \ldots, x_n) &= 0 \\ g_2(x_1, x_2, \ldots, x_n) &= 0 \\ &\cdot \\ &\cdot \\ &\cdot \\ g_m(x_1, x_2, \ldots, x_n) &= 0 \end{aligned} \tag{10}$$

Then the points at which f attains an extreme value are found among the points $(x_1, x_2, \ldots, x_n)$ for which there exist constants $\lambda_1, \lambda_2, \ldots, \lambda_m$ (each λ_i is called a Lagrange multiplier), such that

$$
\begin{aligned}
\frac{\partial f}{\partial x_1} - \lambda_1 \frac{\partial g_1}{\partial x_1} - \lambda_2 \frac{\partial g_2}{\partial x_1} - \cdots - \lambda_m \frac{\partial g_m}{\partial x_1} &= 0 \\
\frac{\partial f}{\partial x_2} - \lambda_1 \frac{\partial g_1}{\partial x_2} - \lambda_2 \frac{\partial g_2}{\partial x_2} - \cdots - \lambda_m \frac{\partial g_m}{\partial x_2} &= 0 \\
&\vdots \\
\frac{\partial f}{\partial x_n} - \lambda_1 \frac{\partial g_1}{\partial x_n} - \lambda_2 \frac{\partial g_2}{\partial x_n} - \cdots - \lambda_m \frac{\partial g_m}{\partial x_n} &= 0 \\
g_1(x_1, x_2, \ldots, x_n) &= 0 \\
g_2(x_1, x_2, \ldots, x_n) &= 0 \\
&\vdots \\
g_m(x_1, x_2, \ldots, x_n) &= 0
\end{aligned}
\tag{11}
$$

The proof of this theorem is omitted. Suppose we consider a special case of this theorem, when the function f of two variables x and y is subject to a single constraint $g(x, y) = 0$. Then the theorem asserts that the points (x, y) at which the function f has an extreme value are among the points (x, y) for which there exists a real number λ, such that

$$
\left\{
\begin{aligned}
\frac{\partial f}{\partial x} - \lambda \frac{\partial g}{\partial x} &= 0 \\
\frac{\partial f}{\partial y} - \lambda \frac{\partial g}{\partial y} &= 0 \\
g(x, y) &= 0
\end{aligned}
\right.
\tag{12}
$$

EXAMPLE 2. Find the extreme values of

$$f(x, y) = xy \tag{13}$$

subject to the constraint

$$g(x, y) = x^2 + y^2 - 1 = 0 \tag{14}$$

SOLUTION. Since

$$\frac{\partial f}{\partial x} = y, \qquad \frac{\partial f}{\partial y} = x, \qquad \frac{\partial g}{\partial x} = 2x, \qquad \frac{\partial g}{\partial y} = 2y$$

Equations (12) become

$$\begin{cases} y - \lambda(2x) = 0 \\ x - \lambda(2y) = 0 \\ x^2 + y^2 - 1 = 0 \end{cases} \tag{15}$$

We solve the first equation in (15) for y and substitute in the second equation to obtain

$$y = \lambda(2x) \tag{16}$$

and

$$x(1 - 4\lambda^2) = 0 \tag{17}$$

From (17), we note that either $x = 0$ or $1 - 4\lambda^2 = 0$. If $x = 0$, then from (16) we must have $y = 0$. But $x = 0$, $y = 0$ does not satisfy (14). Hence, $1 - 4\lambda^2 = 0$, or

$$\lambda = \pm \tfrac{1}{2} \tag{18}$$

Substituting these values of λ in (16), we get

$$y = x \quad \text{or} \quad y = -x \tag{19}$$

For $y = \pm x$, we have from the third equation of (15)

$$\begin{aligned} x^2 + y^2 &= 1 \\ x^2 + (\pm x)^2 &= 1 \\ x^2 + x^2 &= 1 \\ 2x^2 &= 1 \end{aligned}$$

Then $x = \pm 1/\sqrt{2}$ and $y = \pm x = \pm 1/\sqrt{2}$. Thus, the extreme values of f occur among

$$\left(\frac{1}{\sqrt{2}}, \frac{1}{\sqrt{2}}\right), \left(\frac{1}{\sqrt{2}}, -\frac{1}{\sqrt{2}}\right), \left(-\frac{1}{\sqrt{2}}, \frac{1}{\sqrt{2}}\right), \left(-\frac{1}{\sqrt{2}}, -\frac{1}{\sqrt{2}}\right)$$

We now compute the values of f at these points.

$$f\left(\frac{1}{\sqrt{2}}, \frac{1}{\sqrt{2}}\right) = \left(\frac{1}{\sqrt{2}}\right)\left(\frac{1}{\sqrt{2}}\right) = \frac{1}{2}$$

$$f\left(\frac{1}{\sqrt{2}}, -\frac{1}{\sqrt{2}}\right) = \left(\frac{1}{\sqrt{2}}\right)\left(-\frac{1}{\sqrt{2}}\right) = -\frac{1}{2}$$

$$f\left(-\frac{1}{\sqrt{2}}, \frac{1}{\sqrt{2}}\right) = \left(-\frac{1}{\sqrt{2}}\right)\left(\frac{1}{\sqrt{2}}\right) = -\frac{1}{2}$$

$$f\left(-\frac{1}{\sqrt{2}}, -\frac{1}{\sqrt{2}}\right) = \left(-\frac{1}{\sqrt{2}}\right)\left(-\frac{1}{\sqrt{2}}\right) = \frac{1}{2}$$

Thus the maximum value of f is $\frac{1}{2}$, which is attained at the points $(1/\sqrt{2}, 1/\sqrt{2})$ and $(-1/\sqrt{2}, -1/\sqrt{2})$. The minimum value of f is $-\frac{1}{2}$, attained at the points $(1/\sqrt{2}, -1/\sqrt{2})$ and $(-1/\sqrt{2}, 1/\sqrt{2})$.

EXERCISES 8.7

1. State Theorem 1 for a function f of three variables x, y, and z, subject to the constraint $g(x, y, z) = 0$.

2. State Theorem 1 for a function f of three variables x, y, and z, subject to two constraints $g_1(x, y, z) = 0$ and $g_2(x, y, z) = 0$.

3. Find the maximum and minimum values of $f(x, y) = x + y$ subject to $x^2 + y^2 = 1$.

4. Find the maximum and minimum values of $f(x, y) = x^2 - y^2$ subject to $x^2 + y^2 = 1$.

5. By Lagrange's method, find the maximum and minimum values of $f(x, y) = x^2 + y^2$ subject to $x + y = 1$.

6. Do Example 1, using Lagrange's method [*Hint*: Use Problem 1.]

In Problems 7 through 11, use Lagrange's method.

7. Do Problem 10 of Exercises 8.5

8. Do Problem 11 of Exercises 8.5

9. Do Problem 13 of Exercises 8.5

10. Do Problem 14 of Exercises 8.5

11. Find the minimum value of

$$f(x, y, z) = x^2 + y^2 + z^2$$

subject to

$$x + 2y + z = 1 \quad \text{and} \quad 2x - y - 3z = 4$$

8.8 DOUBLE INTEGRALS

In Chapter 4 we saw that if $y = f(x)$ is a continuous function and nonnegative $[f(x) \geq 0]$ on $[a, b]$ then $\int_a^b f(x)dx$ represents the area of the region bounded by the curve $y = f(x)$, the x-axis, and the lines $x = a$ and $x = b$. Now we look for the extension of the integral that will represent volume. It turns out that finding the volume of a solid involves performing integration twice, so the process is called the *double integral.* The rigorous definition of the double integral is beyond the scope of this book. However, we shall give an intuitive description for the fact that we need to integrate twice while considering the volume of a solid.

Let D be a region in the xy-plane which is bounded by two functions $y = g(x)$ and $y = h(x)$, and let $z = f(x, y)$ be the surface that lies above the region D. We are interested in finding the volume under $z = f(x, y)$ bounded by D (see Figure 1).

We imagine a slice cut from the solid by a plane perpendicular to the x-axis at $x = c$. This cross-sectional area is given by the integral

$$\int_{h(c)}^{g(c)} f(c, y)\, dy \tag{1}$$

If $x = a$ and $x = b$ are the lines that bound the region D, then we can find the cross-sectional area given by (1) for every $x \in [a, b]$. Thus, we obtain the area function

$$A(x) = \int_{h(x)}^{g(x)} f(x, y)\, dy$$

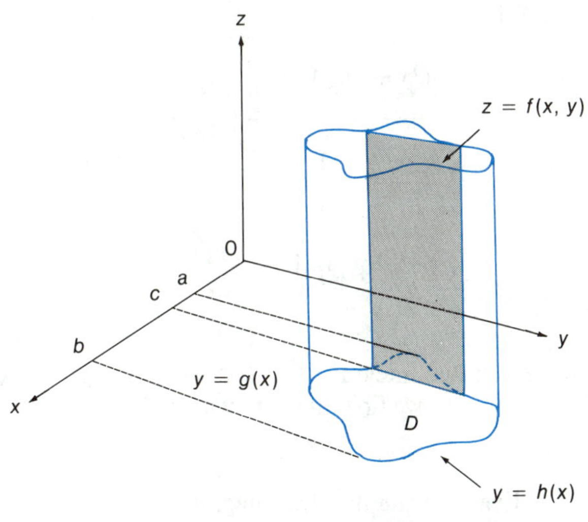

Figure 1

where x is held fixed and the limits of integration depend upon where the cutting plane is taken. In other words, the y-limits are functions of x, the functions that represent the boundary curves for the region D. Finally, we see that the volume V of the solid under $z = f(x, y)$ bounded by D is the sum of all the areas $A(x)$. This sum leads to another integration. Therefore, we have

$$V = \int_a^b A(x)\,dx = \int_a^b \left[\int_{h(x)}^{g(x)} f(x, y)\,dy\right] dx \tag{2}$$

The expression on the right-hand side of (2) is called the *iterated integral.* We usually delete the brackets in the expression for the iterated integral, and write

$$\int_a^b \left[\int_{h(x)}^{g(x)} f(x, y)\,dy\right] dx = \int_a^b \int_{h(x)}^{g(x)} f(x, y)\,dy\,dx \tag{3}$$

To evaluate a given iterated (double) integral, we proceed as follows:

(a) Integrate $\int f(x, y)\,dy$ with respect to y with x held fixed, and evaluate the resulting integral between the limits $y = h(x)$ and $y = g(x)$.

(b) Then integrate the result obtained in (a) with respect to x between the limits $x = a$ and $x = b$.

EXAMPLE 1. Evaluate the interated integral

$$I = \int_2^4 \int_1^{2x+1} (3x + y)\,dy\,dx$$

SOLUTION.

$$\begin{aligned} I &= \int_2^4 \left[3xy + \frac{y^2}{2}\right]_1^{2x+1} dx \\ &= \int_2^4 \left\{\left[3x(2x+1) + \frac{(2x+1)^2}{2}\right] - \left[3x + \frac{1}{2}\right]\right\} dx \\ &= \int_2^4 (8x^2 + 2x)dx = \left[\frac{8x^3}{3} + x^2\right]_2^4 \\ &= \left[\frac{8(4)^3}{3} + (4)^2\right] - \left[\frac{8(2)^3}{3} + (2)^2\right] = \frac{484}{3} \end{aligned}$$

To evaluate the iterated integral $\int_c^d \int_{g(y)}^{h(y)} f(x, y)\,dx\,dy$, we merely interchange the role of x and y. In other words, we first integrate with respect to x and then with respect to y.

EXAMPLE 2. Evaluate the iterated integral

$$I = \int_0^2 \int_0^y (x + e^{y^2})\,dx\,dy$$

SOLUTION.

$$I = \int_0^2 \left(\left[\frac{x^2}{2} + xe^{y^2}\right]_0^y\right) dy = \int_0^2 \left(\frac{y^2}{2} + ye^{y^2}\right) dy$$

$$= \left[\frac{y^3}{6} + \frac{e^{y^2}}{2}\right]_0^2 = \left(\frac{8}{6} + \frac{e^4}{2}\right) - \left(0 + \frac{1}{2}\right) = \frac{5}{6} + \frac{e^4}{2}$$

EXAMPLE 3. Find the volume of the solid whose base is the triangle in the xy-plane bounded by the x-axis, the line $y = x$, and the line $x = 1$, and whose top is in the plane $z = 2x + y + 1$.

SOLUTION. For any x between 0 and 1, y varies from $y = 0$ to $y = x$ (see Figure 2). Hence, the volume V of the solid is given by

$$V = \int_0^1 \int_0^x (2x + y + 1)\, dy\, dx = \int_0^1 \left\{\left[2xy + \frac{y^2}{2} + y\right]_0^x\right\} dx$$

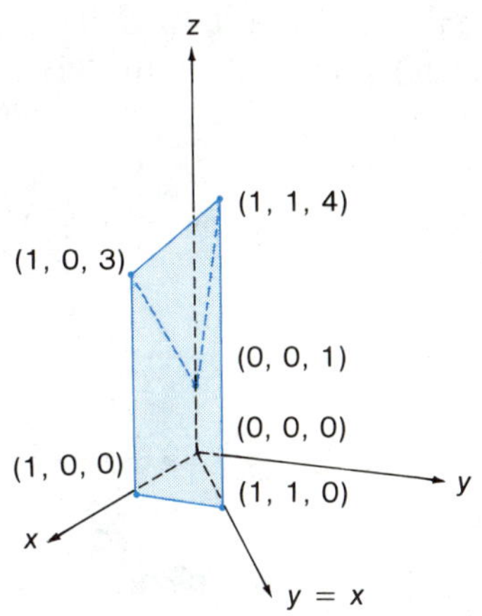

Figure 2

$$= \int_0^1 \left(2x^2 + \frac{x^2}{2} + x\right) dx = \int_0^1 \left(\frac{5}{2}x^2 + x\right) dx$$

$$= \left[\frac{5}{6}x^3 + \frac{x^2}{2}\right]_0^1 = \frac{5}{6} + \frac{1}{2} = \frac{4}{3}$$

EXAMPLE 4. Find the volume of the solid whose base D in the xy-plane is bounded by the curves $y = x^2$ and $y = x$, and whose top is the surface $z = x^2y^3$.

SOLUTION. We first sketch the region D in the xy-plane (see Figure 3).

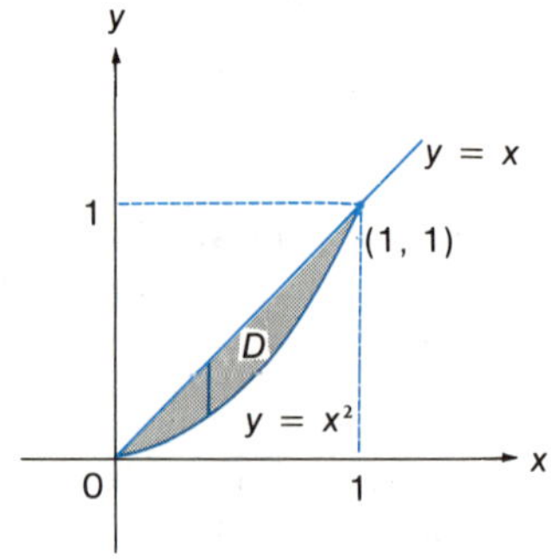

Figure 3

Now suppose x between 0 and 1 is held fixed. Then y varies from $y = x^2$ to $y = x$ in the region D. Consequently, the limits for the inner integral are x^2 and x, while the limits for the outer integral are 0 and 1. Hence, the required volume V is given by

$$V = \int_0^1 \int_{x^2}^{x} x^2 y^3 \, dy \, dx = \int_0^1 \left\{ \left[x^2 \frac{y^4}{4} \right]_{x^2}^{x} \right\} dx$$

$$= \int_0^1 \left(\frac{x^6}{4} - \frac{x^{10}}{4} \right) dx = \frac{1}{4} \int_0^1 (x^6 - x^{10}) dx$$

$$= \frac{1}{4} \left(\frac{x^7}{7} - \frac{x^{11}}{11} \right) \Big|_0^1 = \frac{1}{4} \left(\frac{1}{7} - \frac{1}{11} \right) = \frac{1}{77}$$

We could also compute the required volume by first integrating with respect to x and then with respect to y. For this method, we observe that if y is held fixed between 0 and 1, then x varies from $x = y$ to $x = y^{1/2}$ (see Figure 4). Hence,

$$V = \int_0^1 \int_y^{y^{1/2}} x^2 y^3 \, dx \, dy = \frac{1}{3} \int_0^1 \left[y^3 x^3 \right]_y^{y^{1/2}} dy$$

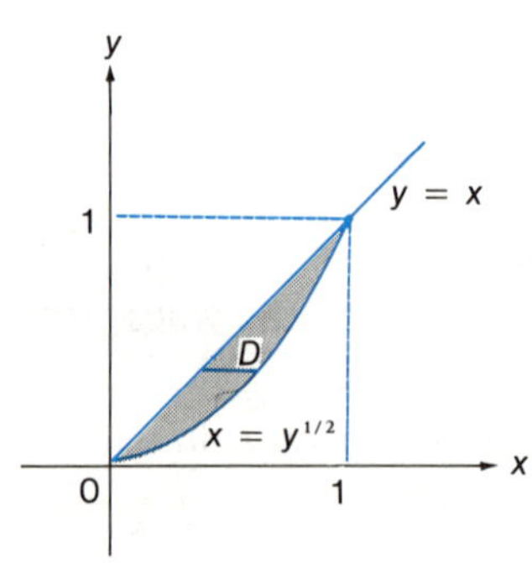

Figure 4

$$= \frac{1}{3}\int_0^1 (y^{9/2} - y^6)\,dy = \frac{1}{3}\left[\frac{y^{11/2}}{11/2} - \frac{y^7}{7}\right]_0^1$$

$$= \frac{1}{3}\left[\frac{2}{11} - \frac{1}{7}\right] = \frac{1}{77}$$

EXERCISES 8.8

In Problems 1 through 24, evaluate the (double) iterated integrals.

1. $\int_0^2\int_0^1 (1 + x)\,dx\,dy$

2. $\int_1^3\int_0^2 3xy^2\,dx\,dy$

3. $\int_0^2\int_1^3 3xy^2\,dy\,dx$

4. $\int_0^3\int_0^3 (x + y^2)\,dx\,dy$

5. $\int_1^2\int_0^1 (y + x^2)\,dx\,dy$

6. $\int_0^3\int_3^4 yx^2\,dx\,dy$

7. $\int_0^1\int_0^1 (x^2 + y)\,dy\,dx$

8. $\int_0^1\int_0^1 (x^2 + y)\,dx\,dy$

9. $\int_0^1\int_2^4 (4 + y)\,dx\,dy$

10. $\int_2^4\int_0^1 (4 + y)\,dy\,dx$

11. $\int_0^1\int_0^1 e^{2x+y}\,dx\,dy$

12. $\int_0^1\int_0^1 e^{2x+y}\,dy\,dx$

13. $\int_0^2\int_1^x (2x + y + 1)\,dy\,dx$

14. $\int_0^1\int_0^{x^2} (xy + e^{x^3})\,dy\,dx$

15. $\int_{-1}^1\int_y^{y^2} (y^2 + 2xy - x^2)\,dx\,dy$

16. $\int_0^2\int_0^{1-x^2} y\,dy\,dx$

17. $\int_1^4\int_0^x \sqrt{x - y}\,dy\,dx$

18. $\int_0^2\int_0^x e^{x^2}\,dy\,dx$

19. $\int_0^1\int_0^{8x} \sqrt{x + y}\,dy\,dx$

20. $\int_1^{\ln 8}\int_0^{\ln y} e^{x+y}\,dx\,dy$

21. $\int_0^2\int_{\sqrt{y}}^1 dx\,dy$

22. $\int_{-1}^2\int_{x^2+1}^{3x+4} dy\,dx$

23. $\int_0^1\int_{y^2}^y \sqrt{\frac{x}{y}}\,dx\,dy$

24. $\int_0^\pi\int_0^{x^2} \frac{y}{x}\,dy\,dx$

25. Find the volume of the solid whose base is the triangle in the xy-plane bounded by the x-axis, the line $y = x$, and the line $x = 1$, and whose top is the plane $z = x + y + 2$.

26. Find the volume of the solid whose base is the region in the xy-plane that is bounded by $x = 0$, $y = 0$, and $y = 1 - x$, and whose top is bounded by the surface $z = x^2 + y^2$.

27. Find the volume of the solid under the surface $z = xy$ bounded by $y = 3x$ and $y = 4 - x^2$ and the plane $z = 0$.

28. Find the volume of the solid bounded by the surface $z = 0$, $z = x$, and $y^2 + x = 2$.

29. Find the volume of the solid bounded by the planes $y = x$, $y = 2 - x$, $y = 0$, $z = 0$, and $x + y + z = 3$.

30. Find the volume of the solid bounded by the surfaces $z = 0$, $x = 0$, $y^2 + x = 4$, and $z - y = 2$.

CHAPTER 8 REVIEW

Part 1 (ORAL)

Define or discuss the following.

1. A real-valued function of two variables
2. Three-dimensional coordinate system
3. Graph of $z = f(x, y)$
4. Partial derivative
5. Mixed partial derivatives
6. Regression line
7. Double integral
8. Total differential
9. Relative maximum and relative minimum
10. Lagrange's method of undetermined multipliers

Part 2 (WRITTEN)

1. Plot the following points.

   (a) $(1, 0, 1)$ (b) $(0, 0, 2)$ (c) $(-1, 2, 3)$

2. Calculate the distance between the points.

(a) $(1, 2, 3)$, $(-1, 0, 1)$ (b) $(-2, 1, -3)$, $(3, -1, -4)$

3. (a) What is the equation of the xy-plane?

(b) What is the equation of the yz-plane?

(c) What is the equation of the xz-plane?

(d) What is the equation of the plane containing the points $(0, 0, 1)$, $(1, 0, 0)$, and $(0, 2, 0,)$?

4. Sketch the graph of the plane $2x + 3y + 4z = 24$.

5. Sketch the graph of $x^2 + y^2 + z^2 = 16$ by drawing several cross sections.

In Problems 6 through 15, find f_x, f_y, f_{xx}, f_{yy}, and f_{yx}.

6. $f(x, y) = 2x - 5y^2$

7. $f(x, y) = xy^2 + 7$

8. $f(x, y) = 8x - 5x^2y + 7xy^3$

9. $f(x, y) = \sqrt{x^2 + y^2}$

10. $f(x, y) = \dfrac{x - y}{x + y}$

11. $f(x, y) = e^x e^y - xe^{y^2+2y}$

12. $f(x, y) = \ln x^2y^3$

13. $f(x, y) = \ln(x^2 + y^3)$

14. $f(x, y) = xye^{x+y}$

15. $f(x, y) = x^2y^3 \ln xy$

In Problems 16 through 19, find $f_x(1, 2)$, $f_y(-1, 3)$, and $f_{xy}(2, 3)$.

16. $f(x, y) = 3x - 4y^2$

17. $f(x, y) = 4xy$

18. $f(x, y) = \dfrac{x + y}{x - y}$

19. $f(x, y) = xe^y$

20. The manufacturing cost of a widget is given by

$$M(x, y) = 4xy^2 + 10x^2 + 5y^2 + 5$$

where x is the cost of labor and y is the cost of the material. Find and interpret the following:

(a) $\left(\dfrac{\partial M}{\partial x}\right)_{(3,2)}$ (b) $\left(\dfrac{\partial M}{\partial y}\right)_{(4,5)}$

21. The period T of a pendulum of length l is given by

$$T = 2\pi\sqrt{\frac{l}{g}}$$

Find the error in g due to errors of 3% in l and 1% in T.

In Problems 22 through 25, find any relative maxima or relative minima for the given function:

22. $f(x, y) = x^2 - xy + y^2 + 2x + 2y + 7$

23. $f(x, y) = 8x + 8y - 10x^2 - 4y^2 + 4xy + 40$

24. $f(x, y) = x^2 + y^2 + 2x + 2y - xy - 6$

25. $f(x, y) = 4x + 4y + 2xy - 5x^2 - 2y^2 + 4$

26. If $u = x^2 + y^2 + z^2$ and $x + y + z = 30$, find the values of x, y, and z for which u is a minimum.

27. Find the points on the surface $z^2 = xy + 1$ whose distance from the origin is minimum.

28. Find the dimensions of a rectangular tank (without the top) having a volume of 32 cubic units and minimum surface area.

29. Find the dimensions of a rectangular box (without the top) having maximum volume and a surface area of 108 sq. in.

30. Find the minimum value of

$$D = \sqrt{x^2 + y^2 + z^2}$$

subject to the constraint

$$x + y + z = 30$$

31. Find the maximum and minimum distances from the origin to the curve

$$5x^2 + 6xy + 5y^2 = 8$$

32. Discuss the maxima and minima of

$$f(x, y) = x^2 + y^2 + z^2$$

subject to constraints

$$\begin{cases} x + y - z = 1 \\ 2x - 3y + 4z = 2 \end{cases}$$

33. The Sleezy-Smut Store in Dimlit City sells three magazines A, B, and C. The daily profit (in cents) for the sale of x, y, and z number of magazines of A, B, and C, respectively, is given by

$$P(x, y, z) = 50x + 30y + 10z + 2xy + xz + 3yz - 8000$$

However, the market in Dimlit City can absorb only 100 magazines. How many magazines of each kind should the store sell per day in order to realize the maximum profit?

34. Find the line of regression for the following data:

x	1	3	5	6	9
y	6.7	6	5.8	5.1	4

Appendices

APPENDIX 1. SETS

The theory of sets plays a central role in the foundations of modern mathematics. Although we shall not plunge deeply into the formal theory of sets, a few of its basic concepts will be presented to give precision to our ideas.

A *set* is any well-defined collection of objects. Each object in a set is called an *element* of the set. The main characteristic of a set is that it is well defined. That is, given any particular object, it must be clear whether the object is an element of the set or not. For example, the collection of all intelligent students at your university is considered ill defined, because it is unlikely that we all can agree on its members. On the other hand, the collection of all those students who maintained a grade-point average of 3.5 or higher at your university during the current academic year is well defined.

We shall denote sets by uppercase letters, such as $A, B, C, \ldots$. We write in symbols

$$a \in A, \quad b \notin A$$

(read "a is an element of A, b is not an element of A") to indicate that a is in the set A and b is not.

A set may have only one element, such as the set consisting of the current president of the United States of America; several elements, such as the set whose elements are the states of the U.S.A., an infinite number of elements, such as the set of counting numbers $1, 2, 3, \ldots$; or no elements at all, such as the set of all women presidents of the U.S.A. as of 1980.

The set that has no elements is called the *empty set* or void set and is denoted by $\varnothing$ or $\{\quad\}$. The reader should note that $\{0\}$ is not the empty set. This set has one element, zero.

Set descriptions are usually included within braces. If the set does not have "too" many elements, we can tabulate all its elements. For example, $A = \{a, b, c\}$ denotes the set A whose elements are a, b, and c and no others.

Besides tabulation, a set is often specified by stating a condition or property which is satisfied by the elements of the set and is written in the form

$$A = \{x \mid x \text{ has the property } P\}$$

which is read "A is the set of all elements x such that x has property P." For example, $E = \{x \mid x \text{ is an even counting number}\}$ is read "E is the set of all x such that x is an even counting number." We note that $2 \in E$, $4 \in E$, $6 \in E$, and so on; and $1 \notin E$, $3 \notin E$.

If every element of a set A is also an element of a set B, then A is called a *subset* of B and we write $A \subseteq B$ or $B \supseteq A$, which may also be read as A is contained in B or B contains A. It is obvious that $A \subseteq A$. The symbol $A \nsubseteq B$ means that A is not a subset of B.

Definition 1. Two sets A and B are said to be *equal*, written $A = B$, if $A \subseteq B$ and $B \subseteq A$.

Thus the two sets A and B are equal if they contain precisely the same elements. The symbol $A \neq B$ means that A is not equal to B. If $A \subseteq B$ and $A \neq B$, then A is said to be a proper subset of B and is written $A \subset B$. In this situation there is at least one element of B that is not in A.

EXAMPLE 1. List all subsets of the set

$$A = \{a, b, c\}$$

SOLUTION. There are eight subsets of A. They are

$$\{a\},\ \{b\},\ \{c\},\ \{a, b\},\ \{b, c\},\ \{a, c\},\ \{a, b, c\},\ \varnothing$$

We note that the set consisting of a single element a is written as $\{a\}$. It is important to note the difference between a and $\{a\}$. We also note that the empty set $\varnothing$ is not written $\{\varnothing\}$. In fact, $\varnothing \neq \{\varnothing\}$, since the empty set $\varnothing$ has no elements, whereas the set $\{\varnothing\}$ has one element, $\varnothing$. The set $\varnothing$ is a subset of every set.

When discussing sets, we always assume that they are subsets of a larger set U, called the *universe of discourse* or the *universal set*. Ideas involving the universal set, its subsets, and certain operations on sets can be visualized by using plane geometric figures called *Venn diagrams*. Figure 1 shows such a diagram. Here all points of the rectangle and its interior represent the universal set U. It also shows some of the subsets of U, denoted by circles and their interiors.

We now define some operations on the subsets of a given universal set U.

Definition 2. The *union* of two subsets A and B of U is the set of all elements of U that belong to either A or B or to both.

The symbol $\cup$ is used to denote the union of sets. Thus $A \cup B$ (read "A union B") is the set of all elements that are either in A or in B or in both. Figure 2 is a Venn diagram in which the shaded region depicts $A \cup B$.

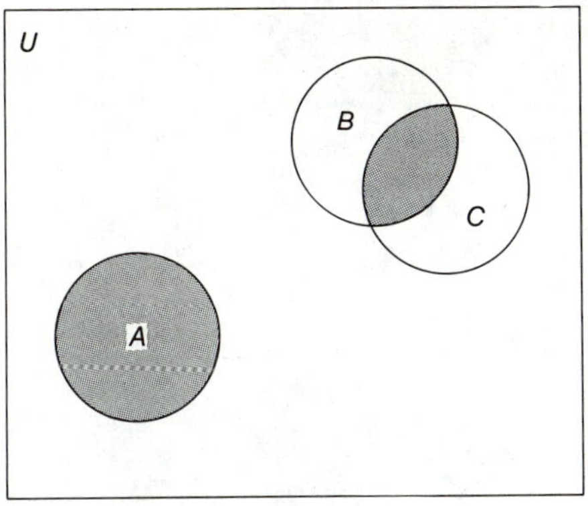

Figure 1

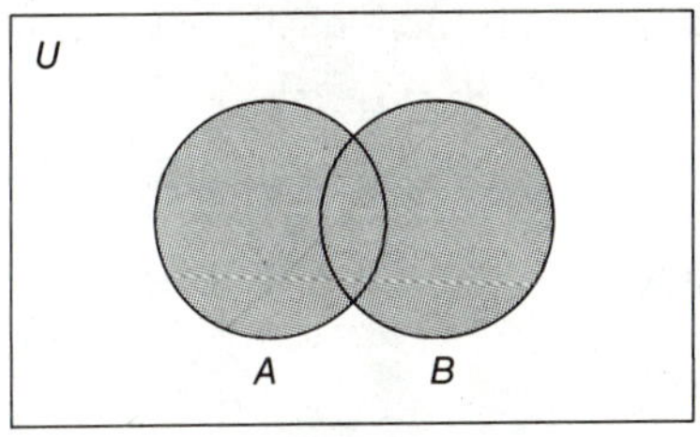

Figure 2

EXAMPLE 2. Let $A = \{2, 3, 4, 6, 8\}$ and $B = \{1, 3, 5, 6, 7\}$. Find $A \cup B$.

SOLUTION. $A \cup B = \{1, 2, 3, 4, 5, 6, 7, 8\}$.

Observe that the elements 3 and 6 appear in both sets, and yet when we form the union, we list each of these elements only once. (Why?)

A second operation of interest is given in Definition 3.

Definition 3. The *intersection* of two subsets A and B of U is the set of all elements of U that belong to both A and B.

The symbol $\cap$ is used to denote the intersection of sets. Thus $A \cap B$ (read "A intersection B") denotes the set of all elements that are in both A and B.

In the Venn diagram of Figure 3, $A \cap B$ is represented by the shaded region.

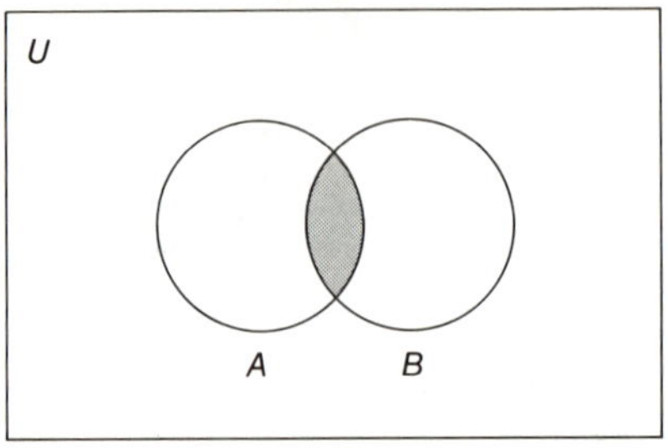

Figure 3

EXAMPLE 3. For the sets A and B of Example 2, find $A \cap B$.

SOLUTION. $A \cap B = \{3, 6\}$.

If $A \cap B = \varnothing$; that is, if A and B have no elements in common, then A and B are said to be disjoint sets.

The operations of union and intersection can be performed on three or more sets by grouping them in pairs.

EXAMPLE 4. Let A and B be the sets of Example 2 and let $C = \{2, 3, 9\}$. Find

(a) $(A \cup B) \cup C$ (b) $(A \cup B) \cap C$

(c) $(A \cap B) \cup C$ (d) $(A \cap B) \cap C$

SOLUTION.

(a) $$(A \cup B) \cup C = \{1, 2, 3, 4, 5, 6, 7, 8\} \cup \{2, 3, 9\} = \{1, 2, 3, 4, 5, 6, 7, 8, 9\}$$

(b) $$(A \cup B) \cap C = \{1, 2, 3, 4, 5, 6, 7, 8\} \cap \{2, 3, 9\} = \{2, 3\}$$

(c) $$(A \cap B) \cup C = \{3, 6\} \cup \{2, 3, 9\} = \{2, 3, 6, 9\}$$

(d) $$(A \cap B) \cap C = \{3, 6\} \cap \{2, 3, 9\} = \{3\}$$

APPENDIX 2. BASIC ALGEBRAIC CONCEPTS

In this appendix we shall list some of the basic formulas for factoring and briefly discuss: the interval notation, inequalities and absolute value.

2.1 FORMULAS FOR FACTORING

$$x^2+2xy+y^2 = (x+y)(x+y) = (x+y)^2 \tag{1}$$

$$x^2-2xy+y^2 = (x-y)(x-y) = (x-y)^2 \tag{2}$$

$$x^2-y^2 = (x+y)(x-y) \tag{3}$$

$$x^3+3x^2y+3xy^2+y^3 = (x+y)^3 \tag{4}$$

$$x^3-3x^2y+3xy^2-y^3 = (x-y)^3 \tag{5}$$

$$x^3+y^3 = (x+y)(x^2-xy+y^2) \tag{6}$$

$$x^3-y^3 = (x-y)(x^2+xy+y^2) \tag{7}$$

$$x^2+(a+b)x+ab = (x+a)(x+b) \tag{8}$$

$$acx^2+(ad+bc)x+bd = (ax+b)(cx+d) \tag{9}$$

EXAMPLE 1. Factor the following polynomials:

(a) a^4+6a^2+9 (b) $4a^4-20a^2b^2+25b^4$

SOLUTION.

(a) By use of (1), with $x = a^2$ and $y = 3$:

$$a^4+6a^2+9 = (a^2)^2+2(a^2)(3)+(3)^2 = (a^2+3)^2$$

(b) By use of (2), with $x = 2a^2$ and $y = 5b^2$,

$$\begin{aligned} 4a^4-20a^2b^2+25b^4 &= (2a^2)^2-2(2a^2)(5b^2)+(5b^2)^2 \\ &= (2a^2-5b^2)^2 \end{aligned}$$

The formula

$$x^2-y^2 = (x+y)(x-y)$$

is one of the most important ones and is called the formula for the *difference of two squares*.

EXAMPLE 2. Factor the following expressions:

(a) $4a^2 - 9b^2$ (b) $16a^4 - 1$

(c) $(u - v)^2 - (a + b)^2$ (d) $a^4 + 4$

SOLUTION.

(a) By formula 3, with $x = 2a$ and $y = 3b$,

$$4a^2 - 9b^2 = (2a)^2 - (3b)^2 = (2a + 3b)(2a - 3b)$$

(b) By formula 3, with $x = 4a^2$ and $y = 1$

$$16a^4 - 1 = (4a^2)^2 - (1) = (4a^2 + 1)(4a^2 - 1)$$

and again by formula (3)

$$4a^2 - 1 = (2a)^2 - (1)^2 = (2a + 1)(2a - 1)$$

Hence

$$16a^4 - 1 = (4a^2 + 1)(2a + 1)(2a - 1)$$

(c) By formula (3), with $x = u - v$ and $y = a + b$

$$\begin{aligned} (u - v)^2 - (a + b)^2 &= [(u - v) + (a + b)][(u - v) - (a + b)] \\ &= (u - v + a + b)(u - v - a - b) \end{aligned}$$

(d) Although $a^4 + 4$ is not given in the form of difference of two squares, we can convert it to that form. Since $(a^2 + 2)^2 = a^4 + 4a^2 + 4$ and

$$a^4 + 4 = a^4 + 4a^2 + 4 - 4a^2$$

we have

$$\begin{aligned} a^4 + 4 &= (a^2 + 2)^2 - (2a)^2 \\ &= (a^2 + 2 + 2a)(a^2 + 2 - 2a) \\ &= (a^2 + 2a + 2)(a^2 - 2a + 2) \end{aligned}$$

The technique in Example 2(d) is sometimes called the *method of completing squares*. The formulas

$$x^3 + y^3 = (x + y)(x^2 - xy + y^2)$$

and

$$x^3 - y^3 = (x - y)(x^2 + xy + y^2)$$

are called formulas for the *sum of cubes* and *difference of cubes*, respectively.

EXAMPLE 3. Factor the following expressions:

(a) $8a^3 + 27b^3$ (b) $a^3 - 64b^3$ (c) $a^6 - b^6$

SOLUTION.

(a) By (4), with $x = 2a$ and $y = 3b$

$$\begin{aligned} 8a^3 + 27b^3 &= (2a)^3 + (3b)^3 \\ &= (2a + 3b)[(2a)^2 - (2a)(3b) + (3b)^2] \\ &= (2a + 3b)(4a^2 - 6ab + 9b^2) \end{aligned}$$

(b) By (5), with $x = a$ and $y = 4b$

$$\begin{aligned} a^3 - 64b^3 &= (a)^3 - (4b)^3 \\ &= (a - 4b)[a^2 + a(4b) + (4b)^2] \\ &= (a - 4b)(a^2 + 4ab + 16b^2) \end{aligned}$$

(c) We first use the difference-of-two-squares formula:

$$\begin{aligned} a^6 - b^6 &= (a^3)^2 - (b^3)^2 \\ &= (a^3 + b^3)(a^3 - b^3) \end{aligned}$$

Now, using (4) and (5), we have

$$\begin{aligned} a^6 - b^6 &= (a^3 + b^3)(a^3 - b^3) \\ &= (a + b)(a^2 - ab + b^2)(a - b)(a^2 + ab + b^2) \end{aligned}$$

Suppose that we are given a polynomial $ax^2 + bx + c$, where a, b, and c are integers. If this polynomial is factorable, then its factorization must be of the form $(dx + e)(fx + g) = dfx^2 + (dg + ef)x + eg$, where d, e, f, and g are integers, and we must have $df = a, eg = c, dg + ef = b$. By trying out various possibilities (there are only finitely many), we get the required factorization of $ax^2 + bx + c$.

EXAMPLE 4. Factor the following polynomials:

(a) $x^2 - 6x + 8$ (b) $2x^2 + 7x + 6$ (c) $6x^2 - xy - 2y^2$

SOLUTION. (a) If we write

$$x^2 - 6x + 8 = (dx + e)(fx + g)$$

then $df = 1$ and $eg = 8$. Trying various possibilities we have the factorization $x^2 - 6x + 8 = (x - 2)(x - 4)$.

(b) If $(2x^2 + 7x + 6) = (dx + e)(fx + g)$, we have $df = 2$ and $eg = 6$. Suppose that we take $d = 1$ and $f = 2$. Thus our factors will look like $(x \quad)(2x \quad)$. Now to find e and g (since $e \cdot g = 6$), we have the following possibilities: $e = 1$, $g = 6$; $e = -1$, $g = -6$; $e = 2$, $g = 3$; $e = -2$, $g = -3$; $e = 3$, $g = 2$; $e = -3$, $g = -2$; $e = 6$, $g = 1$; and $e = -6$, $g = -1$. By trying out these possibilities, we find that $e = 2$ and $g = 3$ is the only one that gives us the middle term $7x$. Consequently,

$$2x^2 + 7x + 6 = (x + 2)(2x + 3)$$

(c) If a factorization with two first-degree polynomials exists, it must be of the form $(dx + ey)(fx + gy)$. Trying out different possibilities, we have

$$6x^2 - xy - 2y^2 = (2x + y)(3x - 2y)$$

Sometimes, the terms in a sum can be grouped together in a suitable fashion to factorize a polynomial by using distributive law.

EXAMPLE 5. Factor the following polynomials:

(a) $x(x + 2y) + 3y(x + 2y)$ (b) $x^3 + 2x^2 - x - 2$

SOLUTION.

(a) By using the distributive law $a \cdot c + b \cdot c = (a + b)c$, with $a = x$, $b = 3y$ *and* $c = x + 2y$, we have the factorization of the expression:

$$x(x + 2y) + 3y(x + 2y) = (x + 3y)(x + 2y)$$

(b) By grouping the first two terms and the last two terms, we have

$$x^3 + 2x^2 - x - 2 = (x^3 + 2x^2) - (x + 2)$$

$$= x^2(x+2) - 1(x+2)$$
$$= (x^2 - 1)(x+2)$$
$$= (x-1)(x+1)(x+2)$$

2.2 THE NUMBER LINE

We may associate the set of real numbers with points on a line in such a way that to each real number x there corresponds one and only one point on the line, and conversely. Such an association is called *one-to-one correspondence.* To do this, we draw a straight line (imagine the line to be extended indefinitely in both directions) and call this line the *number line.* We choose any two points on this line and label them 0 and 1. By agreement we choose 1 to the right of 0. The fixed line segment from 0 to 1 is called the *unit segment*, and its length is chosen as the *unit length.* By reproducing this unit of length successively on both sides of 0 and 1, we get the familiar graphical representation of the set of integers $Z = \{\ldots, -3, -2, -1, 0, 1, 2, 3, \ldots\}$ on the number line (see Figure 1).

Figure 1

We now locate the points corresponding to the rational numbers. Let $x = a/b$, $b \neq 0$ be a positive rational number. We divide the line segment from 0 to a into b equal parts (this can be done with a compass and straightedge) (see Figure 2); then the endpoint of the first subdivision is the number $x = a/b$. The negative rational numbers

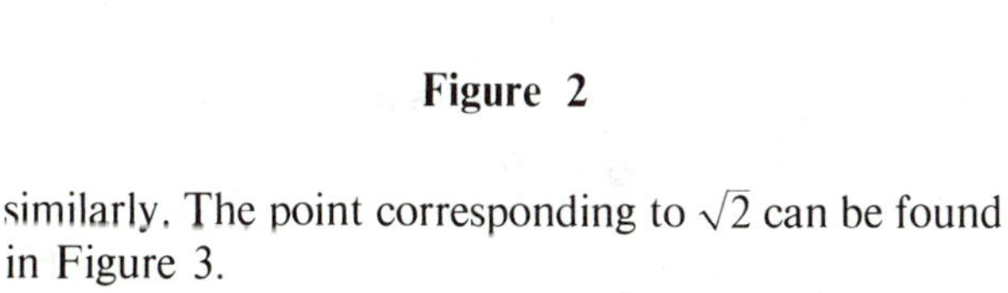

Figure 2

can be obtained similarly. The point corresponding to $\sqrt{2}$ can be found by drawing an arc as indicated in Figure 3.

The number x that is associated with a point, say P on the number line, is called the *coordinate* of P. Thus a number line is also called a *coordinate line.* On the corrdinate line (which we usually draw horizontally) the positive real number are represented by points to the right of zero, the negative real numbers are represented by points to the left of zero. Thus a number line is assigned a direction. The *positive direction* is the direction of increasing coordinates, and the *negative direction* is the direction of

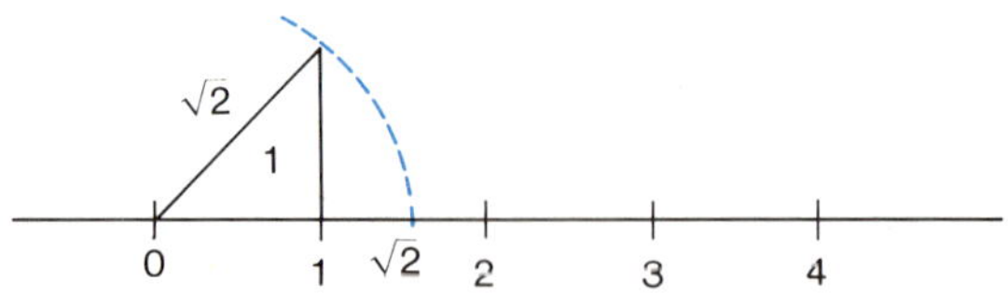

Figure 3

decreasing coordinates. The positive direction is noted by drawing an arrowhead on the number line as shown in Figure 1. With these assignments, it is clear that if a is the coordinate of the point A and b is the coordinate of the point B, then $a < b$ (read a is less than b) means A is to the left of B. Similarly, we say b is greater than a and write $b > a$ if B is to the right of A. The notation $a \geq b$ means that $a > b$ or $a = b$. Similarly, $a \leq b$ means that $a < b$ or $a = b$.

Now instead of saying that a is the coordinate of A on the number line, we usually identify A with a.

An expression involving $<$, $\leq$, $>$ or $\geq$ is called an inequality. We list here some of the important propeties of inequalities for $<$, similar inequalities hold for $\leq$, $>$, and $\geq$. Let a, b, and c be real numbers:

PROPERTY I. For each pair of real numbers a and b, exactly one of the following relations holds:

$$a < b, \qquad a > b, \qquad a = b$$

PROPERTY II. If $a > b$ and $b > c$, then $a > c$.

PROPERTY III. If $a > b$, then $a + c > b + c$.

PROPERTY IV. If $a > b$ and $c > 0$, then $ac > bc$.

PROPERTY V. If $a > b$ and $c < 0$, then $ac < bc$.

The compound inequality $a < c < b$ means that $a < c$ and simultaneously $c < b$. For example, $-2 < 3 < 6$ is correct. However, $3 < x < 1$ is incorrect use of this notation, since $3 < x < 1$ suggests that $3 < x$ and simultaneously that $x < 1$. We thus conclude that $3 < 1$, which is false. In other words, $a < c < b$ means that c is between a and b on the number line and $a < b$.

The sets described below are called *intervals*. For $a < b$, we have:

1. The open interval from a to b, denoted (a, b), is defined by

$$(a, b) = \{x \mid a < x < b\}$$

See Figure 4.

Figure 4

2. The closed interval from a to b, denoted $[a, b]$, is defined by

$$[a, b] = \{x \mid a \leq x \leq b\}$$

See Figure 5.

Figure 5

The numbers a and b are called the *endpoints* of the interval. A square bracket indicates that the endpoint is included in the interval, whereas a parenthesis means that the endpoint does not belong to the interval.

3. Half-open intervals:

$$[a, b) = \{x \mid a \leq x < b\}$$

$$(a, b] = \{x \mid a < x \leq b\}$$

4. One unbounded interval is

$$[a, \infty) = \{x \mid a \leq x\}$$

See Figure 6.

Others are

$$(a, \infty) = \{x \mid a < x\}$$

$$(-\infty, a) = \{x \mid x < a\}$$

Figure 6

$$(-\infty, a] = \{x \mid x \leq a\}$$

$$(-\infty, \infty) = R$$

We remark that the symbol ∞ (or $-\infty$) is not a number. It is simply a part of the notation.

2.3 EQUATIONS

Before we discuss the solutions of inequalities, let us briefly recall the processes involved in the solutions of equations.

Determining the solutions of an equation is called *solving* the equation. Sometimes, it is possible to solve an equation by inspection. In general, however, we first use the concept of equivalent equation.

Definition 1. Equations that have the same solution set are called *equivalent equation*.

EXAMPLE 1. The equations

$$3x - 4 = 8$$

$$3x = 12$$

and

$$x = 4$$

are equivalent equations because the solution set of each of these equations is $\{4\}$.

(b) The equations

$$x = 1$$

and

$$x^2 = x$$

are not equivalent, since the solution set of the first equation is {1}, whereas the solution set of the second equation is {0, 1}.

To solve an equation, we convert the given equation into an equivalent equation whose solution is apparent. Two simple properties allow us to do this.

THE ADDITION PROPERTY. If and equation $a = b$ is true, then $a + c = b + c$ is true for any number c.

Let us illustrate this property by a simple example:

$$x + 3 = 7 \tag{1}$$

We add -3 to both sides of (1) and have

$$x + 3 + (-3) = 7 + (-3) \tag{2}$$

$$x = 4 \tag{3}$$

According to the addition property, if $x + 3 = 7$ is true, then $x = 4$ is true. Thus any solution of $x + 3 = 7$ is also a solution of $x = 4$. Obviously, the only solution of $x = 4$ is {4}. It is also the solution of $x + 3 = 7$. We can check this by substituting $x = 4$ in $x + 3 = 7$. We could also reverse the above processes. That is, starting from (3) and adding 3 to both sides of (3), we get (1). Thus any x that makes (3) true also makes (1) true. Hence equations (1) and (3) are equivalent.

Remark: Whenever the steps in an argument are reversible, we get equivalent equation.

So, the use of the addition property always produces an equation that is equivalent to the original equation. Therefore, checking by substitution is not necessary.

We next consider

THE MULTIPLICATION PROPERTY. If an equation $a = b$ is true, then $a \cdot b = b \cdot c$ is true for any number c.

We observe that the steps in this argument are reversible if $c \neq 0$. But if $c = 0$, then $1/c$ does not exist. Hence we get equivalent equations only if we multiply by a nonzero number.

EXAMPLE 2. Solve for x:

$$3x - 1 = 8 \tag{4}$$

$$\frac{2}{x} = \frac{3}{x} \tag{5}$$

SOLUTION. For Equation (4) we have

$$3x - 1 = 8$$

$$3x - 1 + 1 = 8 + 1 \qquad \text{addition property}$$

$$3x = 9$$

$$\frac{1}{3}(3x) = \frac{1}{3}(9) \qquad \text{multiplication property}$$

$$x = 3$$

Since we only used the addition property and multiplication by a nonzero number $\frac{1}{3}$, the process of solution produced the equivalent equations: $3x - 1 = 8$, $3x = 9$, and $x = 3$.

The solution set in this case is $\{3\}$.

Next we consider Equation (5). Suppose that we multiply both sides of

$$\frac{2}{x} = \frac{3}{x}$$

by x^2; we get

$$\frac{2}{x} \cdot x^2 = \frac{3}{x} \cdot x^2$$

or

$$2x = 3x$$

or

$$2x + (-2x) = 3x + (-2x)$$

or

$$x = 0 \tag{6}$$

So a possible solution for (5) is $x = 0$. However, if we substitute $x = 0$ in the original equation, we get

$$\frac{2}{0} = \frac{3}{0}$$

which is undefined. Thus $x = 0$ is not a solution of (5). Therefore, equations (5) and (6) are not equivalent, and the solution set for (5) is the empty set $\varnothing$.

So, if we multiply both sides of an equation by a variable, we may not get equivalent equations. However, the solution set of the original equation is a subset of the new equation. Therefore, we *must check* possible solutions by substituting in the original equation.

In many applications of algebra and calculus we are faced with problems that are stated in words. Such problems are called *word problems.* In many cases we can introduce variables for the unknown and transform the given information into an equation (or equations) that we can solve. In this section we shall be concerned with only those types of problems which can be solved by using a linear equation in one variable. Thus, after determining the unknown numbers from the statement of the problem, it is necessary to represent those unknowns by the use of only one variable.

For solving word problems, the following strategy may be helpful:

1. *Read the problem carefully*, getting well in mind the knowns and the unknowns.
2. *Draw a picture* (if possible).
3. *Introduce a variable to represent the unknowns.*
4. *If there are several unknowns, use the information in the problem to write each unknown in terms of one variable only.*
5. *From the information given in the problem, write down the equation that relates the unknowns and the known numbers.*
6. *Solve the equation.*
7. *Check to see if the solution of the equation gives a solution of the original problem.*

EXAMPLE 3. The length of a rectangle is 3 feet less than four times its width. Find the dimensions of the rectangle if its perimeter is 64 feet.

SOLUTION. Let x be the width of the rectangle. Then the length of the rectangle is $4x - 3$. See Figure 1. The perimeter of the rectangle, which is the total distance around the rectangle, is $[x + (4x - 3) + x + (4x - 3)]$. The perimeter is given to be 64. Therefore,

$$x + (4x - 3) + x + (4x - 3) = 64$$

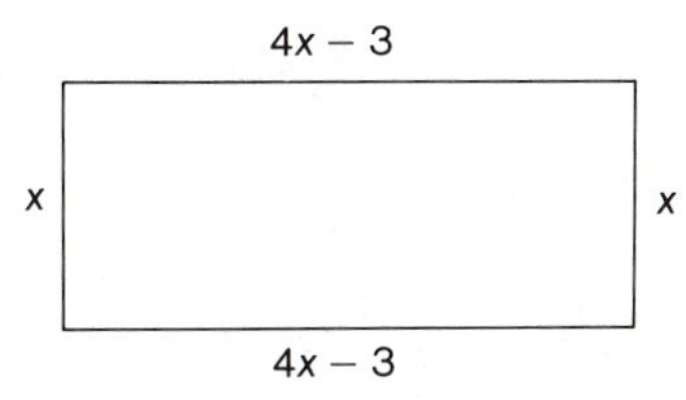

Figure 1

Solving this equation, we have

$$10x - 6 = 64$$

or

$$10x = 70$$

or

$$x = 7$$

Hence the width of the rectangle is 7 feet, and its length $= 4x - 3 = 4(7) - 3 = 28 - 3 = 25$ feet.

CHECK: The perimeter is $7 + 25 + 7 + 25 = 64$ feet.

A third principle for solving equations is based on the following property of real numbers.

PRINCIPLE OF ZERO FACTORS: Let a and b be real numbers, then $a \cdot b = 0$ if and only if $a = 0$ or $b = 0$.

The principle of zero factors is illustrated in the following example.

EXAMPLE 4. Find the solution set of

$$x^2 - x - 6 = 0 \tag{7}$$

SOLUTION. We factor the left-hand side of (7) and obtain

$$(x - 3)(x + 2) = 0 \tag{8}$$

By applying the principle of zero factors, it follows that (8) is a true statement if and only if

$$x - 3 = 0 \tag{9}$$

or

$$x + 2 = 0 \tag{10}$$

From (9) we get $x = 3$, and from (10) we obtain $x = -2$. Therefore, the solution set of (7) is $\{-2, 3\}$. These solutions can be checked by substituting for x, -2 and 3 in the original equation.

The reader may recall that equation (7) can also be solved by the quadratic formula:

QUADRATIC FORMULA.

If $ax^2 + bx + c = 0$ $(a \neq 0)$ has solutions, they are given by the formulas:

$$x = \frac{-b + \sqrt{b^2 - 4ac}}{2a} \quad \text{and} \quad x = \frac{-b - \sqrt{b^2 - 4ac}}{2a}$$

2.4 INEQUALITIES

The processes of solving inequalities are similar to those of solving equations. We use the properties of inequalities to replace a given inequality by a chain of equivalent inequalities such that the solution set of the last inequality is obvious.

Let us recall some of the properties that will be useful for solving a given inequality. Let $a, b, c \in R$ (set of real numbers)

EXAMPLE 1. Solve the inequality

$$2x + 5 < x + 8 \tag{1}$$

SOLUTION. By Property III,

$$2x + 5 - 5 < x + 8 - 5$$

or

$$2x < x + 3$$

Again, by Property III,

$$2x - x < x + 3 - x$$

or

$$x < 3$$

Thus the solution set of the inequality is $\{x \mid x < 3\}$. This solution set can also be written as an interval $(-\infty, 3)$. The graph of the solution set is given in Figure 1.

Figure 1

EXAMPLE 2. Solve the compound inequality

$$3 \leq 4x - 7 < 9 \tag{2}$$

SOLUTION. Adding 7 to each member of the inequality (2), we have

$$10 \leq 4x < 16 \tag{3}$$

Now multiplying each member of (3) by $\frac{1}{4}$, we have

$$\frac{10}{4} \leq x < \frac{16}{4}$$

or

$$\frac{5}{2} \leq x < 4 \tag{4}$$

Since each of these steps is reversible, inequality (4) is equivalent to (2). Thus the solution set of (2) is $\{x \mid \frac{5}{2} \leq x < 4\}$, which is the interval $[\frac{5}{2}, 4)$.

EXAMPLE 3. Solve the inequality

$$x^2 + x < 2 \tag{5}$$

SOLUTION. The following inequalities are equivalent:

$$x^2 + x < 2$$

$$x^2 + x - 2 < 0$$

$$(x + 2)(x - 1) < 0 \tag{6}$$

Therefore, x is in the solution of inequality (5) if and only if the product $(x + 2)(x - 1)$

is negative. The product $(x + 2)(x - 1)$ will be negative if the two factors have opposite signs (one positive and the other negative). Now let us consider the sign of $x + 2$. The factor $x + 2 > 0$ if $x > -2$ and $x + 2 < 0$ if $x < -2$. We illustrate this fact in Figure 2.

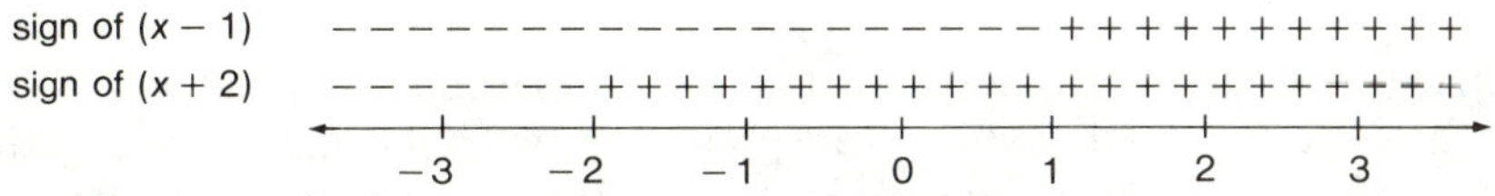

Figure 2

Similarly, the factor $x - 1$ is positive if $x > 1$ and negative is $x < 1$. We again illustrate it in Figure 2. Now, from Figure 2, we observe that the two factors have opposite sign in the interval $(-2, 1)$. Consequently, the solution set of (5) is the interval $(-2, 1)$.

EXAMPLE 4. Solve the inequality

$$\frac{x}{x-4} > 5, \qquad x \neq 4 \tag{7}$$

SOLUTION. Since $x \neq 4$, we have $x - 4 \neq 0$. Hence $(x - 4)^2 > 0$. Therefore, multiplying both members of the inequality (7) by the positive number $(x - 4)^2$, we have the inequality

$$x(x - 4) > 5(x - 4)^2 \tag{8}$$

which gives the following equivalent inequalities:

$$x^2 - 4x > 5(x^2 - 8x + 16)$$

$$0 > 4x^2 - 36x + 80$$

$$0 > x^2 - 9x + 20$$

or

$$x^2 - 9x + 20 < 0$$

$$(x - 5)(x - 4) < 0$$

Now as in Example 3, we can show that the last inequality holds in the interval (4, 5). Thus the solution set of (7) is the open interval (4, 5).

EXAMPLE 5. Solve the inequality

$$(x+2)(x-1)(x-3) > 0 \tag{9}$$

SOLUTION. A number x is in the solution set of this inequality if and only if either all three factors $x+2$, $x-1$, and $x-3$ are positive or two of the factors are negative and one factor is positive. As in Example 3, we shall show the sign of each of the factors in Figure 3 and read off the solution set from the figure. We see that all three factors are positive in the interval $(3, +\infty)$ and that two factors are negative and one positive in the interval $(-2, 1)$. Hence the solution set of the inequality (9) is $(-2, 1) \cup (3, +\infty)$.

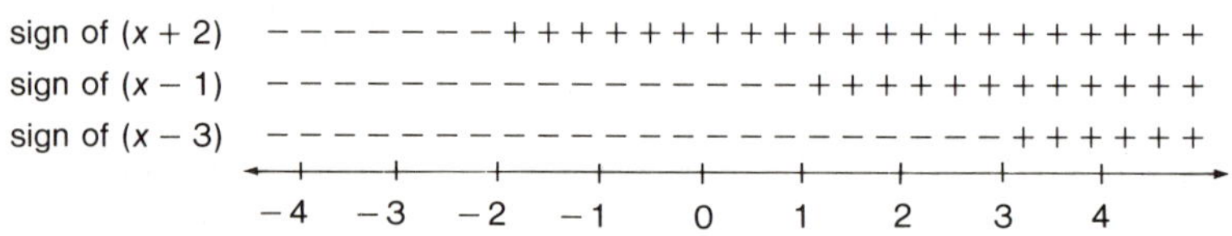

Figure 3

2.5 ABSOLUTE VALUE

We now introduce another useful concept for real numbers.

Definition 1. The *absolute value* of a real number x, denoted $|x|$, is defined as

$$|x| = \begin{cases} x & \text{if } x \geq 0 \\ -x & \text{if } x < 0 \end{cases}$$

For example, $|4| = 4$, $|-4| = -(-4) = 4$, $|0| = 0$, and $|7-13| = |-6| = 6$. The preceding definition gives the following properties. For any real numbers x and y:

$$|x| \geq 0 \quad \text{and} \quad -|x| \leq x \leq |x| \tag{1}$$

$$|x| = 0 \quad \text{if and only if} \quad x = 0 \tag{2}$$

$$|xy| = |x|\,|y| \tag{3}$$

$$|x-y| = |y-x| \tag{4}$$

$$\left|\frac{x}{y}\right| = \frac{|x|}{|y|}; \quad y \neq 0 \tag{5}$$

$$|x+y| \leq |x| + |y| \tag{6}$$

$$|x-y| \geq |x| - |y| \tag{7}$$

We can use the concept of absolute value to assign a distance between two points on a line. We have

Definition 2. Let x and y be the coordinates of two points P and Q, respectively, on a number line; then $d(P, Q)$, the *length* of the line segment from P to Q, is defined by

$$d(P, Q) = |x - y|$$

The quantity $d(P, Q)$ is also called the *distance* between P and Q. It is easy to see from Definition 2 that $d(P, Q) = d(Q, P)$ and $d(P, Q)$ is always nonnegative. Suppose that P has coordinate x and Q is the origin with corrdinate 0; then

$$d(P, Q) = |x - 0| = |x|$$

Thus, geometrically, the absolute value of x is the distance from the origin to the point corresponding to x.

EXAMPLE 1. Let $A = \{x \mid |x-2| \leq 6\}$. Give a geometric description of the set A.

SOLUTION. Let P and Q be points with coordinates x and 2, respectively. Then $|x-2| \leq 6$ states that the distance between P and Q is less than or equal to 6; that is, $d(P, Q) \leq 6$. Consequently, the point P lies in the closed interval with center at 2 and length 12. Hence the set $A = [-4, 8]$. See Figure 1.

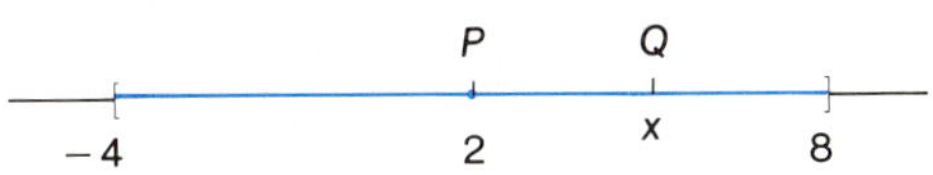

Figure 1

The following two theorems are used to solve inequalities involving absolute value.

THEOREM 1. If $x, a \in R$ and $a > 0$, then

$$|x| < a \quad \text{if and only if} \quad -a < x < a$$

Geometric interpretation: $|x| < a$ means that the distance between x and 0 is less than a units. In other words, x is within a units of 0 (see Figure 2). Using interval notation, $x \in (-a, a)$, which is equivalent to the inequality $-a < x < a$; that is,

$$\{x \mid |x| < a\} = \{x \mid -a < x < a\}$$

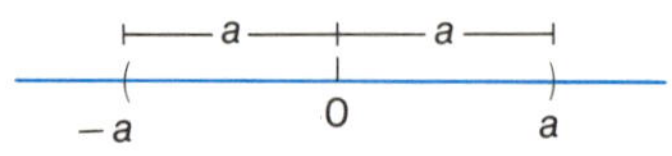

Figure 2

THEOREM 2. If $x, a \in R$ and $a > 0$, then $|x| > a$ if and only if $x > a$ or $x < -a$.

In set notation, Theorem 2 means

$$\{x \mid |x| > a\} = \{x \mid x > a\} \cup \{x \mid x < -a\}$$

The reader is encouraged to provide a geometric interpretation for Theorem 2.

Theorems 1 and 2 are valid if $<$ and $>$ are replaced by $\leq$ and $\geq$, respectively.

EXAMPLE 2. Solve the inequality:

$$|2x - 3| < 7 \tag{8}$$

SOLUTION. By Theorem 3, the inequality (8) is equivalent to

$$-7 < 2x - 3 < 7 \tag{9}$$

or equivalently,

$$-7 + 3 < 2x < 7 + 3$$

or

$$-4 < 2x < 10$$

or

$$-2 < x < 5$$

Therefore, the solution set of (8) is $\{x \mid -2 < x < 5\} = (-2, 5)$.

EXAMPLE 3. Solve the inequality

$$|3x + 4| \geq 5 \tag{10}$$

and sketch the graph of the solution set.

SOLUTION. From Theorem 2, we have

$$3x + 4 \leq -5 \quad \text{or} \quad 3x + 4 \geq 5$$

so

$$3x \leq -9 \quad \text{or} \quad 3x \geq 1$$

$$x \leq -3 \quad \text{or} \quad x \geq \frac{1}{3}$$

Hence

$$\{x \mid |3x + 4| \geq 5\} = \{x \mid x \leq -3\} \cup \left\{x \mid x \geq \frac{1}{3}\right\}$$

$$= (-\infty, -3] \cup \left[\frac{1}{3}, +\infty\right)$$

The graph of the solution set is given in Figure 3.

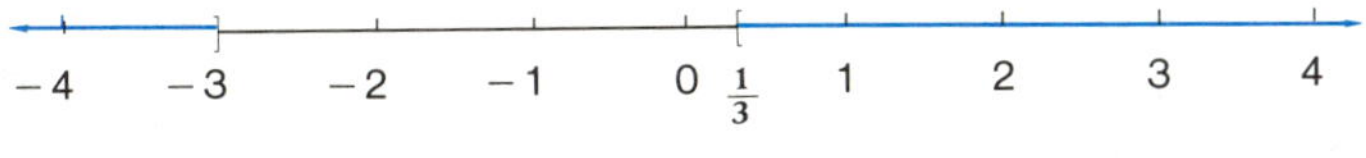

Figure 3

APPENDIX 3. SIGMA NOTATION

In Appendix 4 we shall prove some formulas such as

$$1 + 3 + 5 + \cdots + (2n - 1) = n^2 \tag{1}$$

$$1 + 2 + 3 + \cdots + n = \frac{n(n+1)}{2} \tag{2}$$

$$1^2 + 2^2 + 3^2 + \cdots + n^2 = \frac{n(n+1)(2n+1)}{6} \tag{3}$$

Each of the formulas is true for any natural number n. The three dots indicate the missing terms in the sum. We now introduce a short notation for sums.

The symbol Σ is the capital sigma in the Greek alphabet and corresponds to our English S. Thus it reminds us of the word sum. The symbol

$$\sum_{i=1}^{5} x_i$$

(read "the sum of x_i as i assumes values from 1 to 5") means

$$\sum_{i=1}^{5} x_i = x_1 + x_2 + x_3 + x_4 + x_5$$

The sum need not start at 1 and end at 5. Further, any letter can be used instead of i. The following examples indicate various possibilities.

$$\sum_{k=1}^{3} k^2 = 1^2 + 2^2 + 3^2 = 1 + 4 + 9 = 14$$

$$\sum_{j=2}^{4} j(j+3) = 2\cdot 5 + 3\cdot 6 + 4\cdot 7 = 10 + 18 + 28 = 56$$

$$\sum_{i=1}^{6} 1 = 1 + 1 + 1 + 1 + 1 + 1 = 6$$

$$\sum_{i=4}^{7} x_i = x_4 + x_5 + x_6 + x_7$$

$$\sum_{i=1}^{4} 3y_i = 3y_1 + 3y_2 + 3y_3 + 3y_4$$

$$\sum_{n=2}^{6} na_n = 2a_2 + 3a_3 + 4a_4 + 5a_5 + 6a_6$$

EXAMPLE 1. Show that

(a) $\displaystyle\sum_{i=1}^{4} x_i + \sum_{i=5}^{7} x_i = \sum_{i=1}^{7} x_i$

(b) $\sum_{i=1}^{4}(x_i + y_i) = \sum_{i=1}^{4} x_i + \sum_{i=1}^{4} y_i$

(c) $\sum_{i=1}^{3}(x_i + y_i)^2 = \sum_{i=1}^{3} x_i^2 + 2\sum_{i=1}^{3} x_i y_i + \sum_{i=1}^{3} y_i^2$

SOLUTION.

(a)
$$\begin{aligned}\sum_{i=1}^{4} x_i + \sum_{i=5}^{7} x_i &= (x_1 + x_2 + x_3 + x_4) + (x_5 + x_6 + x_7)\\ &= x_1 + x_2 + x_3 + x_4 + x_5 + x_6 + x_7\\ &= \sum_{i=1}^{7} x_i\end{aligned}$$

(b)
$$\begin{aligned}\sum_{i=1}^{4}(x_i + y_i) &= (x_1 + y_1) + (x_2 + y_2) + (x_3 + y_3) + (x_4 + y_4)\\ &= (x_1 + x_2 + x_3 + x_4) + (y_1 + y_2 + y_3 + y_4)\\ &= \sum_{i=1}^{4} x_i + \sum_{i=1}^{4} y_i\end{aligned}$$

(c)
$$\begin{aligned}\sum_{i=1}^{3}(x_i + y_i)^2 &= (x_1 + y_1)^2 + (x_2 + y_2)^2 + (x_3 + y_3)^2\\ &= (x_1^2 + 2x_1y_1 + y_1^2) + (x_2^2 + 2x_2y_2 + y_2^2) + (x_3^2 + 2x_3y_3 y_3^2)\\ &= (x_1^2 + x_2^2 + x_3^2) + 2(x_1y_1 + x_2y_2 + x_3y_3) + (y_1^2 + y_2^2 + y_3^2)\\ &= \sum_{i=1}^{3} x_i^2 + 2\sum_{i=1}^{3} x_i y_i + \sum_{i=1}^{3} y_i^2\end{aligned}$$

EXAMPLE 2. Use the Σ notation to rewrite formulas (1), (2), and (3).

SOLUTION. With the Σ notation, (1), (2), and (3) become

$$\sum_{k=1}^{n}(2k - 1) = n^2 \tag{1*}$$

$$\sum_{i=1}^{n} i = \frac{n(n + 1)}{2} \tag{2*}$$

$$\sum_{j=1}^{n} j^2 = \frac{n(n + 1)(2n + 1)}{6} \tag{3*}$$

APPENDIX 4. MATHEMATICAL INDUCTION

Suppose we wish to find the sum of the first n odd positive integers. Direct computation shows that

$$\begin{aligned}
&\text{if } n = 1, && 1 = 1 && = 1^2 \\
&\text{if } n = 2, && 1 + 3 = 4 && = 2^2 \\
&\text{if } n = 3, && 1 + 3 + 5 = 9 && = 3^2 \\
&\text{if } n = 4, && 1 + 3 + 5 + 7 = 16 && = 4^2 \\
&\text{if } n = 5, && 1 + 3 + 5 + 7 + 9 = 25 && = 5^2 \\
&\text{if } n = 6, && 1 + 3 + 5 + 7 + 9 + 11 = 36 && = 6^2
\end{aligned}$$

These cases lead us to believe that the sum is always equal to the square of the number of terms in the sum. We observe that 1 is the first term in the sum, and it is easy to see that the nth term of the sum is $2n - 1$. Thus we can express our assertion symbolically as

$$1 + 3 + 5 + \cdots + (2n - 1) = n^2 \tag{1}$$

or using the Σ notation, we claim that

$$\sum_{k=1}^{n} (2k - 1) = n^2 \tag{2}$$

By direct computation, we have seen that Equation (1) [or (2)] is true for $n = 1, 2, 3, 4, 5,$ and 6. Does this mean that Equation (1) is true for any positive integer n? Can we settle this by continuing our numerical computation? Certainly not! No matter how many cases we check, we can never prove that (1) is always true, because there are infinitely many cases and no amount of computation can settle them all. Thus we need some *logical* argument that will prove the (1) is always true.

We use the symbol $p(n)$ (read "p of n") to denote some proposition which depends on the positive integer n. For example, $p(n)$ might denote the proposition that (1) is true. The proof that $p(n)$ is true for all positive integers n can be given in two steps and these two steps form the

Principle of Mathematical Induction. For a given proposition $p(n)$ let us assume that we can prove the following conditions.

1. The proposition is true for $n = 1$.

2. If $p(n)$ is true for $n = k$, then it is also true for $n = k + 1$.

Then the proposition is true for all positive integers n.

Note that the principle of mathematical induction can be used to prove a general proposition for positive integers only. Also the result to be proved should be surmised beforehand.

EXAMPLE 1. Prove that Equation (1) holds for every positive integer n.

SOLUTION.

Step 1. We have already seen that Equation (1) is true for $n = 1$.

Step 2. We now assume that $p(n)$ is true for $n = k$:

$$1 + 3 + 5 + \cdots + (2k - 1) = k^2 \tag{3}$$

To obtain the sum of the first $k + 1$ odd integers, we merely add the next odd one, $2k + 1$, to both sides of (3).

$$\begin{aligned}(1 + 3 + 5 + \cdots + (2k - 1)) + (2k + 1) &= k^2 + (2k + 1)\\ &= (k + 1)^2 \qquad (4)\end{aligned}$$

Equation (4) is precisely Equation (1) when $n = k + 1$. Hence we have shown that if $p(n)$ is true for $n = k$, then it is also true for $n = k + 1$.

By the principle of mathematical induction this completes the proof that Equation (1) is true for every positive integer n.

EXAMPLE 2. Prove that for every positive integer n,

$$1^2 + 2^2 + 3^2 + \cdots + n^2 = \frac{n(n + 1)(2n + 1)}{6} \tag{5}$$

SOLUTION.

Step 1. For $n = 1$, the assertion of Equation (5) is

$$1^2 = \frac{1(1 + 1)(2 + 1)}{6}$$

$$1 = 1$$

which is certainly true.

Step 2. We assume that (5) is true for $n = k$:

$$1^2 + 2^2 + 3^2 + \cdots + k^2 = \frac{k(k+1)(2k+1)}{6} \tag{6}$$

To obtain the left-hand side of (5) for $n = k + 1$, we must add $(k+1)^2$ to the left-hand side of (6). This gives

$$\begin{aligned} 1^2 + 2^2 + 3^2 + \cdots + k^2 + (k+1)^2 &= \frac{k(k+1)(2k+1)}{6} + (k+1)^2 \\ &= \tfrac{1}{6}[k(k+1)(2k+1) + 6(k+1)^2] \qquad (7) \\ &= \tfrac{1}{6}(k+1)[k(2k+1) + 6(k+1)] \\ &= \tfrac{1}{6}(k+1)(2k^2 + 7k + 6) \\ &= \tfrac{1}{6}(k+1)(k+2)(2k+3) \\ &= \tfrac{1}{6}(k+1)[(k+1)+1][2(k+1)+1] \end{aligned}$$

This is Equation (5) when $n = k + 1$. Hence by the principle of mathematical induction, formula (5) is true for every positive integer n.

In the same way one can prove that

$$1 + 2 + 3 + \cdots + n = \frac{n(n+1)}{2} \tag{8}$$

Table A Squares and Square Roots

x	x^2	$\sqrt{x}$	x	x^2	$\sqrt{x}$
1	1	1.000	51	2,601	7.141
2	4	1.414	52	2,704	7.211
3	9	1.732	53	2,809	7.280
4	16	2.000	54	2,916	7.348
5	25	2.236	55	3,025	7.416
6	36	2.449	56	3,136	7.483
7	49	2.646	57	3,249	7.550
8	64	2.828	58	3,364	7.616
9	81	3.000	59	3,481	7.681
10	100	3.162	60	3,600	7.746
11	121	3.317	61	3,721	7.810
12	144	3.464	62	3,844	7.874
13	169	3.606	63	3,969	7.937
14	196	3.742	64	4,096	8.000
15	225	3.873	65	4,225	8.062
16	256	4.000	66	4,356	8.124
17	289	4.123	67	4,489	8.185
18	324	4.243	68	4,624	8.246
19	361	4.359	69	4,761	8.307
20	400	4.472	70	4,900	8.367
21	441	4.583	71	5,041	8.426
22	484	4.690	72	5,184	8.485
23	529	4.796	73	5,329	8.544
24	576	4.899	74	5,476	8.602
25	625	5.000	75	5,625	8.660
26	676	5.099	76	5,776	8.718
27	729	5.196	77	5,929	8.775
28	784	5.292	78	6,084	8.832
29	841	5,385	79	6,241	8.888
30	900	5.477	80	6,400	8.944
31	961	5.568	81	6,561	9.000
32	1,024	5.657	82	6,724	9.055
33	1,089	5.745	83	6,889	9.110
34	1,156	5.831	84	7,056	9.165
35	1,225	5.916	85	7,225	9.220
36	1,296	6.000	86	7,396	9.274
37	1,369	6.083	87	7,569	9.327
38	1,444	6.164	88	7,744	9.381
39	1,521	6.245	89	7,921	9.434
40	1,600	6.325	90	8,100	9.487
41	1,681	6.403	91	8,281	9.539
42	1,764	6.481	92	8,464	9.592
43	1,849	6.557	93	8,649	9.644
44	1,936	6.633	94	8,836	9.695
45	2,025	6.708	95	9,025	9.747
46	2,116	6.782	96	9,216	9.798
47	2,209	6.856	97	9,409	9.849
48	2,304	6.928	98	9,604	9.899
49	2,401	7.000	99	9,801	9.950
50	2,500	7.071	100	10,000	10.000

Table B Exponential Functions

x	e^x	e^{-x}	x	e^x	e^{-x}
0.00	1.0000	1.0000	1.5	4.4817	0.2231
0.01	1.0101	0.9901	1.6	4.9530	0.2019
0.02	1.0202	0.9802	1.7	5.4739	0.1827
0.03	1.0305	0.9705	1.8	6.0496	0.1653
0.04	1.0408	0.9608	1.9	6.6859	0.1496
0.05	1.0513	0.9512	2.0	7.3891	0.1353
0.06	1.0618	0.9418	2.1	8.1662	0.1225
0.07	1.0725	0.9324	2.2	9.0250	0.1108
0.08	1.0833	0.9331	2.3	9.9742	0.1003
0.09	1.0942	0.9139	2.4	11.0230	0.0907
0.10	1.1052	0.9048	2.5	12.182	0.0821
0.11	1.1163	0.8958	2.6	13.464	0.0743
0.12	1.1275	0.8869	2.7	14.880	0.0672
0.13	1.1388	0.8781	2.8	16.445	0.0608
0.14	1.1503	0.8694	2.9	18.174	0.0550
0.15	1.1618	0.8607	3.0	20.086	0.0498
0.16	1.1735	0.8521	3.1	22.198	0.0450
0.17	1.1853	0.8437	3.2	24.533	0.0408
0.18	1.1972	0.8353	3.3	27.113	0.0369
0.19	1.2092	0.8270	3.4	29.964	0.0334
0.20	1.2214	0.8187	3.5	33.115	0.0302
0.21	1.2337	0.8106	3.6	36.598	0.0273
0.22	1.2461	0.8025	3.7	40.447	0.0247
0.23	1.2586	0.7945	3.8	44.701	0.0224
0.24	1.2712	0.7866	3.9	49.402	0.0202
0.25	1.2840	0.7788	4.0	54.598	0.0183
0.30	1.3499	0.7408	4.1	60.340	0.0166
0.35	1.4191	0.7047	4.2	66.686	0.0150
0.40	1.4918	0.6703	4.3	73.700	0.0136
0.45	1.5683	0.6376	4.4	81.451	0.0123
0.50	1.6487	0.6065	4.5	90.017	0.0111
0.55	1.7333	0.5769	4.6	99.484	0.0101
0.60	1.8221	0.5488	4.7	109.950	0.0091
0.65	1.9155	0.5220	4.8	121.510	0.0082
0.70	2.0138	0.4966	4.9	134.290	0.0074
0.75	2.1170	0.4724	5.0	148.41	0.0067
0.80	2.2255	0.4493	5.5	244.69	0.0041
0.85	2.3396	0.4274	6.0	403.43	0.0025
0.90	2.4596	0.4066	6.5	665.14	0.0015
0.95	2.5857	0.3867	7.0	1096.60	0.0009
1.0	2.7183	0.3679	7.5	1808.0	0.0006
1.1	3.0042	0.3329	8.0	2981.0	0.0003
1.2	3.3201	0.3012	8.5	4914.8	0.0002
1.3	3.6693	0.2725	9.0	8103.1	0.0001
1.4	4.0552	0.2466	10.0	22026.0	0.00005

Table C Common Logarithms

x	0	1	2	3	4	5	6	7	8	9
1.0	.0000	.0043	.0086	.0128	.0170	.0212	.0253	.0294	.0334	.0374
1.1	.0414	.0453	.0492	.0531	.0569	.0607	.0645	.0682	.0719	.0755
1.2	.0792	.0828	.0864	.0899	.0934	.0969	.1004	.1038	.1072	.1106
1.3	.1139	.1173	.1206	.1239	.1271	.1303	.1335	.1367	.1399	.1430
1.4	.1461	.1492	.1523	.1553	.1584	.1614	.1644	.1673	.1703	.1732
1.5	.1761	.1790	.1818	.1847	.1875	.1903	.1931	.1959	.1987	.2014
1.6	.2041	.2068	.2095	.2122	.2148	.2175	.2201	.2227	.2253	.2279
1.7	.2304	.2330	.2355	.2380	.2405	.2430	.2455	.2480	.2504	.2529
1.8	.2553	.2577	.2601	.2625	.2648	.2672	.2695	.2718	.2742	.2765
1.9	.2788	.2810	.2833	.2856	.2878	.2900	.2923	.2945	.2967	.2989
2.0	.3010	.3032	.3054	.3075	.3096	.3118	.3139	.3160	.3181	.3201
2.1	.3222	.3243	.3263	.3284	.3304	.3324	.3345	.3365	.3385	.3404
2.2	.3424	.3444	.3464	.3483	.3502	.3522	.3541	.3560	.3579	.3598
2.3	.3617	.3636	.3655	.3674	.3692	.3711	.3729	.3747	.3766	.3784
2.4	.3802	.3820	.3838	.3856	.3874	.3892	.3909	.3927	.3945	.3962
2.5	.3979	.3997	.4014	.4031	.4048	.4065	.4082	.4099	.4116	.4133
2.6	.4150	.4166	.4183	.4200	.4216	.4232	.4249	.4265	.4281	.4298
2.7	.4314	.4330	.4346	.4362	.4378	.4393	.4409	.4425	.4440	.4456
2.8	.4472	.4487	.4502	.4518	.4533	.4548	.4564	.4579	.4594	.4609
2.9	.4624	.4639	.4654	.4669	.4683	.4698	.4713	.4728	.4742	.4757
3.0	.4771	.4786	.4800	.4814	.4829	.4843	.4857	.4871	.4886	.4900
3.1	.4914	.4928	.4942	.4955	.4969	.4983	.4997	.5011	.5024	.5038
3.2	.5051	.5065	.5079	.5092	.5105	.5119	.5132	.5145	.5159	.5172
3.3	.5185	.5198	.5211	.5224	.5237	.5250	.5263	.5276	.5289	.5302
3.4	.5315	.5328	.5340	.5353	.5366	.5378	.5391	.5403	.5416	.5428
3.5	.5441	.5453	.5465	.5478	.5490	.5502	.5514	.5527	.5539	.5551
3.6	.5563	.5575	.5587	.5599	.5611	.5623	.5635	.5647	.5658	.5670
3.7	.5682	.5694	.5705	.5717	.5729	.5740	.5752	.5763	.5775	.5786
3.8	.5798	.5809	.5821	.5832	.5843	.5855	.5866	.5877	.5888	.5899
3.9	.5911	.5922	.5933	.5944	.5955	.5966	.5977	.5988	.5999	.6010
4.0	.6021	.6031	.6042	.6053	.6064	.6075	.6085	.6096	.6107	.6117
4.1	.6128	.6138	.6149	.6160	.6170	.6180	.6191	.6201	.6212	.6222
4.2	.6232	.6243	.6253	.6263	.6274	.6284	.6294	.6304	.6314	.6325
4.3	.6335	.6345	.6355	.6365	.6375	.6385	.6395	.6405	.6415	.6425
4.4	.6435	.6444	.6454	.6464	.6474	.6484	.6493	.6503	.6513	.6522
4.5	.6532	.6542	.6551	.6561	.6571	.6580	.6590	.6599	.6609	.6618
4.6	.6628	.6637	.6646	.6656	.6665	.6675	.6684	.6693	.6702	.6712
4.7	.6721	.6730	.6739	.6749	.6758	.6767	.6776	.6785	.6794	.6803
4.8	.6812	.6821	.6830	.6839	.6848	.6857	.6866	.6875	.6884	.6893
4.9	.6902	.6911	.6920	.6928	.6937	.6946	.6955	.6964	.6972	.6981
5.0	.6990	.6998	.7007	.7016	.7024	.7033	.7042	.7050	.7059	.7067
5.1	.7076	.7084	.7093	.7101	.7110	.7118	.7126	.7135	.7143	.7152
5.2	.7160	.7168	.7177	.7185	.7193	.7202	.7210	.7218	.7226	.7235
5.3	.7243	.7251	.7259	.7267	.7275	.7284	.7292	.7300	.7308	.7316
5.4	.7324	.7332	.7340	.7348	.7356	.7364	.7372	.7380	.7388	.7396
x	0	1	2	3	4	5	6	7	8	9

Table C (continued)

x	0	1	2	3	4	5	6	7	8	9
5.5	.7404	.7412	.7419	.7427	.7435	.7443	.7451	.7459	.7466	.7474
5.6	.7482	.7490	.7497	.7505	.7513	.7520	.7528	.7536	.7543	.7551
5.7	.7559	.7566	.7574	.7582	.7589	.7597	.7604	.7612	.7619	.7627
5.8	.7634	.7642	.7649	.7657	.7664	.7672	.7679	.7686	.7694	.7701
5.9	.7709	.7716	.7723	.7731	.7738	.7745	.7752	.7760	.7767	.7774
6.0	.7782	.7789	.7796	.7803	.7810	.7818	.7825	.7832	.7839	.7846
6.1	.7853	.7860	.7868	.7875	.7882	.7889	.7896	.7903	.7910	.7917
6.2	.7924	.7931	.7938	.7945	.7952	.7959	.7966	.7973	.7980	.7987
6.3	.7993	.8000	.8007	.8014	.8021	.8028	.8035	.8041	.8048	.8055
6.4	.8062	.8069	.8075	.8082	.8089	.8096	.8102	.8019	.8116	.8122
6.5	.8129	.8136	.8142	.8149	.8156	.8162	.8169	.8176	.8182	.8189
6.6	.8195	.8202	.8209	.8215	.8222	.8228	.8235	.8241	.8248	.8254
6.7	.8261	.8267	.8274	.8280	.8287	.8293	.8299	.8306	.8312	.8319
6.8	.8325	.8331	.8338	.8344	.8351	.8357	.8363	.8370	.8376	.8382
6.9	.8388	.8395	.8401	.8407	.8414	.8420	.8426	.8432	.8439	.8445
7.0	.8451	.8457	.8463	.8470	.8476	.8482	.8488	.8494	.8500	.8506
7.1	.8513	.8519	.8525	.8531	.8537	.8543	.8549	.8555	.8561	.8567
7.2	.8573	.8579	.8585	.8591	.8597	.8603	.8609	.8615	.8621	.8627
7.3	.8633	.8639	.8645	.8651	.8657	.8663	.8669	.8675	.8681	.8686
7.4	.8692	.8698	.8704	.8710	.8716	.8722	.8727	.8733	.8739	.8745
7.5	.8751	.8756	.8762	.8768	.8774	.8779	.8785	.8791	.8797	.8802
7.6	.8808	.8814	.8820	.8825	.8831	.8837	.8842	.8848	.8854	.8859
7.7	.8865	.8871	.8876	.8882	.8887	.8893	.8899	.8904	.8910	.8915
7.8	.8921	.8927	.8932	.8938	.8943	.8949	.8954	.8960	.8965	.8971
7.9	.8976	.8982	.8987	.8993	.8998	.9004	.9009	.9015	.9020	.9025
8.0	.9031	.9036	.9042	.9047	.9053	.9058	.9063	.9069	.9074	.9079
8.1	.9085	.9090	.9096	.9101	.9106	.9112	.9117	.9122	.9128	.9133
8.2	.9138	.9143	.9149	.9154	.9159	.9165	.9170	.9175	.9180	.9186
8.3	.9191	.9196	.9201	.9206	.9212	.9217	.9222	.9227	.9232	.9238
8.4	.9243	.9248	.9253	.9258	.9263	.9269	.9274	.9279	.9284	.9289
8.5	.9294	.9299	.9304	.9309	.9315	.9320	.9325	.9330	.9335	.9340
8.6	.9345	.9350	.9355	.9360	.9365	.9370	.9375	.9380	.9385	.9390
8.7	.9395	.9400	.9405	.9410	.9415	.9420	.9425	.9430	.9435	.9440
8.8	.9445	.9450	.9455	.9460	.9465	.9469	.9474	.9479	.9484	.9489
8.9	.9494	.9499	.9504	.9509	.9513	.9518	.9523	.9528	.9533	.9538
9.0	.9542	.9547	.9552	.9557	.9562	.9566	.9571	.9576	.9581	.9586
9.1	.9590	.9595	.9600	.9605	.9609	.9614	.9619	.9624	.9628	.9633
9.2	.9638	.9643	.9647	.9652	.9657	.9661	.9666	.9671	.9675	.9680
9.3	.9685	.9689	.9694	.9699	.9703	.9708	.9713	.9717	.9722	.9727
9.4	.9731	.9736	.9741	.9745	.9750	.9754	.9759	.9763	.9768	.9773
9.5	.9777	.9782	.9786	.9791	.9795	.9800	.9805	.9809	.9814	.9818
9.6	.9823	.9827	.9832	.9836	.9841	.9845	.9850	.9854	.9859	.9863
9.7	.9868	.9872	.9877	.9881	.9886	.9890	.9894	.9899	.9903	.9908
9.8	.9912	.9917	.9921	.9926	.9930	.9934	.9939	.9943	.9948	.9952
9.9	.9956	.9961	.9965	.9969	.9974	.9978	.9983	.9987	.9991	.9996
x	0	1	2	3	4	5	6	7	8	9

Table D Natural Logarithms of Numbers

x	$\ln x$	x	$\ln x$	x	$\ln x$
	*	4.5	1.5041	9.0	2.1972
0.1	7.6974	4.6	1.5261	9.1	2.2083
0.2	8.3906	4.7	1.5476	9.2	2.2192
0.3	8.7960	4.8	1.5686	9.3	2.2300
0.4	9.0837	4.9	1.5892	9.4	2.2407
0.5	9.3069	5.0	1.6094	9.5	2.2513
0.6	9.4892	5.1	1.6292	9.6	2.2618
0.7	9.6433	5.2	1.6487	9.7	2.2721
0.8	9.7769	5.3	1.6677	9.8	2.2824
0.9	9.8946	5.4	1.6864	9.9	2.2925
1.0	0.0000	5.5	1.7047	10	2.3026
1.1	0.0953	5.6	1.7228	11	2.3979
1.2	0.1823	5.7	1.7405	12	2.4849
1.3	0.2624	5.8	1.7579	13	2.5649
1.4	0.3365	5.9	1.7750	14	2.6391
1.5	0.4055	6.0	1.7918	15	2.7081
1.6	0.4700	6.1	1.8083	16	2.7726
1.7	0.5306	6.2	1.8245	17	2.8332
1.8	0.5878	6.3	1.8405	18	2.8904
1.9	0.6419	6.4	1.8563	19	2.9444
2.0	0.6931	6.5	1.8718	20	2.9957
2.1	0.7419	6.6	1.8871	25	3.2189
2.2	0.7885	6.7	1.9021	30	3.4012
2.3	0.8329	6.8	1.9169	35	3.5553
2.4	0.8755	6.9	1.9315	40	3.6889
2.5	0.9163	7.0	1.9459	45	3.8067
2.6	0.9555	7.1	1.9601	50	3.9120
2.7	0.9933	7.2	1.9741	55	4.0073
2.8	1.0296	7.3	1.9879	60	4.0943
2.9	1.0647	7.4	2.0015	65	4.1744
3.0	1.0986	7.5	2.0149	70	4.2485
3.1	1.1314	7.6	2.0281	75	4.3175
3.2	1.1632	7.7	2.0412	80	4.3820
3.3	1.1939	7.8	2.0541	85	4.4427
3.4	1.2238	7.9	2.0669	90	4.4998
3.5	1.2528	8.0	2.0794	100	4.6052
3.6	1.2809	8.1	2.0919	110	4.7005
3.7	1.3083	8.2	2.1041	120	4.7875
3.8	1.3350	8.3	2.1163	130	4.8676
3.9	1.3610	8.4	2.1282	140	4.9416
4.0	1.3863	8.5	2.1401	150	5.0106
4.1	1.4110	8.6	2.1518	160	5.0752
4.2	1.4351	8.7	2.1633	170	5.1358
4.3	1.4586	8.8	2.1748	180	5.1930
4.4	1.4816	8.9	2.1861	190	5.2470

*Subtract 10 for $x < 1$. Thus $\ln 0.3 = 8.7960 - 10 = -1.2040$.

Table E A Brief Table of Integrals (An arbitrary constant is to be added to the formula for each indefinite integral.)

Elementary Integrals

1. $\int cf(u)\,du = c\int f(u)\,du$

2. $\int (f(u) + g(u))\,du = \int f(u)\,du + \int g(u)\,du$

3. $\int u^n\,du = \dfrac{n^{n+1}}{n+1}, \quad n \neq -1$

4. $\int \dfrac{du}{u} = \ln|u|$

5. $\int e^u\,du = e^u$

6. $\int a^u\,du = \dfrac{a^u}{\ln a}, \quad a > 0$

7. $\int \dfrac{du}{a^2 - u^2} = \dfrac{1}{2a}\ln \dfrac{a+u}{a-u}$

8. $\int \dfrac{du}{u^2 - a^2} = \dfrac{1}{2a}\ln \dfrac{u-a}{u+a}$

9. $\int \dfrac{du}{\sqrt{u^2 - a^2}} = \ln|u + \sqrt{u^2 - a^2}|, \quad u^2 \geq a^2$

10. $\int \dfrac{du}{\sqrt{u^2 + a^2}} = \ln(u + \sqrt{u^2 + a^2})$

Integrand Containing $a + bu$

11. $\int \dfrac{u\,du}{a+bu} = \dfrac{1}{b^2}(a + bu - a\ln|a+bu|)$

12. $\int \dfrac{u\,du}{(a+bu)^2} = \dfrac{1}{b^2}\;\dfrac{a}{a+bu} + \ln|a+bu|$

13. $\int \dfrac{u^2\,du}{(a+bu)^2} = \dfrac{1}{b^3}\; a + bu - \dfrac{a^2}{a+bu} - 2a\ln|a+bu|$

14. $\int \dfrac{u\,du}{(a+bu)^3} = \dfrac{1}{b^2}\; -\dfrac{1}{a+bu} + \dfrac{a}{2(a+bu)^2}$

15. $\int \dfrac{du}{u(a+bu)} = \dfrac{1}{a}\ln \dfrac{u}{a+bu}$

16. $\int \dfrac{du}{u^2(a+bu)} = -\dfrac{1}{au} + \dfrac{b}{a^2}\ln \dfrac{a+bu}{u}$

17. $\int \dfrac{du}{u(a+bu)^2} = \dfrac{1}{a(a+bu)} - \dfrac{1}{a^2}\ln \dfrac{a+bu}{u}$

18. $$\int u^m(a+bu)^n\,du = \frac{u^{m+1}(a+bu)^n}{m+n+1} + \frac{an}{m+n+1}\int u^m(a+bu)^{n-1}\,du$$

Integrand Containing $\sqrt{a+bu}$

19. $$\int u\sqrt{a+bu}\,du = -\frac{2(2a-3bu)(a+bu)^{3/2}}{15b^2}$$

20. $$\int u^n\sqrt{a+bu}\,du = \frac{2u^n(a+bu)^{3/2}}{b(2n+3)} - \frac{2an}{b(2n+3)}\int u^{n-1}\sqrt{a+bu}\,du$$

21. $$\int \frac{u\,du}{\sqrt{a+bu}} = \frac{2(bu-2a)\sqrt{a+bu}}{3b^2}$$

22. $$\int \frac{u^n\,du}{\sqrt{a+bu}} = \frac{2u^n\sqrt{a+bu}}{b(2n+1)} - \frac{2an}{b(2n+1)}\int \frac{u^{n-1}\,du}{\sqrt{a+bu}}$$

23. $$\int \frac{du}{u\sqrt{a+bu}} = \frac{1}{\sqrt{a}}\ln\frac{\sqrt{a+bu}-\sqrt{a}}{\sqrt{a+bu}+\sqrt{a}}, \quad a>0$$

24. $$\int \frac{du}{u^n\sqrt{a+bu}} = -\frac{\sqrt{a+bu}}{a(n-1)u^{n-1}} - \frac{b(2n-3)}{2a(n-1)}\int \frac{du}{u^{n-1}\sqrt{a+bu}}$$

25. $$\int \frac{\sqrt{a+bu}}{u}\,du = 2\sqrt{a+bu} + a\int \frac{du}{u\sqrt{a+bu}}$$

Integrand Containing $\sqrt{a^2-u^2}$

26. $$\int \frac{\sqrt{a^2-u^2}}{u}\,du = \sqrt{a^2-u^2} - a\ln\frac{a+\sqrt{a^2-u^2}}{u}$$

27. $$\int \frac{du}{u\sqrt{a^2-u^2}} = -\frac{1}{a}\ln\frac{a+\sqrt{a^2-u^2}}{u}$$

28. $$\int \frac{du}{u^2\sqrt{a^2-u^2}} = -\frac{\sqrt{a^2-u^2}}{a^2u}$$

29. $$\int \frac{du}{(a^2-u^2)^{3/2}} = \frac{u}{a^2\sqrt{a^2-u^2}}$$

Integrand Containing $\sqrt{u^2 \pm a^2}$

See also formulas (8), (9), and (10).

30. $$\int \sqrt{u^2\pm a^2}\,du = \frac{u}{2}\sqrt{u^2\pm a^2} \pm \frac{a^2}{2}\ln\left|u+\sqrt{u^2\pm a^2}\right|$$

31. $$\int \sqrt{(u^2\pm a^2)^n}\,du = \frac{u\sqrt{(u^2\pm a^2)^n}}{n+1} \pm \frac{na^2}{n+1}\int \sqrt{(u^2\pm a^2)^{n-2}}\,du, \quad n \neq -1$$

32. $$\int \frac{du}{(u^2\pm a^2)^{3/2}} = \frac{\pm u}{a^2\sqrt{u^2\pm a^2}}$$

33. $$\int \frac{u^2\,du}{\sqrt{u^2\pm a^2}} = \frac{u}{2}\sqrt{u^2\pm a^2} \mp \frac{a^2}{2}\ln\left|u+\sqrt{u^2\pm a^2}\right|$$

34. $$\int \frac{du}{u\sqrt{u^2+a^2}} = -\frac{1}{a}\ln\frac{a+\sqrt{u^2+a^2}}{u}$$

35. $\displaystyle\int \frac{du}{u^2\sqrt{u^2 \pm a^2}} = \mp \frac{\sqrt{u^2 \pm a^2}}{a^2u}$

36. $\displaystyle\int \frac{du}{u^3\sqrt{u^2 + a^2}} = -\frac{\sqrt{u^2 + a^2}}{2a^2u^2} + \frac{1}{2a^3} \ln \frac{a + \sqrt{u^2 + a^2}}{u}$

37. $\displaystyle\int \frac{\sqrt{u^2 + a^2}}{u}\, du = \sqrt{u^2 + a^2} - a \ln \frac{a + \sqrt{u^2 + a^2}}{u}$

38. $\displaystyle\int \frac{\sqrt{u^2 \pm a^2}}{u^2}\, du = -\frac{\sqrt{u^2 \pm a^2}}{u} + \ln \left| u + \sqrt{u^2 \pm a^2} \right|$

Integrand Containing the Exponential Function or the Logarithmic Function

See also formulas (5) and (6).

39. $\displaystyle\int ue^u\, du = (u - 1)e^u$

40. $\displaystyle\int u^ne^n\, du = u^ne^n - n\int u^{n-1}e^u\, du$

41. $\displaystyle\int \frac{e^u}{u^u} du = -\frac{e^u}{(n-1)u^{n-1}} + \frac{1}{n-1}\int \frac{e^u\, du}{u^{n-1}}, \quad n \neq 1$

42. $\displaystyle\int e^u \ln u\, du = e^u \ln u - \int \frac{e^u}{u} du, \quad u > 0$

43. $\displaystyle\int \ln u\, du = u \ln u - u$

44. $\displaystyle\int u^n \ln u\, du = u^{n+1} \frac{\ln u}{n+1} - \frac{1}{(n+1)^2}, \quad n \neq -1$

45. $\displaystyle\int \frac{du}{u \ln u} = \ln |\ln u|$

46. $\displaystyle\int u^n(\ln u)^m\, du = \frac{u^{n+1}}{n+1}(\ln u)^m - \frac{m}{n+1}\int u^n(\ln u)^{m-1}\, du, \quad n \neq -1$

Some Definite Integrals

47. $\displaystyle\int_0^\infty x^{n-1}e^{-x}\, dx = \int_0^1 \ln \frac{1}{x}^{n-1} dx = \Gamma(n), \quad n > 0,$

and $\Gamma(n) = (n-1)!$, if n is a positive integer.

48. $\displaystyle\int_0^1 x^{m-1}(1-x)^{n-1}\, dx = \int_0^\infty \frac{x^{m-1}}{(1+x)^{m+n}} = \frac{\Gamma(m)\Gamma(n)}{\Gamma(m+n)}, \quad m, n > 0$

49. $\displaystyle\int_0^\infty e^{-a^2x^2}\, dx = \frac{\sqrt{\pi}}{2a}$

50. $\displaystyle\frac{1}{\sqrt{2\pi}} \int_{-\infty}^{\infty} e^{-x^2/2}\, dx = 1$

Answers To Selected Problems

EXERCISES 1.1

3. x-coordinate is zero. **5.** Line parallel to y-axis at $x = -2$.
7. $\sqrt{34}$, $(\frac{5}{2}, \frac{11}{2})$ **9.** $\sqrt{13}$, $(-2, -\frac{9}{2})$ **11.** $\sqrt{2}$, $(-\frac{5}{2}, \frac{1}{2})$
13. 4, $(-1, 5)$ **21.** isosceles **23.** equilateral
25. (a) collinear, (b) not collinear, (c) collinear
27. $17\sqrt{2}$ **29.** $d(D, M) = 1000$, $d(M, P) = 1500$, $d(D, P) = 500\sqrt{13}$ **31.** $a = 1 \pm 2\sqrt{21}$

EXERCISES 1.2

1. Straight line through (0, 0) **3.** Line **5.** Line **7.** Line
9. Parabola opening upward **11.** Parabola opening downward
13. Parabola opening downward **15.** Parabola opening downward
17. Parabola opening to the right
19. (with figure)

$y = \frac{1}{x}$

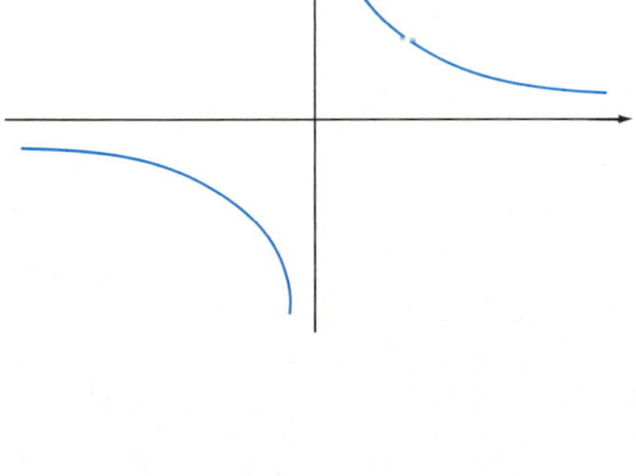

21. $C(2, 3)$; $r = 5$ **23.** $C(-2, -3)$; $r = \sqrt{11}$ **27.** $(x - \frac{1}{2})^2 + y^2 = \frac{1}{4}$
29. $(x + \frac{1}{4})^2 + (y - \frac{3}{8})^2 = \frac{397}{64}$ **31.** $(x + 2)^2 + (y - 3)^2 = 25$
33. $(x + 2)^2 + (y + 3)^2 = 16$ **35.** $(x - 1)^2 + (y - 2)^2 = 4$
37. $(x - 2)^2 + (y - 5)^2 = 26$ **39.** $|x| = |y|$ **41.** $y^2 = 4(x - 1)$
43. (a) $y = 7x$, (c) 42, (d) 8 **45.** 3605

EXERCISES 1.3

1. $(y - 3) = 2(x - 2)$ **3.** $(y + 1) = -7(x - 4)$ **5.** $y = \frac{2}{3}(x + 1)$
7. $y = -(x + 4)$ **9.** $y - 1 = \frac{4}{3}(x - \sqrt{2})$ **11.** $y + 4 = 2(x - 3)$
13. $y = -2$ **15.** $x = 2$ **17.** slope $= -\frac{1}{2}$; y-intercept $= 2$; x-intercept $= 4$
19. $\frac{3}{2}$; $\frac{7}{2}$; $-\frac{7}{3}$ **21.** No slope, no y-intercept, 5 **24.** $\dfrac{x}{\frac{5}{3}} + \dfrac{y}{\frac{5}{4}} = 1$ **26.** 7
28. $\frac{1}{2}$ **30.** $y = -4$ **32.** (a) $\frac{2}{7}$, (b) $-\frac{7}{2}$ **34.** (a) $-\frac{1}{2}$, (b) 2
36. (a) no slope, (b) 0 **38.** (a) $(y - 1) = \frac{2}{7}(x - 1)$, (b) $(y - 1) = -\frac{1}{2}(x - 1)$, (c) $x = 1$, (d) $y = 1$ **40.** (a) $c = 0$, (b) $c = -3$, (c) $c = -6$, (d) $c = -4$. **49.** $y - 4 = \frac{3}{2}(x - \frac{1}{2})$ **51.** $(-\frac{5}{3}, -\frac{7}{3})$
53. $(-\frac{5}{7}, -\frac{1}{7})$ **55.** $(-\frac{3}{16}, -\frac{7}{16})$ **57.** $(-\frac{2}{11}, -\frac{10}{11})$ **59.** $(2, \frac{5}{4})$

EXERCISES 1.4

1. $f(0) = 6$, $f(a) = 6$, $f(a + h) = 6$ **3.** $f(a) = 2a - 1$, $f(a + h) = 2(a + h) - 1$ **5.** $f(0) = 0$, $f(1) = 0$, $f(a) = a - a^2$, $f(a^2) = a^2 - a^4$, $f(a + h) = (a + h) - (a + h)^2$
7. $f(0) = 1$, $f(1) = 0$, $f(-3) = 2$, $f(5)$ is undefined, $f(a^2) = \sqrt{1 - a^2}$
9. $f(0)$ and $f(-1)$ are undefined. **11.** $f(-2) = 2(-2) + 1 = -3$; $f(0) = (0)^2 = 0$; $f(4) = (4)^2 = 16$ **13.** R **15.** R **17.** R
19. $(-\infty, -2) \cup (0, \infty)$ **21.** $x^2 - 6x + 8 \geq 0$; $(-\infty, 2] \cup [4, \infty)$
33. (a) 350, (b) $\frac{1}{2}$, (c) 600, (d) 250
35. (a) $25t + 6p - 24 = 0$, where t is measured from 8 P.M. (b) $53\frac{1}{3}$, $p(6) = 45$, $p(8) = 36\frac{2}{3}$, (c) $-\frac{25}{6}$

EXERCISES 1.5

1. Domain of $f = \{x \mid x \in R\}$ = domain of g
$(f + g)(x) = 2x + 5 + x^2$, $D(f + g) = \{x \mid x \in R\}$
$(f - g)(x) = 2x + 5 - x^2$, $D(f - g) = \{x \mid x \in R\}$
$(f \cdot g)(x) = x^2(2x + 5)$, $D(f \cdot g) = \{x \mid x \in R\}$
$\left(\dfrac{f}{g}\right)(x) = \dfrac{2x + 5}{x^2}$, $D\left(\dfrac{f}{g}\right) = \{x \mid x \in R, x \neq 0\}$

3. Domain $f = \{x \mid |x| \geq 2\}$, domain $g = \{x \mid x \in R\}$,
$(f + g)(x) = \sqrt{x^2 - 4} + x^2 + 2x - 3$, $(f - g)(x) = \sqrt{x^2 - 4} - x^2 - 2x + 3$,
$(fg)(x) = (x^2 + 2x - 3)\sqrt{x^2 - 4}$, domain $f + g$, $f - g$, $fg = \{x \mid |x| \geq 2\}$,
domain $(f/g) = \{x \mid |x| \geq 2, x \neq -3\}$

4. $(f+g)(x) = \dfrac{x^2+3x-1}{(x+1)(2x-1)}$, domain $f+g$, $f-g$, $fg = \{x \mid x \neq -1, \frac{1}{2}\}$
domain $f/g = \{x \mid x \neq -1, 0, \frac{1}{2}\}$

5. Domain $f = \{x \mid x > 0\}$, $g = \{x \mid x \neq 0\}$,
$(f+g)(x) = \dfrac{1}{\sqrt{x}} + \dfrac{1}{x^2}$, $(f-g)(x) = \dfrac{1}{\sqrt{x}} - \dfrac{1}{x^2}$, $(f\cdot g)(x) = \dfrac{1}{x^2\sqrt{x}}$, domain
$f+g$, $f-g$, $f\cdot g = \{x \mid x > 0\}$, $\left(\dfrac{f}{g}\right)(x) = x^{3/2}$, domain $\left(\dfrac{f}{g}\right) = \{x \mid x > 0\}$

7. $(f\circ g)(x) = 2+3x$, $(g\circ f)(x) = 6+3x$

9. $(f\circ g)(x) = x+4$, $(g\circ f)(x) = \sqrt{x^2+4}$

10. $(f\circ g)(x) = \sqrt[3]{2x^3+2}$, $(g\circ f)(x) = 2x+3$

12. $(f\circ g)(x) = \dfrac{x}{1+2x^2}$, $(g\circ f)(x) = \dfrac{x^2+2}{x}$

13. $(f\circ g)(x) = \dfrac{\sqrt{3x^4+8}}{x^2}$, $(g\circ f)(x) = \dfrac{2}{2x^2+3}$

14. $(f\circ g)(x) = \dfrac{1}{\sqrt{x-2}}$, domain $f\circ g = \{x \mid x > 2\}$,
$(g\circ f)(x) = \dfrac{\sqrt{x}-2x}{x}$, domain $g\circ f = \{x \mid x > 0\}$

15. (a) 16 (b) -52 (c) -2 (d) 7 (e) 56 (f) 242

16. (a) Undefined (b) $2(\sqrt{5}-\sqrt{10})$ (c) Undefined (d) Undefined
(e) Undefined (f) Undefined

17. (c) $(f\circ f)(x) = \sqrt{4+(\sqrt{4+x})}$, domain $f\circ f$ = domain f
(d) $(f\circ f)(x) = \dfrac{2+x}{5+2x}$, domain $f\circ f = \{x \mid x \neq -2, -\frac{5}{2}\}$
(e) $(f\circ f)(x) = \dfrac{8x^4+9x^2+2}{2x^3+x}$, domain $f\circ f$ = domain f

21. $2x^2+9$ **23.** $4x^2+20x+25$

EXERCISES 1.6

2. -21 **3.** -7 **5.** 24 **7.** 0 **9.** $\frac{16}{7}$ **11.** $-\frac{2}{7}$ **12.** $\frac{25}{67}$
13. -4 **19.** $\frac{3}{2}$ **20.** $\sqrt[3]{\frac{1}{3}}$ **21.** Does not exist **22.** $\dfrac{1}{2\sqrt{x}}$ **25.** 1
27. -3 **29.** $\dfrac{1}{2\sqrt{x}}$ **31.** $\dfrac{1}{\sqrt{2x+3}}$ **33.** $3x^2$ **35.** $2x-2$
37. 0

EXERCISES 1.7

1. Discont. **3.** Discont. **5.** Discont. **7.** Cont. **9.** Discont.
11. Cont. **13.** Discont. at -4; Cont. at 4. **15.** Cont. **17.** Discont.
19. Cont. **21.** Cont. **23.** $f(1) = 3$ **25.** $f(-1) = -2$

EXERCISES 2.1

1. $1, R$ **3.** $10x, R$ **5.** $6x, R$ **7.** $-\frac{2}{x^2}, x \neq 0$ **9.** $1-\frac{1}{x^2}, x \neq 0$
11. $-\frac{1}{2x^{3/2}}, x > 0$ **13.** $\frac{x}{\sqrt{x^2+1}}, R$ **15.** 0 **17.** -1 **19.** 0
23. $y = 12x - 16$ **25.** $y = -\frac{1}{4}x + 1$

EXERCISES 2.2

1. 7 **3.** $-10x$ **5.** $x^2 + 3x$ **7.** $-16x^3 - 9x^2$ **9.** $3\sqrt{x}$
11. $\frac{1}{2\sqrt{x}}$ **13.** $-\frac{8}{x^5}$ **15.** $\frac{3}{x^2}$ **17.** $505x^{100} + 17$ **19.** $-\frac{1}{x^2}$
21. $-\frac{1}{x^2} + \frac{1}{2\sqrt{x}}$ **23.** $-\frac{2}{x^3}$ **25.** $\frac{3}{2\sqrt{t}} - 27t^2$ **27.** $25 - \frac{2}{s^3}$
29. $2x + \frac{1}{\sqrt{x}} - \frac{1}{x^2}$ **31.** $24x^5 - 20x^4 + 4x^3$ **33.** $7x^{5/2} - 10x^{3/2}$
35. $-\frac{3}{(x-1)^2}$ **37.** $\frac{12}{(4+x)^2}$ **39.** $-\frac{16}{(2+5x)^2}$
43. $(x^5 - 3x^2)(4 - x^3)(10x - 1) + (x^5 - 3x^2)(-3x^2)(5x^2 - x) + (5x^4 - 6x)(4 - x^3)(5x^2 - x)$ **45.** $12x - y - 16 = 0$
47. $x + 4y - 4 = 0$ **49.** $y = -4, 4x - y - 4 = 0$ **53.** $(5, 5)$

EXERCISES 2.3

1. $6(3x+1)$ **3.** $-(x+1)^{-2}$ **5.** $4(x-5)^3$ **7.** $6(4x^2+x)^5(8x+1)$
9. $12(5x^3 + 6x^2 + 9)^{11}(15x^2 + 12x)$ **11.** $40x(4x-1)^4 + 2(4x-1)^5$
13. $\frac{15x}{2\sqrt{3x+1}} + 5\sqrt{3x+1}$ **15.** $\frac{4x^2 + x + 10}{\sqrt{x^2+5}}$ **17.** $\frac{8x^2 - x}{(2x^2 - x)^{5/2}}$
19. $2x - \frac{x}{(x^2+3)^{3/2}}$

EXERCISES 2.4

1. (a) 5 (b) 8 (c) 4, 12 **3.** (a) 96 (b) 128, 64 (c) 256
5. (a) 4 (b) 2 (c) 2, 2 **7.** (a) $\frac{\sqrt{10}}{4}$ (b) 2500, 625 **9.** \$1.99 = approximate profit on the sale of 11th pizza **13.** (a) $1, \frac{3}{4}, \frac{1}{2}$ (b) $\frac{9}{4}$ miles
15. $80\pi, 200\pi, 800\pi$ **17.** (a) $\frac{1}{125\pi}$, (b) $\frac{4}{5}$ **19.** $\frac{8}{25\pi}$ **21.** $-\frac{2}{59}$
23. $-\frac{11}{45}$ **25.** $\frac{2}{127}$ **27.** $\frac{11}{63}$
29. 0.8333, 0.1667 **31.** 0.6567, 0.3433

EXERCISES 2.5

1. $-\dfrac{x}{y}$ **3.** $-\dfrac{2}{y(x-2)^2}$ **5.** $\dfrac{9x^2+10x}{2y}$ **7.** $\dfrac{x(a^2-y^2)}{y(x^2+a^2)}$

9. $\dfrac{x(2y^3+3xy^2-x^3)}{y(y^3-2x^3-3x^2y)}$ **11.** $\dfrac{\sqrt{1-y^2}(8\sqrt{1-x^2}+x)}{\sqrt{1-x^2}(8\sqrt{1-y^2}-y)}$ **13.** $\dfrac{6x^2-4xy-1}{1+2x^2}$

15. $y = 2x-1$ **17.** $y = -x+5$ **21.** $\dfrac{85}{\sqrt{13}}$

EXERCISES 2.6

1. $6x^2,\ 12x$ **3.** $3x^2+3\sqrt{x},\ 6x+\dfrac{3}{2\sqrt{x}}$ **5.** $-\dfrac{1}{x^2},\ \dfrac{2}{x^3}$

7. $-2(x-1)^{-2},\ 4(x-1)^{-3}$ **9.** $\dfrac{\sqrt{2}}{2}\left[\dfrac{1}{\sqrt{x}}-\dfrac{1}{2x^{3/2}}\right],\ \dfrac{\sqrt{2}}{4}\left[-\dfrac{1}{x^{3/2}}+\dfrac{3}{2x^{5/2}}\right]$

11. $\dfrac{4}{(x^2+1)^2},\ \dfrac{4(1-3x^2)}{(x^2+1)^3}$ **13.** $-20-32t,\ -32$

15. $6t^2-24t+24,\ 12t^2-24$

EXERCISES 3.1

1. f increasing on $(-\infty, \infty)$ **3.** f increasing on $(-\infty, \infty)$ **5.** f increasing on $(-\infty, 0)$; decreasing on $(0, \infty)$ **7.** f increasing on $(-3, \infty)$; decreasing on $(-\infty, -3)$ **9.** f increasing on $(\frac{3}{2}, \infty)$; decreasing on $(-\infty, \frac{3}{2})$
11. f increasing on $(-\infty, -1) \cup (3, \infty)$; decreasing on $(-1, 3)$
13. f increasing on $(-\infty, -1) \cup (1, \infty)$; decreasing on $(-1, 1)$.
15. f increasing on $(-\frac{1}{2}, 0) \cup (\frac{1}{2}, \infty)$; decreasing on $(-\infty, -\frac{1}{2}) \cup (0, \frac{1}{2})$.
17. R increasing on $[0, 2)$; decreasing on $(2, 4]$. **19.** decreasing.
21. (a) (1) $p < 20$, (2) $p > 20$; (b) $\dfrac{p}{p-40}$, (1) $p < 20$, (2) $p > 20$.

EXERCISES 3.2

3. Abs. max. 9; Abs. min. -27. **5.** Abs. max. 124; Abs. min. -2
7. Abs. max. 79; Abs. min. -11. **9.** Abs. max. 71; Abs. min. -1
11. Abs. max. 61; Abs. min. -11 **13.** No abs. max. or min.
15. No abs. max.; Abs min. $\frac{17}{16}$ **17.** Abs. max. 144; Abs min. 0.
19. Abs. max. 32; Abs. min. 0. **21.** Rel. min. $(3, 2)$. **23.** Rel. min. $(-4, -3)$.
25. Rel. max. $(-\frac{3}{2}, \frac{17}{4})$ **27.** Rel. max. $(1, 2)$; Rel. min. $(7, -7)$
29. Rel. min. $(-6, 3)$; Rel. max. $(-2, 7)$ **31.** Rel. min. $(-1, -2)$ and $(1, -2)$; Rel. max. $(0, -1)$ **33.** No rel. max. or min. **35.** No rel. max. or min.
37. Rel. max. $(0, \frac{8}{3})$ **39.** Rel. max. $(\frac{1}{5}, (\frac{1}{5})^{1/3}(\frac{4}{5})^{4/3})$; Rel. min. $(1, 0)$
41. (a) $a = -3$, $b = -24$ (b) Rel. max. at $x = -2$; Rel. min. at $x = 4$

EXERCISES 3.3

1. 14, 14 **3.** 16, 32 **5.** (a) $x = 4$, $y = 4$; min. = 128; (b) No
7. $\frac{P}{4} \times \frac{P}{4}$ **9.** $4\sqrt[3]{2} \times 4\sqrt[3]{2} \times (4)^{2/3}$ **11.** $28 \times 70 \times 10$
13. $\frac{20\pi}{4+\pi}$, $\frac{80}{4+\pi}$ **15.** $12 \times 12 \times 2$ **17.** $10 \times 10 \times 10$ **19.** No
21. $x = 100$; 120,000 **23.** $(10)^{-5/2}$ **25.** (a) $\frac{5}{3}$, (b) 250, (c) 1000
27. (b) 49,900, (c) \$37,349,641 **29.** $R'(x) = C'(x)$

EXERCISES 3.4

1. -5 **3.** $\frac{1}{3}$ **5.** 0 **7.** does not exist **9.** does not exist
11. does not exist **13.** does not exist **15.** does not exist **17.** does not exist **19.** does not exist **21.** does not exist **23.** $y = 1$; $x = -1$; neither **25.** $y = 6$; $x = -1$; $x = -3$; neither **27.** $y = 3$; $x = 1$; neither **29.** none **31.** Increasing on $(\frac{3}{2}, \infty)$; decreasing on $(-\infty, \frac{3}{2})$; no inflection point; concave up for every x; relative minimum is $(\frac{3}{2}, \frac{7}{4})$
32. Increasing on $(-\infty, \frac{3}{4})$; decreasing on $(\frac{3}{4}, \infty)$; no inflection point; concave down for every x; relative maximum is $(\frac{3}{4}, -\frac{7}{8})$ **35.** Increasing on $(0, \infty)$; decreasing on $(-\infty, 0)$; concave down in $(-\infty, 0) \cup (0, \infty)$; absolute minimum at $(0, 0)$ **36.** Decreasing on $(-\infty, \infty)$; concave down in $(0, \infty)$; concave up in $(-\infty, 0)$; $x = 0$ is a point of inflection
37. Increasing on $(-\infty, -6) \cup (0, \infty)$; decreasing on $(-6, 0)$; $x = -3$ is a point of inflection; concave up in $(-3, \infty)$; concave down in $(-\infty, -3)$; $x = 0$ is a relative minimum, $x = -6$ is a relative maximum
39. Increasing on $(-\infty, -1) \cup (-\frac{1}{3}, \infty)$; decreasing on $(-1, -\frac{1}{3})$; concave up in $(-\frac{2}{3}, \infty)$; concave down in $(-\infty, -\frac{2}{3})$; $x = -\frac{2}{3}$ is a point of inflection; $x = -\frac{1}{3}$ is a relative minimum; $x = -1$ is a relative maximum
42. Decreasing on $(-\infty, \infty)$; $x = 1$ is a point of inflection; concave up in $(-\infty, 1)$; concave down in $(1, \infty)$ **44.** Decreasing on $(-\infty, 0) \cup (0, \infty)$; concave up in $(0, \infty)$; concave down in $(-\infty, 0)$ **47.** Concave up in $(0, \infty)$; concave down in $(-\infty, 0)$; decreasing for every $x \neq 0$
50. Increasing on $(-3, 3)$; decreasing on $(-\infty, -3) \cup (3, \infty)$; $x = 6$ point of inflection; concave up in $(6, \infty)$; concave down in $(-\infty, -3) \cup (-3, 6)$; $x = 3$ is a relative maximum **51.** $3x + y - 1 = 0$; $18x + 27y - 11 = 0$; $9x + 27y - 17 = 0$; $60x - 432y - 3 = 0$; and $3x - 27y = 0$ **52.** -8
55. Concave down at $x = 100$; has an inflection point at $x = 200$; and is concave up for $x = 300, 400$

EXERCISES 3.5

2. 44.5, 50.75, 6.25 **3.** 6.5, 6.75, 0.25 **6.** -0.03125, -0.0286, 0.00265 **8.** 11.1364 **11.** 15.9687 **15.** 0.5031 **16.** $\frac{8}{5}\pi$ sq. in.
17. 2.4π sq. in., 1.7% **18.** ≈ 0.007 ft. **20.** 100π cu. cm.
21. (b) 7 (c) 0.01 (d) 0.875% **22.** (b) $\left(\frac{4}{5}x - 35 - \frac{x^2}{500}\right) dx$ (c) 0 **23.** ≈ 5444 **25.** ≈ 500; actual = 459

EXERCISES 4.1

1.

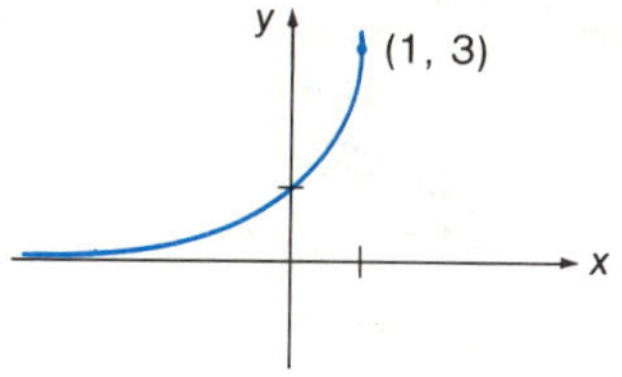

2.

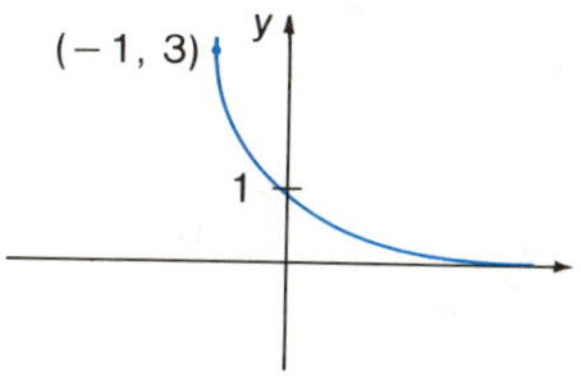

5.

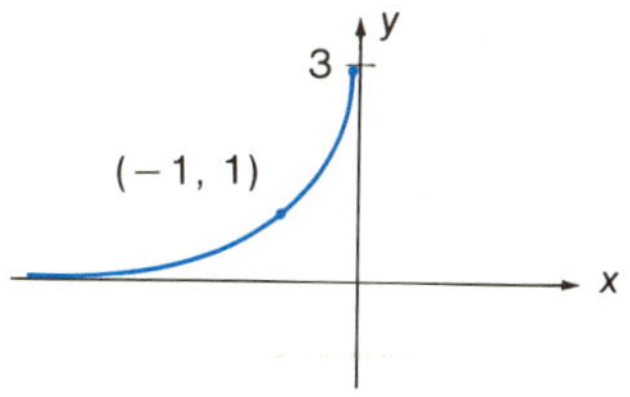

7.

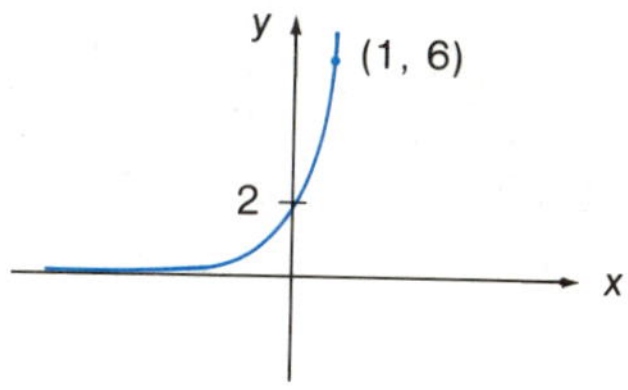

8.

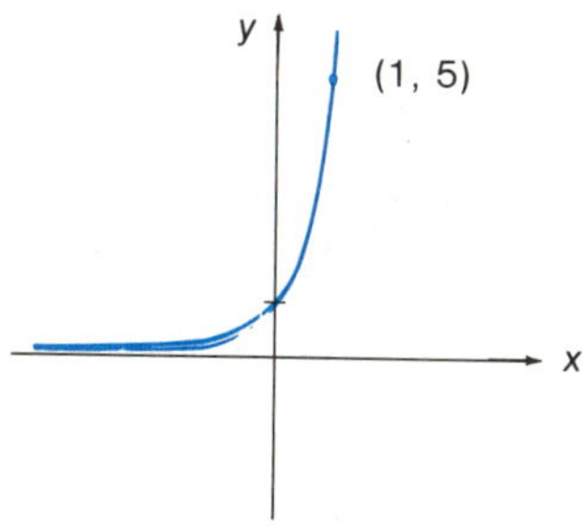

9.

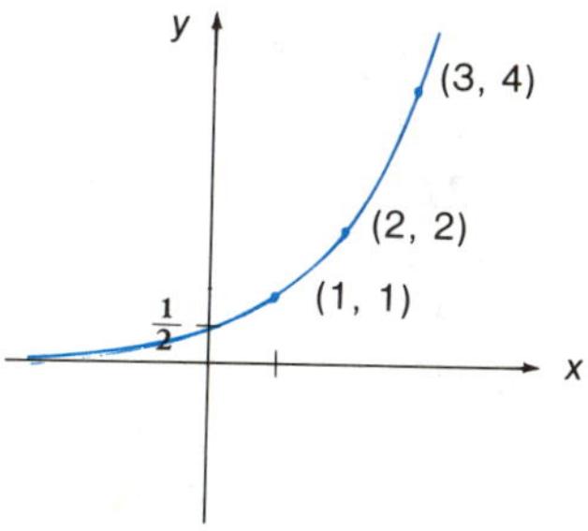

10.

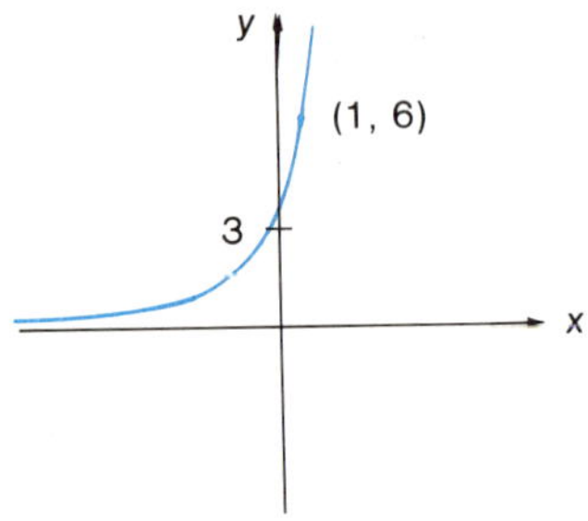

11.

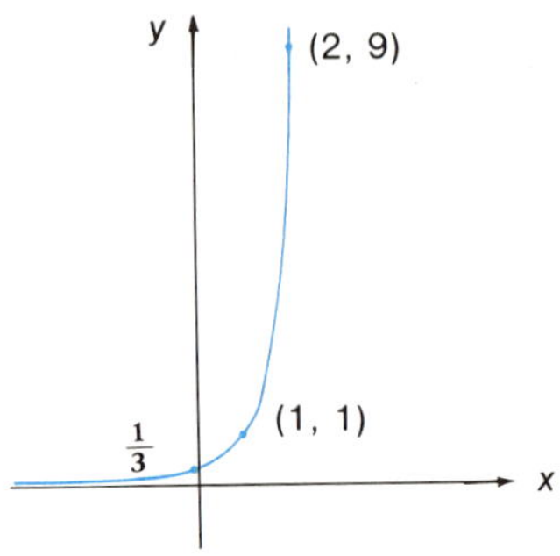

12.

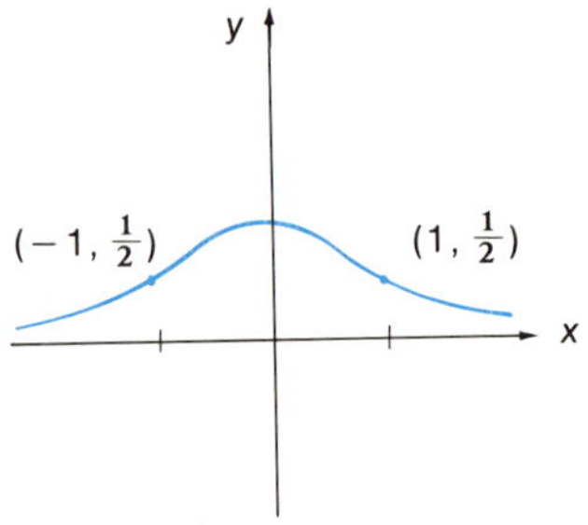

15.

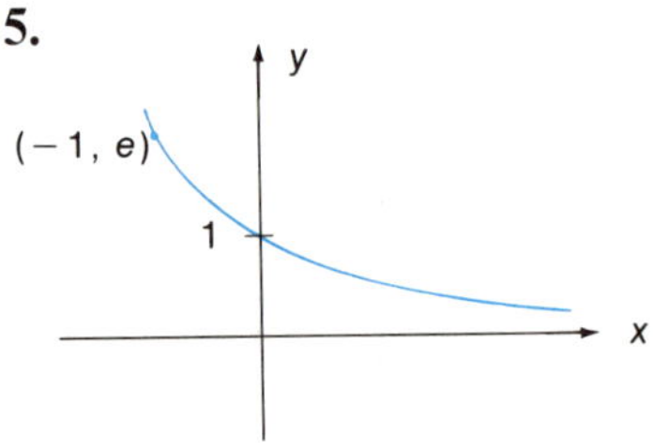

19. 3 **21.** $\frac{1}{2}$ **23.** $\frac{1}{5}$ **24.** 4 **27.** (a) 100; (b) 90.483; (c) 36.787 **29.** \$104,937; \$118,394; \$106,767 **31.** \$8880 **33.** \$9661.95 **35.** \$9684.86 **37.** \$9695.96 **39.** (a) \$5247.80; (b) \$5220.81

EXERCISES 4.2

1. $\log_2 32 = 5$ **3.** $\log_{1/2} \frac{1}{16} = 4$ **5.** $\log_{100} 1000 = \frac{3}{2}$
7. $\log_{10} 0.00001 = -5$ **9.** $\log_a y = x$ **11.** $100 = (10)^2$ **13.** $\frac{4}{9} = (\frac{2}{3})^2$
15. $1 = 6^0$ **17.** $5 = e^x$ **19.** 81 **21.** 32 **23.** $\pm\frac{1}{4}$ **25.** $-\frac{1}{5}$
27. $\log_2 21$ **29.** $\log_{10} \frac{15}{8}$ **31.** 1.5441 **33.** 0.1761 **35.** 0.4226
37. 0.8271 **41.** $\log_b x + \log_b y + \log_b z$ **43.** $2 \log_b x + 3 \log_b y + 4 \log_b z$
45. $\frac{1}{2} \log_b x + \frac{1}{3} \log_b y + \frac{1}{4} \log_b z$ **47.** $\frac{1}{2} \log_b x + \frac{1}{4} \log_b y + \frac{1}{4} \log_b z$
49. $\dfrac{\log 9}{\log 2}$ **51.** $1 + \dfrac{\log 27}{\log 4}$ **53.** $\dfrac{2 \log 5 - \log 3}{3 \log 5 - \log 3}$ **55.** $150{,}000\sqrt{3}$
57. $\dfrac{1}{0.36} \ln 2$

EXERCISES 4.3

1. $4e^{4x}$ **3.** $-\dfrac{4}{x^3} e^{2/x^2}$ **5.** $x^2e^{2x}(3 + 2x)$ **7.** $-4xe^{-2x^2}$ **9.** $\dfrac{2x}{x^2 + 4}$

11. $e^{x \ln x}(1 + \ln x)$ **13.** $\dfrac{4xe^{2x^2+4}}{e^{2x^2+4}+1}$ **15.** $5^x \ln 5$ **17.** $\frac{1}{2}(10)^x \ln 10$
19. $2e^{2x} + 2e(2x)^{e-1}$ **21.** $-\frac{1}{3}(5x^2 + 7x + 3)^{-4/3}(x^2 + 2x + 4)^{-2/3}(3x^2 + 34x + 22)$
23. $\dfrac{1}{2}\left(\dfrac{x+2}{1-2x}\right)^{-1/2} \cdot \dfrac{5-4x}{(1-2x)^2}$ **25.** 208 **29.** $y\left[\dfrac{30x}{3x^2+1} + \dfrac{3x}{x^2+1} - \dfrac{20x}{x^2+4}\right]$
31. $x^x[1 + \ln x]$ **33.** Absolute maximum = 1; no absolute minimum.

EXERCISES 5.1

1. $\frac{1}{2}x + C$ **3.** $\frac{3}{4}x^4 + C$ **5.** $\dfrac{x^3}{3} + x^2 + x + C$ **7.** $-\dfrac{1}{x} + C$
9. $\dfrac{x^3}{3} - x^2 + x + C$ **11.** $\frac{4}{3}x^{3/2} + C$ **13.** $\frac{3}{2}x^{2/3} + C$ **15.** $\dfrac{x^2}{2} - x + C$
17. $\frac{2}{7}x^{7/2} - 4x^{1/2} + C$ **19.** $\frac{2}{3}x^{3/2} - \sqrt{2}x + C$ **21.** $\frac{1}{4}e^{4x} + C$
23. $\frac{3}{2}e^{(2/3)x} + C$ **25.** $\frac{1}{2}e^{2x} + \ln x + C$ **27.** $\pi r^2 + 2r^4 + C$
29. $2u^3 + 12u - \dfrac{6}{u} + C$ **31.** $y = x^3 + 4x - 2$ **33.** $y = \dfrac{5x^2}{2} + 5 \ln x + 2$
35. $y = 2e^{5x} - x^4 - 7$ **37.** $f(x) = x^2 - x + 2$
39. $v(t) = 3t^2 - 5t^{3/2} + \frac{5}{2}$, $s(t) = t^3 - 2t^{5/2} + \frac{5}{2}t + 3$
43. $C(x) = 18x^{2/3} + 10{,}000$ **45.** $P(x) = \dfrac{3x^2}{2} + 5000x - 200$
47. $C(x) = \dfrac{x^3}{6} - \dfrac{x^2}{20} + 50x + 161{,}180$

EXERCISES 5.2

2. $\frac{17}{8}$ **3.** $\frac{51}{2}$ **6.** $\dfrac{3}{8}\sum_{k=1}^{8}\sqrt{\dfrac{48+3k}{8}}$ **7.** 15 **9.** 36 **11.** 2
13. $\frac{1}{3}$ **15.** $\frac{15}{2}$ **17.** 9 **19.** 3

EXERCISES 5.3

1. $\frac{8}{3}$ **2.** 12 **6.** $\frac{4}{3}$ **7.** 30 **9.** $-\frac{32}{3}$ **12.** $\frac{256}{5}$ **14.** 10
15. 36 **18.** $\frac{9}{2}$ **19.** $\frac{1}{3}$ **20.** $\frac{8}{3}$ **21.** $\frac{4}{3}$ **22.** 36
25. $(x-1)^2$ **26.** $2\sqrt{x}$ **28.** $8x^2$ **30.** 0 **33.** 2 **34.** $-\frac{3}{2}, 2$

EXERCISES 5.4

1. $\dfrac{(x+4)^3}{3} + C$ **3.** $\frac{2}{3}(3x+4)^{3/2} + C$ **5.** $-\dfrac{(4-x)^4}{4} + C$
7. $-\dfrac{1}{4(4x-1)} + C$ **9.** $\dfrac{(4x^5+6)^{101}}{2020} + C$ **11.** $\frac{1}{3}(x^2-4)^{3/2} + C$
13. $\dfrac{2}{3}(1 - x^{-1})^{3/2} + C$ **15.** $\dfrac{e^{7x}}{7} + C$ **17.** $\dfrac{1}{2}e^{x^2} + C$ **19.** $e^x - e^{-x} + C$

21. $3e^{-1/x} + C$ **23.** $-e^{-x} + \frac{1}{2}\ln(2x+3) + C$ **25.** $\ln(1+e^x) + C$

27. $\dfrac{(\ln x)^2}{2} + C$ **29.** $\dfrac{6}{5}(x+1)^{5/2} + \dfrac{4}{5}(x+1)^{3/2} + C$ **31.** $\dfrac{x^2}{2} + \ln(x+1) + C$

33. $\frac{1}{2}$ **35.** $2\sqrt{2}(\sqrt{3}-1)$ **37.** $\frac{1}{4}$ **39.** $\frac{1}{2}(e-1)$ **41.** $\frac{5}{2} - \ln 2$

EXERCISES 5.5

1. $\frac{2}{3}x^3 - \frac{1}{2}x^2 + C$ **3.** $\frac{1}{5}x^2(1+x^2)^{5/2} - \frac{2}{35}(1+x^2)^{7/2} + C$ **5.** $xe^x - x + C$

7. $\dfrac{x^2}{2}\ln x - \dfrac{1}{4}x^2 + C$ **9.** $x^5 \ln x - \dfrac{x^5}{5}\Big|_2^4 = (2^{11} - 2^5)\ln 2 - \dfrac{2^{10}}{5} + \dfrac{2^5}{5}$

11. $e^x(2-x) + C$ **13.** $\dfrac{xe^{2x}}{2} - \dfrac{e^{2x}}{4} + C$

15. $\frac{1}{2}(2x+3)\ln(2x+3) - \frac{1}{2}(2x+3) + C$ **17.** $-(2+x)e^{-x} + C$

19. $(\ln x)[\ln(\ln x) - 1] + C$ **21.** 1

EXERCISES 5.6

1. $2\ln(x+1) + 3\ln(x-3) + C$ **3.** $\ln(x-3) - 2\ln(x-2) - 3\ln(x+4) + C$

5. $-5\ln x + 2\ln(x+1) + 4\ln(x-1) + C$

7. $\frac{3}{2}\ln(x+1) - \frac{3}{2}\ln(x-1) - \dfrac{1}{x-1} + C$

9. $\frac{2}{3}\ln(x-2) - \frac{2}{3}\ln(x+1) + \dfrac{1}{x+1} + C$

11. $\frac{11}{32}\ln(x-2) - \dfrac{13}{16(x-2)} + \frac{21}{32}\ln(x+2) + \dfrac{3}{16(x+2)} + C$

13. $\frac{11}{8}\ln(x-1) + \frac{13}{8}\ln(x+3) - \dfrac{1}{2(x-1)} + C$

EXERCISES 5.7

The number of the formula used is set in parentheses.

1. (11) $\frac{1}{64}(3 + 8x - 3\ln|3+8x|) + C$ **3.** (15) $4\ln\left|\dfrac{z}{2+15z}\right| + C$

5. (14) $-\dfrac{1}{y-6} - \dfrac{3}{(y-6)^2} + C$ **7.** (29) $\dfrac{s}{9\sqrt{9-s^2}} + C$

9. (10) $\ln(x + 3 + \sqrt{x^2+6x+13}) + C$

11. (38) $-\dfrac{\sqrt{x^2+4x+5}}{x+2} + \ln|x + 2 + \sqrt{x^2+4x+5}| + C$

13. (39) $\frac{4}{3}(6u-1)e^{6u} + C$

EXERCISES 6.1

1. $t - \dfrac{3t^2}{2} + 4$ **2.** $x^4 + \dfrac{x^3}{3} - \dfrac{10}{3}$ **4.** $\dfrac{2}{3}t^{3/2} + 4t - \dfrac{55}{3}$
6. $2x - \frac{2}{5}x^{5/2} + \frac{406}{5}$ **9.** $-\frac{2}{5}x^{5/2} + 2x^{1/2} + \frac{83}{10}$
10. $\frac{9}{4}x^{4/3} - \frac{9}{5}x^{5/3} + \frac{103}{5}$ **11.** (b) $-2x^2 + 360x - 500$ (c) 90 units
12. $C(x) = \frac{20}{3}x^{3/2} + 350$ **13.** $p(x) = \dfrac{R(x)}{x} = \frac{1}{100}x + 20$
14. (a) 9 min. (b) 48 in. (c) 18 min. **15.** (b) $\approx$ 674 min. **16.** Yes, by 6 ft.
17. (a) $v(t) = \dfrac{t^2}{2} + t + 2$, $a(t) = \dfrac{t^3}{6} + \dfrac{t^2}{2} + 2t + 1$ (b) $v(t) = \dfrac{t^3}{3} - 5t - 1$, $a(t) = \dfrac{t^4}{12} - \dfrac{5}{2}t^2 - t + 4$ **18.** (a) $v(t) = -32t + 288$, $s(t) = -16t^2 + 288t$ (b) 1296 ft. **19.** $3\sqrt{2}$ sec. **20.** $t = 4$ sec.
21. (a) 5333 (b) 9600 **23.** (a) 3600 (b) 45000 (c) 330,000
25. 6353, 77708 **27.** 710, 3181 **29.** 9626, 26164

EXERCISES 6.2

1. 8 **3.** 5 **5.** 4 **7.** $\frac{2}{3}$ **9.** $\frac{1}{3}$ **11.** $\frac{25}{12}$ **13.** $\dfrac{1}{e-1}$
15. $\frac{1}{4}[\ln 5 - \ln 3]$ **17.** $\frac{45}{28}$ **19.** (a) -878.25 (b) -29.275
21. (a) 5 at $x = 3$, (b) $\varrho(0) = 2 + 6\sqrt{3}$, (c) $\dfrac{7 + 8\sqrt{3}}{2}$ (d) $\dfrac{21 + 24\sqrt{3}}{2}$
23. 10

EXERCISES 6.3

1. 18 **3.** $\frac{128}{3}$ **5.** $\frac{992}{3}$ **7.** $\frac{9}{2}$ **9.** $\frac{64}{3}$ **11.** 18
13. (a) (80, 80) (b) 6400 (c) 3200 **15.** (a) (2, 4) (b) $\frac{16}{3}$ (c) $\frac{16}{3}$
17. (a) (50, 150) (b) 1250 (c) $\frac{2500}{3}$ **19.** (a) (8, 394) (b) $\frac{4096}{3}$ (c) $\frac{5152}{3}$

EXERCISES 6.4

1. Yes **3.** Yes **5.** Yes **7.** 44 **9.** $\frac{77}{3}$ **11.** 4
14. Does not exist **15.** $2\sqrt{2}$ **16.** $2\sqrt{6}$ **17.** Does not exist
21. Does not exist **22.** $\sqrt{3}$ **23.** 1 **24.** Does not exist **26.** $\frac{1}{2}$
29. $\frac{1}{2}$ **30.** $-\frac{1}{2}$ **31.** Does not exist **33.** 2 **35.** Does not exist
36. 0 **40.** (a) $50,000

EXERCISES 6.5

7. $\frac{1}{4}$ **9.** $\frac{1}{234}$ **11.** 1 **13.** 4 **15.** $\frac{3}{2}$ **17.** $\dfrac{1}{8 + e^{-4}}$

19. (a) 0.38 (b) 0.21 (c) 0.039 **21.** (a) 0.95 (b) approx. ≈ 1 (c) approx. ≈ 0 **25.** 0.96

EXERCISES 7.1

1. Second order, linear **2.** Second order, nonlinear **3.** First order, nonlinear **4.** Second order, nonlinear **5.** Fourth order, linear **6.** Third order, linear **7.** First order, nonlinear **8.** Third order, nonlinear **9.** Second order, linear **10.** Second order, nonlinear **11.** Second order, nonlinear **12.** Second order, linear **29.** (b) $y = -3e^{-2t}$ **30.** (b) $c = 2$ (c) $k = \ln c$ (d) $y(2) = ce^{-2k}$ **31.** (b) $y = 9e^{2x} - 6e^{3x}$ **32.** (b) $(e^3-1)^{-1}[(e^3-2e)e^{-x} + (2e-1)e^{2x}]$
34. (b) $y = \dfrac{1}{\sqrt{2x - \frac{7}{4}}}$

EXERCISES 7.2

1. $y = ce^x$ **2.** $y = ce^{-2x}$ **3.** $y = ce^{-\sqrt{3}x}$ **4.** $y = ce^{-3x/2}$ **5.** $y = cx^4$
6. $y = ce^{3x}$ **7.** $y = \dfrac{-1}{\ln|x+1| + C}$ **8.** $\dfrac{y^2}{2} - 4y = \dfrac{x^2}{2} + 3x + C$
10. $y = -\frac{1}{2}e^{-2x} + c$ **11.** $y = -\ln|-e^x + c|$ **16.** $y = ce^{-2t^2}$
17. $y = \dfrac{-1}{xe^{-x} + e^{-x} + c}$ **19.** $y = \dfrac{3}{2} + ce^{-2x}$ **20.** $y = -10 + ce^x$
21. $y = \dfrac{x}{3} - \dfrac{1}{9} + ce^{-3x}$ **22.** $y = \dfrac{x^3}{2} + \dfrac{c}{x^3}$ **24.** $y = \dfrac{1}{2}t^2 - \dfrac{1}{2} + ce^{-t^2}$
26. $y = 2x^2 + cx^{-2}$ **28.** $y = te^{-t} + 2e^{-t}$ **29.** $y^2 = \dfrac{t^4}{2} + \dfrac{199}{4}$
35. $m = 1, 3$ **36.** $m = -1, -2$ **37.** $m = \pm 2$ **38.** $m = \pm\sqrt{2}$
40. $m = \dfrac{1 \pm \sqrt{7}}{2}$

EXERCISES 7.3

1. $20{,}000(1 - (\frac{4}{5})^t)$, ≈ 3 **2.** $\approx 38.1°$ **3.** $t \approx 1.35$ hr.
4. (a) 21.8° (b) ≈ 9.4 min. **5.** (a) ≈ -0.0875 (b) ≈ 16 min.
6. (a) 2.96×10^8 (b) 4.28×10^8 (c) Will double in $t \approx 17.7$ years (d) Will triple in $t \approx 28.1$ years **7.** ≈ 1990 **8.** In the year 2022
9. (a) 1.2528 (b) ≈ 2229 **10.** (a) About 65% (b) 74.91%
12. 63.496 ft. **13.** $(7, 4 \ln 2)$

EXERCISES 8.1

1. 1, 5, 13 **2.** $-2, 1, e-1$ **5.** $\{(x, y) \mid x, y \in R\}$
7. $\{(x, y) \mid 1 \le x \le 2 \text{ and } 3 \le y \le 4\}$ **9.** $\{(x, y) \mid (x, y) \ne (0, 0)\}$

11. (a) yz-plane (b) xz-plane (c) At $z = 3$ plane parallel to xy-plane (d) z-axis
(e) Straight line parallel to z-axis going through the point (1, 0, 0)
(f) Plane parallel to xy-plane 2 units below (at $z = -2$)
(g) Straight line through (0, 0, 0) and (1, 1, 1) (h) Point (1, −1, 2)

12. (a) $\sqrt{a^2 + b^2 + c^2}$ (c) 5 (e) 13

EXERCISES 8.2

4.

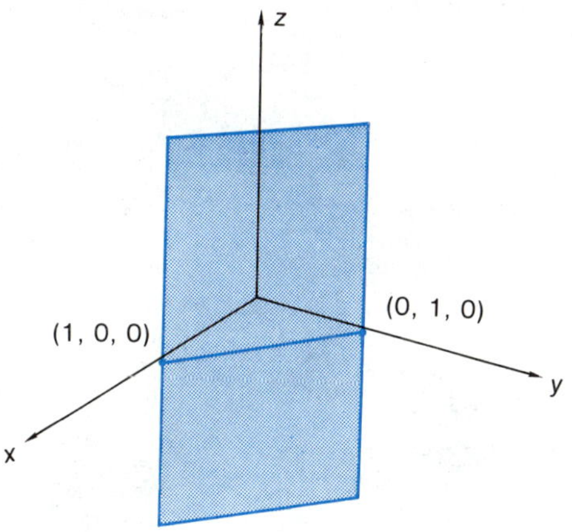

6.

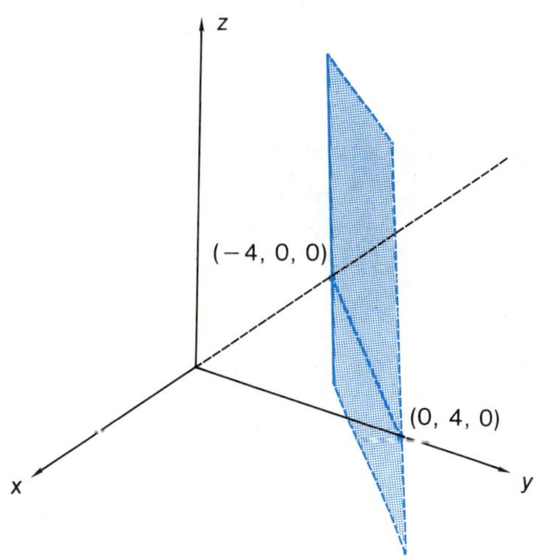

11.

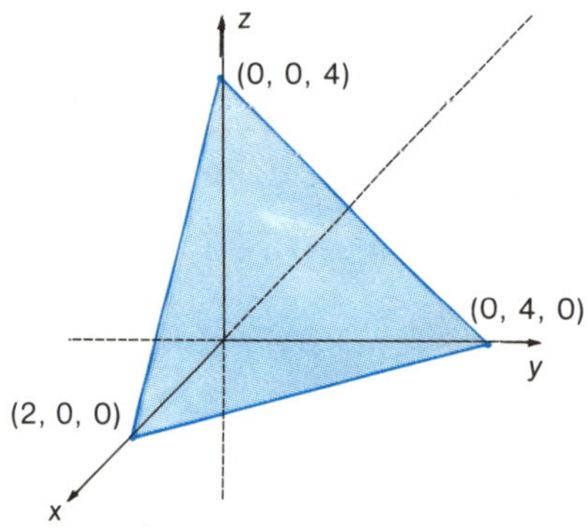

12.

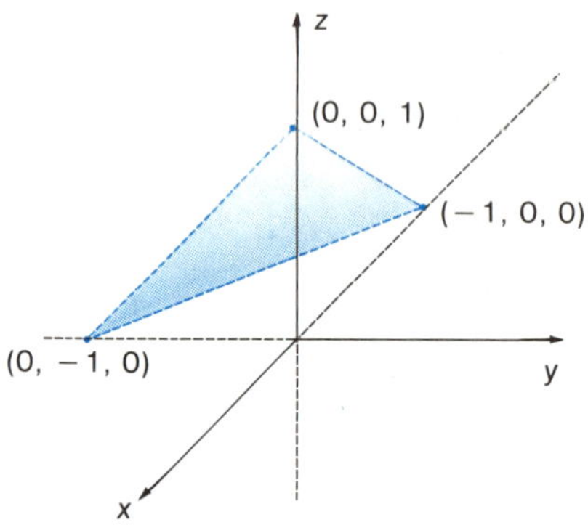

14.

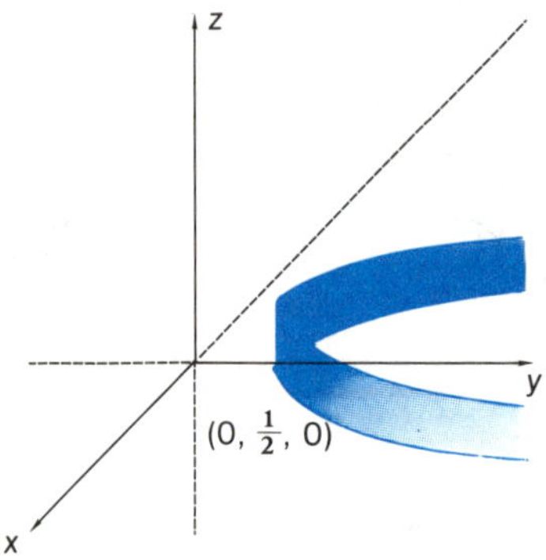

16.

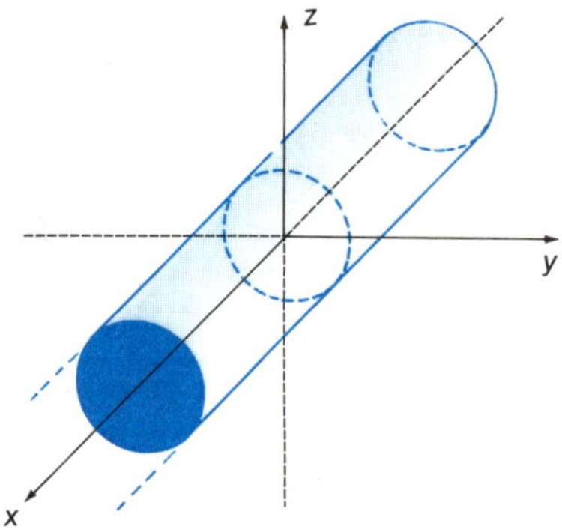

19.

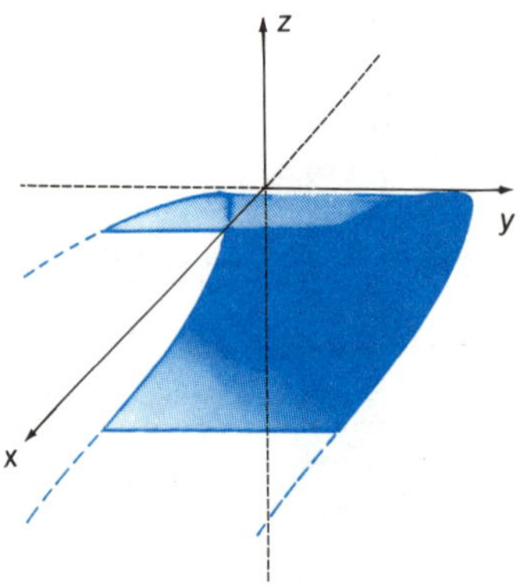

22. Sphere with origin at (0, 0, 0) and radius $r = 4$ **23.** Sphere with origin at (1, 0, 0) and radius 3 **26.** Sphere with origin at $(-3, 1, -2)$ and radius 4

29.

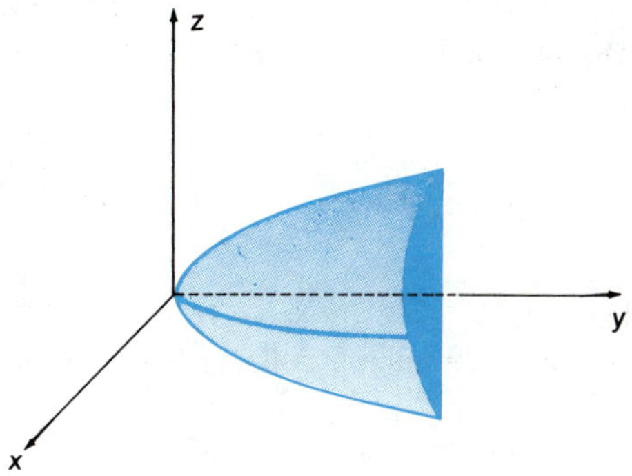

EXERCISES 8.3

1. $f_x = 2,\ f_y = -3$ **3.** $f_x = -5y,\ f_y = -5x$ **5.** $f_x = 2xy - y^2,$
$f_y = x^2 - 2xy$ **7.** $f_x = 3x^2y^3,\ f_y = 3x^3y^2$ **9.** $f_x = \frac{1}{y},\ f_y = -\frac{x}{y^2}$

11. $f_x = \frac{2y}{(x+y)^2},\ f_y = -\frac{2x}{(x+y)^2}$ **13.** $f_x = 2xe^{x^2+2y^2},\ f_y = 4ye^{x^2+2y^2}$

15. $f_x = e^{x+y}\left[\frac{2x}{x^2+y^2} + \ln(x^2+y^2)\right],\ f_y = e^{x+y}\left[\frac{2y}{x^2+y^2} + \ln(x^2+y^2)\right]$

17. $f_{xx}(1,2) = 16,\ f_{xy}(1,2) = 24 = f_{yx}(1,2),\ f_{yy}(1,2) = 12$

21. (a) $f_x = 2xe^{u+v},\ f_y = 3y^2e^{u+v},\ f_u = (x^2+y^3)e^{u+v},\ f_v = (x^2+y^3)e^{u+v}$
(b) $f_{xy} = 0,\ f_{xu} = 2xe^{u+v},\ f_{uv} = (x^2+y^3)e^{u+v}$ (c) $f_{xyu} = 0,\ f_{uyx} = 0,\ f_{uxy} = 0$

EXERCISES 8.4

1. $\left.\frac{\partial P}{\partial x}\right|_{(3,5)} = 37,\ \left.\frac{\partial P}{\partial y}\right|_{(2,3)} = 10$

3. $\frac{\partial P}{\partial x} = \frac{[y(x^2+y^2+y) - 2x]e^{xy}}{(x^2+y^2+y)^2},\ \frac{\partial P}{\partial y} = \frac{[x(x^2+y^2+y) - (2y+1)]e^{xy}}{(x^2+y^2+y)^2}$

5. (a) $f_x = 50y,\ f_y = 50x$ (b) $f_x\big|_{y=5} = 250$ (c) $f_y\big|_{x=2} = 100$

6. (a) $f_x = 2x + 20y,\ f_y = 2y + 20x$ (b) $f_x\big|_{y=5} = 2x + 100$ (c) $f_y\big|_{x=2} = 2y + 40$

7. (a) $(y^2 + e^y)\,dx + x(2y + e^y)\,dy$ (b) $(1 + y/x)\,dx + \ln x\,dy$ **9.** 3%

10. 4% **11.** 1% **12.** 27 **13.** (a) $\frac{\partial C}{\partial x} = 8x + y,\ \frac{\partial C}{\partial y} = x - 26y$

(b) $\frac{\partial C}{\partial x} = \frac{y}{x} + 6xy,\ \frac{\partial C}{\partial y} = \ln x + 3x^2$ (c) $\frac{\partial C}{\partial x} = e^y,\ \frac{\partial C}{\partial y} = xe^y + 1$

(d) $\frac{\partial C}{\partial x} = \ln(16x + e^y) + x\frac{16}{16x + e^y},\ \frac{\partial C}{\partial y} = x\frac{e^y}{16x + e^y}$

EXERCISES 8.5

1. At (0, 0) is a saddle point **2.** At (0, 0) is the relative minimum **5.** At (0, 0) is the relative minimum **7.** None **8.** None **9.** (0, 0, 1) **10.** $4' \times 4' \times 2'$ **11.** $6'' \times 6'' \times 3''$ **14.** $l \approx 3.495$, $w \approx 3.495$, $h \approx 5.241$ **15.** $l \approx 3.319$, $w \approx 3.319$, $h \approx 5.810$ **17.** $y = 10$, $x = 6$ **19.** $x = 9$, $y = 11$

EXERCISES 8.6

1. $y = 0.34x + 4.54$ **3.** $y = -3x + 4$ **5.** $y = -1.581x + 11.459$ **7.** $R(t) = 113.8t + 29$ million dollars, $P(t) = 0.3035t + 0.014$ dollars per share

EXERCISES 8.7

3. f attains maximum value $\sqrt{2}$ at $(1/\sqrt{2}, 1/\sqrt{2})$; f attains minimum value $-\sqrt{2}$ at $(-1/\sqrt{2}, -1/\sqrt{2})$ **4.** f attains maximum value 1 at $(\pm 1, 0)$; f attains minimum value -1 at $(0, \pm 1)$ **5.** $\frac{1}{2}$ is minimum value at $(\frac{1}{2}, \frac{1}{2})$ **11.** $\frac{134}{75}$

EXERCISES 8.8

1. 3 **2.** 52 **5.** $\frac{11}{6}$ **7.** $-\frac{5}{6}$ **10.** 9 **12.** $\frac{1}{2}(e^3 - e^2 - e + 1)$ **13.** $\frac{5}{3}$ **16.** $\frac{23}{15}$ **18.** $\frac{1}{2}(e^4 - 1)$ **21.** $2 - \frac{4}{3}\sqrt{2}$ **22.** $\frac{21}{2}$ **24.** $3 \ln 3 - 2$ **25.** $\frac{3}{2}$ **26.** $\frac{1}{6}$

Index